Real
Analysis

Real
Analysis

William O. Ray

The University of Oklahoma

Prentice Hall
Englewood Cliffs, New Jersey   07632

Library of Congress Cataloging-in-Publication Data

RAY, WILLLIAM O.
    Real analysis.

    Includes index.
    1. Functions of real variables. 2. Functional
analysis. 3. Measure theory. I. Title.
QA331.R36 1988   515   87-14402
ISBN 0-13-762386-0

Editorial/Production supervision and
        interior design: Fay Ahuja
Cover design: Edsal Enterprises
Manufacturing buyer: Paula Benevento

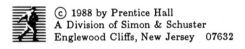
Printed in the United States of America
10 9 8 7 6 5 4 3 2 1

ISBN 0-13-762386-0 01

Prentice-Hall International (UK) Limited, *London*
Prentice-Hall of Australia Pty. Limited, *Sidney*
Prentice-Hall Canada Inc., *Toronto*
Prentice-Hall Hispanoamericana, S.A., *Mexico*
Prentice-Hall of India Private Limited, *New Delhi*
Prentice-Hall of Japan Inc., *Tokyo*
Simon & Schuster Asia Pte. Ltd., *Singapore*
Editora Prentice-Hall do Brasil, Ltda., *Rio de Janeiro*

Typeset using ᵖᶜTEX ™, Personal TEX Incorporated, Mill Valley, CA.

to

Shireen

Contents

Preface

This book evolved from the course in integration and measure theory taught at the University of Oklahoma. Students in this course are primarily beginning graduate work in mathematics or statistics, although, occasionally, a few physicists or engineers enroll. The book presumes a general background in undergraduate mathematics, particularly in undergraduate analysis. A student having completed a course from either R. R. Goldberg's *Methods of Real Analysis* or R. G. Bartle's *The Elements of Real Analysis*, for example, should be amply prepared. In addition, some familiarity with vector spaces is presupposed starting in Chapter 3.

No attempt has been made to write a comprehensive text, although I believe that I have covered all of the major topics. Some topics (especially in measure theory) have been relegated to the Problems, but a deliberate effort has been made to avoid dependence upon the these. I believe that all of the fundamental concepts and techniques are discussed in the body of the text. Although the result may not read as cleanly as it would otherwise, it is my firm belief that this approach is of greater value to the beginning student.

Some of the fundamental material from undergraduate analysis is summarized in Appendix A; all students are encouraged to at least browse through this appendix, as they may find some new material (such as Theorem A.2.4). Much of the material in Chapter 1 is also review, and may either be omitted or covered more hastily. Note, however, that this chapter contains much that is important for the rest of the text, including sets of measure zero and the Cantor set.

Chapter 2 is the heart of the book. In this chapter the Lebesgue integral is defined and its fundamental properties are deduced. The development in this chapter (and elsewhere) owes much to the classic book *Introduction to Real Functions and Orthogonal Expansions* by B. Sz.-Nagy — in particular we approach the integral from the point of view of the Cauchy Principal Value (i.e., the Daniell integral) rather than the more traditional approach through measure theory. It is our hope and belief that this approach is more intuitive and more readily followed by the beginning student. We have nonetheless included all of the standard topics from measure theory in this chapter.

After concluding Chapter 2, any one of Chapter 3 (Banach spaces), Chapter 4 (metric spaces), or Chapter 5 (differentiation), could be covered. However, if they are done out of order, some of the examples and applications may have to be omitted until later. Chapter 6 (Stieltjes integrals) depends on Chapter 5 but is otherwise independent of all the preceding chapters. Chapter 7 contains some of the most important mathematics in the text. In this chapter, the abstract integral is developed (once again, following the Daniell approach). This more abstract integral is then used in the development of Fubini's Theorem, signed measures, and the Radon-Nikodym Theorem. The chapter concludes with a discussion of the Fourier transform. Chapter 8 is on basic functional analysis. In this chapter an effort is made to relate abstract principles such as the Hahn-Banach Theorem back to the more concrete notions in integration theory from which they arose.

Anyone writing a book incurs an ocean of debts. There is, of course, the debt owed to the great mathematicians responsible for the beautiful theories and the debt to earlier expositors on the topic; many of these are referenced in Appendix C. In addition, I have been inspired throughout by the many fine teachers and colleagues who have enriched my mathematical life. The most notable of these are W. A. Kirk, Paul Waltman, Jon Simon, Richard Goldberg and Morris Marx.

A number of persons read the manuscript for this book and made many valuable suggestions. These include Professor Michael McAsey of Bradley University, Professor Thomas Banchoff of Brown University, Professor Peter Colwell of Iowa State University, Professor William C. Fox of SUNY Stony Brook, and Professor Arthur Robinson of Princeton. In addition, I am indebted to my students at the University of Oklahoma who suffered through early versions of the book. The constructive criticisms and comments made by these individuals have greatly enhanced the final product. Of course, I remain fully responsible for any remaining errors or deficiencies.

The manuscript was prepared using TeX. I am greatly indebted in those responsible for developing this remarkable software, especially to Donald Knuth, who developed TeX, and to Lance Carnes, who wrote the personal computer version.

This entire project was instigated by my wife, who typed much of the manuscript and provided invaluable advice on matters of style and programming in TeX. In addition, without her constant encouragement (nagging) this project would never have been completed. As a consequence, although a textbook should properly be dedicated to one's children, this one is dedicated to her.

W.O.R.
Norman, Oklahoma

Real
Analysis

Chapter 1

The Riemann Integral

1.1. CONTINUOUS FUNCTIONS. The notion of a continuous function is one of the cornerstones of analysis. The concept of function evolved during the seventeenth century together with the theories of integration and differentiation. While Descartes had included only algebraically defined curves in his geometry, other nonalgebraic curves, such as the logarithmic and exponential functions, soon became important as the calculus evolved. Eventually, compositions, sums, and infinite series obtained from these algebraic and nonalgebraic sources all came to be thought of as functions. Largely as a consequence of the study of trigonometric series in the late eighteenth and early nineteenth centuries it became clear that the notion of function needed to be refined and clarified. In many ways modern analysis was born at this time as mathematicians began to distinguish between functions, continuous functions, differentiable functions, integrable functions, and so on.

In this section we give a brief summary of the properties of continuous functions of a real variable. At the outset, the student may find our approach unusual; eventually everything will be related back to the more conventional notions given in calculus.

1.1.1. Definitions. Let f be a real-valued function with domain $D \subseteq \Re$, and let J be a set of real numbers. If f is bounded on J and if $J \cap D \neq \emptyset$, we define the *oscillation of f over J* to be the number

$$\omega(f; J) = \sup\{f(x) : x \in J \cap D\} - \inf\{f(x) : x \in J \cap D\}; \qquad (i)$$

in all other cases, we define

$$\omega(f; J) = +\infty.$$

For $x \in D$ *the oscillation of f at x* is the number

$$\omega(f; x) = \inf\left\{\omega(f; J) : J \text{ is an open interval about } x\right\}. \qquad (ii)$$

In general, $\omega(f; x) \geq 0$; if $\omega(f; x) > 0$, then the graph of f must "jump" at $(x, f(x))$. For example, if

$$f(x) = \begin{cases} \sin\left(\frac{1}{x}\right) & x \neq 0 \\ 0 & x = 0, \end{cases}$$

then $\omega(f; 0) = 2$, while $\omega(f; x) = 0$ if $x \neq 0$.

Some elementary properties of the oscillation are summarized in the following proposition. The proof is immediate from the definitions, and so is omitted.

1.1.2. Proposition. *Let f and D be as in Definition* 1.1.1.

(i) *If $J_1 \subseteq J_2$, then $\omega(f; J_1) \leq \omega(f, J_2)$.*

(ii) *If $x, y \in J \cap D$, then $|f(x) - f(y)| \leq \omega(f; J)$.*

(iii) *If f is bounded on J (and $J \cap D \neq \emptyset$), then for each $\epsilon > 0$ there are points $x, y \in J \cap D$ such that $\omega(f; J) \leq f(x) - f(y) + \epsilon$.*

In general we will want to concentrate on the case when $\omega(f; x) = 0$ ("continuous functions"). Before turning to this case, however, we prove the following proposition, which is needed in the next section (Theorem 1.2.14.).

1.1.3. Proposition. *Let f be an arbitrary real-valued function defined on all of $\Re$, and let $\epsilon > 0$ be arbitrary. Then the set*

$$E = \{x : \omega(f; x) \geq \epsilon\}$$

is closed.

Proof. If E is empty, there is nothing to prove, so we assume that $E \neq \emptyset$. Let $\{x_n\}$ be a Cauchy sequence in E with limit x_∞. Fix $\beta > 0$ and select an open interval J about x_∞ so that

$$\omega(f; J) \leq \omega(f; x_\infty) + \beta.$$

For n sufficiently large, $x_n \in J$ and so

$$\epsilon \leq \omega(f; x_n) \leq \omega(f; J) \leq \omega(f; x_\infty) + \beta,$$

i.e., $\omega(f; x_\infty) \geq \epsilon - \beta$. Since $\beta > 0$ was arbitrary, this completes the proof.

1.1.4. Definition. Let f and D be as in Definition 1.1.1; f is *continuous at* $x \in D$ if $\omega(f; x) = 0$. If $\omega(f; x) = 0$ for all $x \in D$, then we say that f *is continuous on D.*

Note that the infimum in 1.1.1(i) must be taken over *open* intervals containing x (see Problem 1.1).

This definition of continuous turns out to be slightly more convenient to apply in integration theory; it is equivalent to the more conventional notions.

1.1.5. Theorem. *Let f and D be as in Definition 1.1.1 and let $x \in D$; then the following are equivalent:*

(i) *f is continuous at x;*

(ii) *for each $\epsilon > 0$ there is a $\delta > 0$ such that if $|x - y| \leq \delta$ and if $y \in D$, then $|f(x) - f(y)| \leq \epsilon$;*

(iii) *if U is an open interval containing $f(x)$, then there is an open interval V about x with $f(V \cap D) \subseteq U$;*

(iv) *if $\{x_n\} \subseteq D$ and $\lim x_n = x$, then $\lim f(x_n) = f(x)$.*

Proof. (i) $\Rightarrow$ (ii). Fix $\epsilon > 0$ and select an open interval J about x with the property that $\omega(f; J) \leq \epsilon$. Choose $\delta > 0$ so that $(x - \delta, x + \delta) \subseteq J$ and apply 1.1.2(ii).

(ii) $\Rightarrow$ (iii). Select $\epsilon > 0$ so that $(f(x) - \epsilon, f(x) + \epsilon) \subseteq U$; apply part (ii).

(iii) $\Rightarrow$ (iv). Given an open interval U about $f(x)$, select an open interval V about x so that $f(V \cap D) \subseteq U$. Select N so large that $n \geq N$ implies that $x_n \in V$; then $n \geq N$ implies that $f(x_n) \in U$, showing that $\lim f(x_n) = f(x)$.

(iv) $\Rightarrow$ (i). Suppose that $\omega(f; x) = \epsilon > 0$; set

$$I_n = \left(x - \frac{1}{n}, x + \frac{1}{n} \right).$$

Since $\omega(f; I_n) \geq \omega(f; x) = \epsilon$, we may select $x_n, y_n \in I_n$ so that

$$f(y_n) - f(x_n) \geq \frac{\epsilon}{2}.$$

But since $\lim x_n = \lim y_n = x$, we must have, by (iv),

$$\lim f(y_n) - f(x_n) = f(x) - f(x) = 0,$$

a contradiction.

The following proposition follows easily from standard results on limits and the above; we leave the proof to the reader.

1.1.6. Proposition. *Suppose that f and g are continuous at x; then each of the following is also continuous at x:*

(i) $f + g$;

(ii) $f \cdot g$;

(iii) $f \wedge g \equiv \min\{f, g\}$;

(iv) $f \vee g \equiv \max\{f, g\}$;

(v) $\alpha \cdot f$, where α is any real number.

If, in addition, $g \neq 0$, then

(vi) f/g is continuous at x.

By themselves, compactness and continuity are each very powerful concepts; when applied together they complement one another and often simplify problems. Indeed, our entire approach to integration theory in Chapter 2 is based on some of the consequences of compactness and continuity, which we now discuss.

1.1.7. Theorem. *Let f be continuous on the compact set K. Then for each $\epsilon > 0$ there is a $\delta > 0$ such that $|f(x) - f(y)| \leq \epsilon$ whenever $x, y \in K$ and $|x - y| \leq \delta$.*

Of course, for each *fixed* $x \in K$ there corresponds a δ with the foregoing property. The point of Theorem 1.1.7 is that δ can be selected *independently* of

x, or *uniformly for $x \in K$*; we will sometimes say that f is *uniformly continuous.* Another way to phrase the theorem is the following:

1.1.8. Corollary. *Let f be continuous on the compact set K. Then for each $\epsilon > 0$ there is a $\delta > 0$ such that if $J \subseteq K$ is any interval of length δ, then $\omega(f; J) \leq \epsilon$.*

Proof of Theorem 1.1.7. For each $x \in K$ select $\delta_x > 0$ such that $|f(x) - f(y)| \leq \epsilon/2$ if $|x - y| \leq \delta_x$ and $y \in K$. Now the intervals $(x - \delta_x, x + \delta_x)$ are an open cover for K, and so admit a finite subcover

$$(x_1 - \delta_{x_1}, x_1 + \delta_{x_1}), \ldots, (x_n - \delta_{x_n}, x_n + \delta_{x_n}).$$

Applying A.3.5, there is a $\delta > 0$ so that if $|x - y| \leq \delta$, then x and y are *both* in the *same* interval $(x_i - \delta_{x_i}, x_i + \delta_{x_i})$.
Consequently,

$$|f(x) - f(y)| \leq |f(x) - f(x_i)| + |f(x_i) - f(y)| \leq \epsilon,$$

completing the proof.

Our definition of the Lebesgue integral in Chapter 2 relies on sequences of continuous functions. Of course, in general, the limit of continuous functions, even on a compact set, need not be continuous (consider $f_n(x) = x^n$ on $[0, 1]$). Sufficient conditions for the limit of continuous functions to be continuous constitute the final results in this section.

1.1.9. Definition. Let $\{f_n\}$ be a sequence of functions with domain D. We will say that $\{f_n\}$ *converges to f pointwise on D* if

$$\lim f_n(x) = f(x)$$

for each x. If corresponding to each $\epsilon > 0$ there is an N so that

$$|f_n(x) - f(x)| \leq \epsilon$$

whenever $x \in D$ and $n \geq N$, then we will say that $\{f_n\}$ *converges uniformly to f on the set D.*

Certainly, uniform convergence implies pointwise convergence; once again, the point of uniformity is that N can be chosen uniformly for x in the set D.

1.1.10. Theorem. *Let $\{f_n\}$ be a sequence of continuous functions defined on a domain D and suppose that $\{f_n\}$ converges uniformly to the function f on D. Then the function f is continuous on the set D.*

Proof. Fix $x \in D$ and $\epsilon > 0$. Choose N so large that $|f_n(u) - f(u)| \leq \epsilon/3$ whenever $u \in D$ and $n \geq N$. Next, choose $\delta > 0$ so that if $|y - x| \leq \delta$ and $y \in D$, then $|f_N(y) - f_N(x)| \leq \epsilon/3$. Then

$$|f(x) - f(y)| \leq |f(x) - f_N(x)| + |f_N(x) - f_N(y)| + |f_N(y) - f(y)|$$
$$\leq \epsilon$$

whenever $|x - y| \leq \delta$ and $y \in D$, completing the proof.

1.1.11. Corollary. *Let $\{f_n\}$ be a sequence of continuous functions defined on a domain D. Suppose that corresponding to each $\epsilon > 0$ there is an N so that*

$$|f_n(x) - f_m(x)| \leq \epsilon$$

whenever $n, m \geq N$ and $x \in D$. Then there is a continuous function f defined on D to which the sequence $\{f_n\}$ converges uniformly.

Proof. For each fixed x, $\{f_n(x)\}$ is a Cauchy sequence and thus $f(x) \equiv \lim f_n(x)$ is well-defined. Moreover, $\{f_n\}$ converges *uniformly* to $f(x)$, for if N is chosen as in the hypothesis, then

$$|f_n(x) - f(x)| = \lim_{m \to \infty} |f_n(x) - f_m(x)| \leq \epsilon$$

for all $n \geq N$ and all $x \in D$. Continuity of f is immediate from Theorem 1.1.10.

    While uniform convergence gives a test for when the limit of continuous functions is continuous, it can often be difficult to decide if a particular sequence converges uniformly. The final theorem in this section gives a test for uniform convergence and constitutes one of the fundamental links in the development of integration in Chapter 2. A sequence of functions $\{f_n\}$ is *nondecreasing* [respectively, *nonincreasing*] if $f_n(x) \leq f_{n+1}(x)$ [respectively, $f_n(x) \geq f_{n+1}(x)$] for all x.

1.1.12. Dini's Theorem. *Let $\{f_n\}$ be a nonincreasing sequence of nonnegative continuous functions, and suppose that $\lim f_n(x) = 0$ for all x in some compact set K. Then $\{f_n\}$ converges uniformly to zero on K.*

Proof. Fix $\epsilon > 0$. For each $x \in K$ select an integer $N(x)$ depending on x so that

$$0 \leq f_n(x) \leq \epsilon/2$$

whenever $n \geq N(x)$. Next select an open interval I_x, again depending on x, about x so that

$$|f_{N(x)}(x) - f_{N(x)}(y)| \leq \epsilon/2$$

whenever $y \in I_x \cap K$. Thus if $y \in I_x \cap K$ and $n \geq N(x)$, then

$$f_n(y) \leq f_{N(x)}(y) \leq f_{N(x)}(x) + \epsilon/2 \leq \epsilon.$$

The intervals $\{I_x\}$ cover K, and so admit a finite subcover

$$I_{x_1}, \ldots, I_{x_p}.$$

Set $N = \max\{N(x_1), \ldots, N(x_p)\}$.

If $y \in K$ and $n \geq N$, then $y \in I_{x_j}$ for some $j = 1, \ldots, p$ and so

$$\begin{aligned}
f_n(y) \leq f_N(y) &\leq f_{N(x_j)}(y) \\
&\leq f_{N(x_i)}(x) + \epsilon/2 \\
&\leq \epsilon,
\end{aligned}$$

showing that $\{f_n\}$ converges uniformly to zero.

PROBLEMS

1.1.   If $f(x) = \mathrm{sgn}(x)$, show that $\omega(f; 0) = 2$ while

$$\inf\{\omega(f; J) : J \text{ is an interval (not necessarily open) about } 0\}$$

is equal to 1. [Recall that

$$\mathrm{sgn}(x) = \begin{cases} 1 & \text{if } x > 0 \\ 0 & \text{if } x = 0 \\ -1 & \text{if } x < 0. \end{cases}$$

1.2.   Prove Proposition 1.1.6.

1.3.   If f is a continuous function defined on the compact set K, show that $f(K)$ is compact.

1.4.   If f is a continuous injective (i.e. one to one) function defined on the compact set K, show that f^{-1} is continuous.

1.5.   *Prove or disprove:* Let $\{f_n\}$ be a nondecreasing sequence of continuous functions defined on the compact set K. If

$$\lim f_n(x) < +\infty$$

for all $x \in K$, then $\{f_n\}$ converges uniformly to a continuous function.

1.6.   If U is an open interval, show that there is a nondecreasing sequence $\{f_n\}$ of continuous functions such that $\lim f_n = \chi_U$. [Recall that if U is a set, then the *characteristic function* of U is

$$\chi_U(t) = \begin{cases} 1 & \text{if } t \in U \\ 0 & \text{otherwise.} \end{cases}$$

1.7.   *Semicontinuous functions.* **Definition.** A function f is *lower semicontinuous* at x if

$$\liminf f(x_n) \leq f(x) \text{ whenever } \lim x_n = x$$

and is *upper semicontinuous* at x if

$$\limsup f(x_n) \geq f(x) \text{ whenever } \lim x_n = x.$$

The function f is *lower (upper) semicontinuous on a set* D if f is lower (upper) semicontinuous at each $x \in D$.

(a) Show that f is lower semicontinuous on $\Re$ if and only if $f^{-1}(t, \infty)$ is an open set for each $t \in \Re$.

(b) Show that f is upper semicontinuous on $\Re$ if and only if $f^{-1}(-\infty, t)$ is an open set for each $t \in \Re$.

(c) Let f be an upper semicontinuous function defined on the compact set K. Show that f attains its maximum on the set K.

(d) Let f be an lower semicontinuous function defined on the compact set K. Show that f attains its minimum on the set K.

(e) For a set $A \subseteq \Re$, show that χ_A is lower semicontinuous if and only if A is open.

(f) For a set $A \subseteq \Re$, show that χ_A is upper semicontinuous if and only if A is closed.

(g) If $\{f_n\}$ is a sequence of lower semicontinuous functions defined on a set $D \subseteq \Re$ and if $f(x) = \sup_n f_n(x) < \infty$ for all $x \in D$, then show that f is lower semicontinuous.

(h) Let f be a lower semicontinuous function defined on a bounded set $D \subseteq \Re$ and suppose that $f(D) \subset [0, 1]$. Fix $\epsilon > 0$. Find collections $\{U_1, \ldots, U_p\}$ of open sets and $\{\alpha_1, \ldots, \alpha_p\}$ of scalars so that

$$0 \le f - \sum_{i=1}^{p} \alpha_i \chi_{U_i} \le \epsilon.$$

(i) Let f be a a lower semicontinuous function defined on a bounded set $D \subseteq \Re$ and suppose that $f(D) \subset [0, 1]$. Show that there is a nondecreasing sequence $\{f_n\}$ of continuous functions defined on D with the property that $f = \lim f_n$. [This provides a partial converse to Problem 1.7(g).]

1.8. Let f be a continuous function defined on the interval $[a, b]$ and suppose that $f([a, b]) \subseteq [a, b]$. Show that there is an $x_0 \in [a, b]$ with the property that $f(x_0) = x_0$. (This is a special case of the *Brouwer Fixed Point Theorem*.)

1.9. *Continuous functions on* $\Re \times \Re$. (See Problem A.16.) **Definition.** Let $f : D \subseteq \Re \times \Re \to \Re$; then f is *continuous at* $(x, y) \in D$ if

$$\lim f(x_n, y_n) = f(x, y)$$

whenever $\{(x_n, y_n)\}$ is a sequence in D with limit (x, y); f is *continuous on* D if f is continuous at each $(x, y) \in D$.

(a) Show that the following are equivalent:

(i) f is continuous at $(x, y) \in D$.

(ii) For each $\epsilon > 0$ there is a $\delta > 0$ such that

$$(u - x)^2 + (v - y)^2 \le \delta \;\Rightarrow\; |f(u, v) - f(x, y)| \le \epsilon.$$

(iii) For each open set $U \subseteq \Re$ about $f(x, y)$ there is an open set $V \subset \Re \times \Re$ with $f(V \cap D) \subseteq U$.

(b) For each fixed x, show that the function $f_x(y) \equiv f(x, y)$ is a continuous function of the single variable y.

(c) For each fixed y, show that the function $f_y(x) \equiv f(x, y)$ is a continuous function of the single variable x.

(d) *Prove or find a counterexample:* f is continuous $\leftrightarrow$ both f_x and f_y are continuous.

(e) If $C \subset \Re \times \Re$ is compact and if f is continuous on C, then f attains its maximum and minimum on C.

1.10. If f is continuous at x, then show that $|f|$ is also continuous at x.

1.11. Suppose that f is continuous at x and that g is continuous at $y = f(x)$. Show that $h = g \circ f$ is continuous at x.

1.2. THE RIEMANN INTEGRAL. This section is devoted to a rigorous, if rather basic treatment of the integral studied in calculus, the Riemann integral. The basic idea (which has historical roots dating back to Archimedes) is to approximate the areas under curves with rectangular areas. We will first define the integral for step functions and then extend our definitions to include more general functions.

1.2.1. Definitions. If I is a bounded interval, then we define the *length of I* to be the real number

$$\ell(I) = \sup(I) - \inf(I).$$

(Note that each of the intervals $[a, b]$, $(a, b]$, $[a, b)$, and (a, b) has the same length.)

If $[a, b]$ is a closed bounded interval, a *partition* of $[a, b]$ is a finite collection

$$\{I_1, \ldots, I_n\}$$

of disjoint intervals whose union is $[a, b]$. By reindexing if needed, the endpoints of the intervals $\{I_1, \ldots, I_n\}$ may be ordered so that

$$a = t_0 \leq t_1 \leq \cdots \leq t_n = b,$$

where I_j has endpoints t_{j-1} and t_j. Usually, it is unimportant whether these endpoints belong to I_j or to an adjacent interval; because of this we will also refer to the endpoints as a partition.

A *step function s* is a function defined on $[a, b]$ of the form

$$s = \sum_1^n \alpha_j \chi_{I_j},$$

where $\{I_1, \ldots, I_n\}$ is a partition of $[a, b]$ and $\{\alpha_1, \ldots, \alpha_n\}$ are real numbers.

The "area under a step function" s is just the area of a finite number of rectangles, and so we may define the *integral from a to b of s* to be

$$\int_a^b s(x)\, dx = \sum_1^n \alpha_j \ell(I_j).$$

Notice that the algebraic representation for a step function s need not be unique; for example,

$$s_1 = \chi_{[0,1]} \quad \text{and} \quad s_2 = 2\chi_{[0,\frac{1}{2}]} - \chi_{[0,1]} + 2\chi_{(\frac{1}{2},1]}$$

both define the same step function. Since each function has the same graph, the "area under the graph" must be the same. The point is that while the definition of the integral appears to depend upon the representation of s, it is in fact *independent* of this representation and depends only upon the graph of the function. We will make free use of this geometric observation.

Before extending the integral to curvilinear functions, we list some basic properties of the integral for step functions.

1.2.2. Proposition. *Let s_1 and s_2 be step functions defined on the interval* $[a, b]$.

(i) If $s_1 \leq s_2$, then $\int_a^b s_1(x)\, dx \leq \int_a^b s_2(x)\, dx$.

(ii) If α and β are real numbers, then $\alpha s_1 + \beta s_2$ is a step function defined on $[a, b]$ and

$$\int_a^b \alpha s_1(x) + \beta s_2(x)\, dx = \alpha \int_a^b s_1(x)\, dx + \beta \int_a^b s_2(x)\, dx.$$

(iii) Both $s_1 \bigwedge s_2$ and $s_1 \bigvee s_2$ are step functions.

(iv) If $a \leq c \leq b$, then

$$\int_a^b s_1(x)\, dx = \int_a^c s_1(x)\, dx + \int_c^b s_1(x)\, dx.$$

Proof. All four conclusions are verified in essentially the same way; we prove only (ii) leaving the remainder to the reader.

Let $s_1 = \sum_1^n \alpha_j \chi_{I_j}$ and $s_2 = \sum_1^m \beta_k \chi_{J_k}$ be representations for s_1 and s_2. Now $I_j \cap J_k$ is always an interval (possibly empty) and the family of nm intervals $\{I_j \cap J_k\}$ satisfies

$$\chi_{I_j} = \sum_{k=1}^m \chi_{(I_j \cap J_k)} \quad \text{and}$$

$$\chi_{J_k} = \sum_{j=1}^n \chi_{(I_j \cap J_k)}.$$

Moreover, the intervals $\{I_j \cap J_k\}$ are all disjoint and their union is the entire interval $[a, b]$ – i.e., the family $\{I_j \cap J_k\}$ is a partition of $[a, b]$. Consequently,

$$s_1 = \sum_{j=1}^n \sum_{k=1}^m \alpha_j \chi_{(I_j \cap J_k)} \quad \text{and} \quad s_2 = \sum_{k=1}^m \sum_{j=1}^n \beta_k \chi_{(I_j \cap J_k)}$$

are representations of the step functions s_1 and s_2. Thus

$$\alpha s_1 + \beta s_2 = \sum_{j=1}^{n}\sum_{k=1}^{m}(\alpha\alpha_j + \beta\beta_j)\chi_{(I_j \cap J_k)}$$

is a step function (changing the order of summation in s_2).

From this it follows at once that

$$\int_a^b \alpha s_1(x) + \beta s_2(x)\,dx = \alpha\sum_{j=1}^{n}\alpha_j\sum_{k=1}^{m}\ell(I_j \cap J_k) + \beta\sum_{k=1}^{m}\beta_k\sum_{j=1}^{n}\ell(I_j \cap J_k)$$

$$= \alpha\int_a^b s_1(x)\,dx + \beta\int_a^b s_2(x)\,dx,$$

completing the proof of *(ii)*.

Next we extend this simple notion of "integrable function" to include a much broader class of functions. The basic idea behind the next approximation is founded on Archimedes' method for computing the volume of a sphere.

1.2.3. Definition.   Let f be a bounded function defined on a closed bounded interval $[a, b]$. Suppose that for every $\epsilon > 0$ there are step functions s_1 and s_2 satisfying

(i) $s_1(x) \le f(x) \le s_2(x)$ for all $x \in [a, b]$; and

(ii) $\int_a^b s_2(x) - s_1(x)\,dx \le \epsilon$.

Then we say that f is *Riemann integrable on* $[a, b]$ and define the Riemann integral of f to be

$$\int_a^b f(x)\,dx = \inf\left\{\int_a^b s_2(x)\,dx : s_2 \ge f \text{ and } s_2 \text{ is a step function}\right\}.$$

[Note also that

$$\int_a^b f(x)\,dx = \sup\left\{\int_a^b s_1(x)\,dx : s_1 \le f \text{ and } s_1 \text{ is a step function}\right\}$$

since, by (ii), the supremum and infimum must agree.]

The functions s_1 and s_2 correspond to the "lower" and "upper" Darboux sums considered in calculus. The definition greatly increases the number of functions which are integrable — including, for example, the continuous functions.

1.2.4 Theorem.   *Let f be a continuous function defined on the closed bounded interval $[a, b]$. Then f is Riemann integrable on $[a, b]$.*

Proof.   The idea of the proof is to show that every continuous function can be uniformly approximated from above and from below by step functions; it is then routine to verify 1.2.3(ii).

Fix $\epsilon > 0$; by Corollary 1.1.8 there is a $\delta > 0$ so that $\omega(f; J) \leq \epsilon/(b-a)$ whenever $J \subseteq [a, b]$ and $\ell(J) \leq \delta$. Choose a partition

$$a = t_0 < t_1 < \cdots < t_n = b$$

of $[a, b]$ so that if $I_k = (t_{k-1}, t_k]$, then $\ell(I_k) \leq \delta$ for each k. Set

$$M_k = \sup\{f(x) : x \in I_k\} \quad \text{and}$$
$$m_k = \inf\{f(x) : x \in I_k\},$$

so that $M_k - m_k = \omega(f; I_k) \leq \epsilon/(b-a)$. Then if

$$s_1 = \sum_1^n m_k \chi_{I_k} \quad \text{and}$$

$$s_2 = \sum_1^n M_k \chi_{I_k},$$

it follows at once that

$$s_1 \leq f \leq s_2 \quad \text{and}$$
$$\int_a^b s_2(x) - s_1(x)\, dx = \sum_1^n (M_k - m_k)\ell(I_k) \leq \epsilon.$$

Since we approximate the Riemann integral for general functions with the Riemann integral for step functions, Proposition 1.2.2 extends readily to include all Riemann integrable functions.

1.2.5. Proposition. *Let f and g be Riemann integrable functions defined on the interval $[a, b]$.*

(i) If $f \leq g$, then $\int_a^b f\, dx \leq \int_a^b g(x)\, dx$.

(ii) If α and β are real numbers, then $\alpha f + \beta g$ is Riemann integrable and

$$\int_a^b \alpha f(x) + \beta g(x)\, dx = \alpha \int_a^b f(x)\, dx + \beta \int_a^b g(x)\, dx.$$

(iii) Both $f \wedge g$ and $f \vee g$ are Riemann integrable.

(iv) If $a \leq c \leq b$, then

$$\int_a^c f(x)\, dx + \int_c^b f(x)\, dx = \int_a^b f(x)\, dx.$$

Proof. Once again, we prove only (ii), leaving the remaining similar parts to the reader. We begin with the case $\alpha, \beta > 0$.

Fix $\epsilon > 0$ and choose step functions s_1, s_2 and σ_1, σ_2 so that

$$s_1 \leq f \leq s_2, \quad \sigma_1 \leq g \leq \sigma_2,$$

$$\int_a^b s_2(x) - s_1(x)\, dx \leq \frac{\epsilon}{2\alpha} \quad \text{and} \quad \int_a^b \sigma_2(x) - \sigma_1(x)\, dx \leq \frac{\epsilon}{2\beta}.$$

Thus

$$\alpha s_1 + \beta \sigma_1 \leq \alpha f + \beta g \leq \alpha s_2 + \beta \sigma_2$$

and

$$\int_a^b \alpha s_2(x) + \beta \sigma_2(x) - (\alpha s_1(x) + \beta \sigma_1(x))\, dx$$

$$\leq \alpha \int_a^b s_2(x) - s_1(x)\, dx + \beta \int_a^b \sigma_2(x) - \sigma_1(x)\, dx$$

$$\leq \epsilon,$$

showing that $\alpha f + \beta g$ is Riemann integrable.

Moreover,

$$\alpha \int_a^b f(x)\, dx + \beta \int_a^b g(x)\, dx \leq \alpha \int_a^b s_2(x)\, dx + \beta \int_a^b \sigma_2(x)\, dx$$

$$\leq \alpha \left(\int_a^b s_1(x)\, dx + \frac{\epsilon}{2\alpha} \right) + \cdots$$

$$\cdots + \beta \left(\int_a^b \sigma_1(x)\, dx + \frac{\epsilon}{2\beta} \right)$$

$$= \int_a^b \alpha s_1(x) + \beta \sigma_1(x)\, dx + \epsilon$$

$$\leq \int_a^b \alpha f(x) + \beta g(x)\, dx + \epsilon.$$

Since ϵ was arbitrary, this shows that

$$\alpha \int_a^b f(x)\, dx + \beta \int_a^b g(x)\, dx \leq \int_a^b \alpha f(x) + \beta g(x)\, dx.$$

A symmetric argument gives the reverse inequality, and thus *(ii)* is proved in the case that $\alpha, \beta \geq 0$.

For the general case, we first observe that if f and g are Riemann integrable, then so are $-f$ and $-g$. Thus, writing

$$\alpha f + \beta g = |\alpha|\, \mathrm{sgn}(\alpha) f + |\beta|\, \mathrm{sgn}(\beta) g$$

gives the conclusion.

Because continuous functions defined on compact intervals have so many nice properties, it is often useful to know that continuous functions are integrable. Indeed, we can bypass step functions altogether with the following theorem.

1.2.6. Theorem. *Let f be a bounded function defined on the interval $[a, b]$. Then f is Riemann integrable if and only if for each $\epsilon > 0$ there are continuous functions g and h satisfying:*

(i) *$g(x) \leq f(x) \leq h(x)$ for all $a \leq x \leq b$; and*
(ii)

$$\int_a^b h(x) - g(x)\, dx \leq \epsilon.$$

Proof. For sufficiency, fix $\epsilon > 0$ and select continuous functions g and h satisfying (i) and

$$\int_a^b h(x) - g(x)\, dx \leq \frac{\epsilon}{3}.$$

Next, select step functions s_1, s_2 and σ_1, σ_2 so that

$$s_1 \leq g \leq s_2, \quad \sigma_1 \leq h \leq \sigma_2,$$

$$\int_a^b s_2(x) - s_1(x)\, dx \leq \frac{\epsilon}{3} \quad \text{and} \quad \int_a^b \sigma_2(x) - \sigma_1(x)\, dx \leq \frac{\epsilon}{3}.$$

Thus $s_1 \leq f \leq \sigma_2$ and

$$\int_a^b \sigma_2(x) - s_1(x)\, dx = \int_a^b \sigma_2(x) - h(x)\, dx + \cdots$$

$$\cdots + \int_a^b h(x) - g(x)\, dx + \int_a^b g(x) - s_1(x)\, dx$$

$$\leq \int_a^b \sigma_2(x) - \sigma_1(x)\, dx + \frac{\epsilon}{3} + \cdots$$

$$\cdots + \int_a^b s_2(x) - s_1(x)\, dx + \frac{\epsilon}{3}$$

$$\leq \epsilon,$$

showing that f is Riemann integrable.

For the converse, we first prove the following lemma.

1.2.7. Lemma. *Let $[a, b]$ and $[c, d]$ be bounded intervals with $[a, b] \subseteq [c, d]$. Then for each $\epsilon > 0$ there are continuous functions g and h with the following properties:*

(i)   *$g \leq \chi_{[a,b]} \leq h$;   and*

(ii)

$$\int_c^d h(x) - g(x)\, dx \le \epsilon.$$

Proof. We can approximate $\chi_{[a,b]}$ with piecewise linear continuous functions. For example, if $a < b$ and $\delta > 0$ is sufficiently small, then the functions

$$g(x) = \begin{cases} 0 & c \le x \le a \\ \delta^{-1}(x-a) & a < x \le a + \delta \\ 1 & a + \delta < x \le b - \delta \\ \delta^{-1}(b-x) & b - \delta < x \le b \\ 0 & b < x \le d \end{cases}$$

and

$$h(x) = \begin{cases} 0 & c \le x \le a - \delta \\ \delta^{-1}(x - a + \delta) & a - \delta < x \le a \\ 1 & a < x \le b \\ \delta^{-1}(b - x + \delta) & b < x \le b + \delta \\ 0 & b + \delta < x \le d \end{cases}$$

will do. (We must take care that $a + \delta < b - \delta$ in order for g to be well-defined; if $a = b$, we may take $g \equiv 0$.) Notice that

$$\int_c^d h(x) - g(x)\, dx = 2\delta.$$

Proof of 1.2.6 completed. If f is Riemann integrable on $[a, b]$, we may select step functions so that $s_1 \le f \le s_2$ and

$$\int_a^b s_2(x) - s_1(x)\, dx \le \frac{\epsilon}{3}.$$

By applying Lemma 1.2.7 finitely many times to the components of s_1 and s_2 we may select continuous functions $g_1, g_2, h_1,$ and h_2 so that

$$g_1 \le s_1 \le h_1 \quad \text{and} \quad g_2 \le s_2 \le h_2$$

$$\int_a^b h_1(x) - g_1(x)\, dx \le \frac{\epsilon}{3} \quad \text{and} \quad \int_a^b h_2(x) - g_2(x)\, dx \le \frac{\epsilon}{3}.$$

Then

$$\int_a^b h_2(x) - g_1(x)\, dx = \int_a^b h_2(x) - s_2(x)\, dx + \cdots$$

$$\cdots + \int_a^b s_2(x) - s_1(x)\, dx + \int_a^b s_1(x) - g_1(x)\, dx$$

$$\leq \int_a^b h_2(x) - g_2(x)\, dx + \frac{\epsilon}{3} + \cdots$$

$$\cdots + \int_a^b h_1(x) - g_1(x)\, dx$$

$$\leq \epsilon,$$

as desired.

If $s = \sum_1^n \alpha_j \chi_{I_j}$ is a step function, then by the triangle inequality

$$\left| \int_a^b s(x)\, dx \right| = \left| \sum_1^n \alpha_j \ell(I_j) \right|$$

$$\leq \sum_1^n |\alpha_j| \ell(I_j)$$

$$= \int_a^b |s(x)|\, dx.$$

The next result extends this elementary observation to arbitrary Riemann-integrable functions.

1.2.8. Theorem. *If f is a Riemann-integrable function defined on $[a, b]$, then so is $|f|$; moreover,*

$$\left| \int_a^b f(x)\, dx \right| \leq \int_a^b |f(x)|\, dx.$$

Proof. By 1.2.5(iii), both of the functions

$$f^+ \equiv f \vee 0 \quad \text{and} \quad f^- \equiv (-f) \vee 0$$

are Riemann integrable. Since $|f| = f^+ + f^-$, this verifies that $|f|$ is Riemann integrable. For the other conclusion

$$\left| \int_a^b f(x)\,dx \right| = \left| \int_a^b f^+(x) - f^-(x)\,dx \right|$$

$$\leq \left| \int_a^b f^+(x)\,dx - \int_a^b f^-(x)\,dx \right|$$

$$\leq \left| \int_a^b f^+(x)\,dx \right| + \left| \int_a^b f^-(x)\,dx \right|$$

$$= \int_a^b f^+(x)\,dx + \int_a^b f^-(x)\,dx$$

$$= \int_a^b |f(x)|\,dx,$$

as desired.

(The functions f^+ and f^- defined in the proof above will be used throughout this text.)

As we have already observed, the value of a step function can be changed at a finite number of points (the endpoints of the partition) without changing the value of the integral; this is because the "rectangles" $(t_i \times \alpha_i)$ have height but no width (and hence zero area). It is not difficult to see that this observation applies to an arbitrary Riemann integrable function; in particular, if f and g are both Riemann-integrable and agree everywhere except at a finite number of points, then

$$\int_a^b f(x)\,dx = \int_a^b g(x)\,dx. \qquad (*)$$

Precisely how much f and g can disagree and still maintain equality in (*) is unclear; the answer is contained in the next definition.

1.2.9. Definition. A set E of real numbers has *measure zero* if for each $\epsilon > 0$, there is a family $\{I_k\}$ of intervals with the property that

$$E \subseteq \cup_1^\infty I_k \quad \text{and} \quad \sum_1^\infty \ell(I_k) \leq \epsilon.$$

Intuitively, sets of measure zero have arbitrarily small "length"; we would therefore expect that "rectangles" constructed over such sets would have zero area. Note, however, that the characteristic function of a set of measure zero need *not* be Riemann integrable (see below). We can show, however, that if f and g are Riemann integrable and if $f = g$ everywhere except on a set of measure zero, then

$$\int_a^b |f(x) - g(x)|\,dx = 0;$$

in particular, the Riemann integral cannot distinguish between two such functions.

We observe that the covering intervals $\{I_k\}$ in Definition 1.2.9 can be taken to be both open and disjoint (Problem 1.12); we will use this observation both later in this section and in Section 2.1.

1.2.10. Examples

(i) Every finite and every countably infinite set has measure zero.

(ii) *(The Cantor Set.)* From the unit interval $[0,1]$, first delete the middle open third $(\frac{1}{3}, \frac{2}{3})$. From the remaining two intervals delete the open middle thirds; from the remaining four intervals, again delete the open middle thirds. Continuing this process, we delete intervals having aggregate length

$$\sum_1^\infty 2^{n-1}3^{-n} = 1.$$

The remaining points, which should intuitively have "zero length" are called the *Cantor set*. This set does indeed have measure zero [see Problem 1.16(f)) and is in addition uncountable [problem 1.16(c)]. The Cantor set consists of all those numbers in $[0,1]$ having a base 3 expansion containing no 1's [Problem 1.16(b)].

As this example suggests, sets of measure zero can have a very complex structure. These sets play a critical role in both integration and differentiation; in particular the Cantor set is a source of many examples and counterexamples in both theories. The student should study this example and the related problems very carefully.

Because sets of measure zero are, intuitively, small, we think of their complements as being large. The phrase *almost everywhere* is commonly used as a synonym for "everywhere except on a set of measure zero." Thus, for example, saying "$f = g$ almost everywhere" means:

"There is a set E of measure zero such that $f(x) = g(x)$ for all $x \notin E$."

This shorthand terminology is virtually universal and will be used throughout this text; "almost everywhere" is generally abbreviated "a.e." Sets of measure zero are also sometimes said to have "content zero" or to be "null sets."

Notice that if $E = \chi_{(\mathcal{Q} \cap [0,1])}$, then E has measure zero and thus $\chi_E = 0$ a.e. However, χ_E is *not* Riemann integrable: If g and h are continuous functions satisfying $g \le \chi_E \le h$, then then $g \le 0$ everywhere and $h \ge 1$ everywhere on $[0,1]$ (since $g \le 0$ and $h \ge 1$ on a dense subset of $[0,1]$). Consequently,

$$\int_0^1 h(x) - g(x)\, dx \ge 1,$$

showing that χ_E is not Riemann integrable.

However, we can easily show that if $f = 0$ a.e. *and* if f is Riemann integrable, then $\int_a^b f(x)\, dx = 0$. As a first step we establish the following:

1.2.11. Proposition.   *If f is Riemann integrable on $[a, b]$ and if $f \geq 0$ a.e. on $[a, b]$, then*

$$\int_a^b f(x)\, dx \geq 0.$$

Proof.   If g is any continuous function with $g \geq f$, then our assumptions imply that $g \geq 0$ everywhere. To see this, suppose, for some x, that $g(x) \equiv \alpha < 0$. Since g is continuous, there is an open interval J about x with

$$\omega(g; J) \leq \left|\frac{\alpha}{2}\right|;$$

in particular, if $t \in J$, then $g(t) \leq \alpha/2 < 0$. Since $f \leq g$, it follows that $f(t) \leq \alpha/2 < 0$ for all $t \in J$, which contradicts $f \geq 0$ a.e.

Now, by 1.2.6,

$$\int_a^b f(x)\, dx = \inf\left\{\int_a^b g(x)\, dx : g \geq f \text{ and } g \text{ is continuous}\right\} \geq 0,$$

completing the proof.

1.2.12. Theorem.   *Let f and g be Riemann-integrable functions defined on the interval $[a, b]$. If $f = g$ a.e. on $[a, b]$, then*

$$\int_a^b |f(x) - g(x)|\, dx = 0.$$

Proof.   Since $|f - g| \geq 0$, it follows that

$$\int_a^b |f(x) - g(x)|\, dx \geq 0.$$

Since $-|f - g| \geq 0$ a.e., Lemma 1.2.11 implies that

$$\int_a^b |f(x) - g(x)|\, dx \leq 0,$$

giving the result.

When combined with Theorem 1.2.8, the above implies that $\int_a^b f(x)\, dx = \int_a^b g(x)\, dx$ whenever $f = g$ a.e. and *both* f and g are Riemann-integrable.

It would be possible at this time to broaden our notion of "integrability" to include functions which agree almost everywhere with a Riemann integrable function f. If f is Riemann integrable and if $f = g$ a.e., we could take the integral of g to be

$$\int_a^b g(x)\,dx = \int_a^b f(x)\,dx$$

whether or not g meets the conditions of Definition 1.2.3. Observe that this definition would be independent of the function f in the following sense: If there were another Riemann integrable function $\hat{f}$ such that $\hat{f} = g$ a.e., then by 1.2.12, $\int_a^b f(x)\,dx = \int_a^b \hat{f}(x)\,dx$. This would extend the integral, in an entirely natural manner, to include the characteristic function of the rationals. A somewhat more involved extension, based on the Cauchy principal value, is the idea behind the Lebesgue integral.

A countable combination of null sets must again be a null set; this fundamental observation will be used throughout the remainder of this text:

1.2.13 Lemma. *Let $\{E_n\}$ be a countable family of sets each having measure zero and set $E = \cup_1^\infty E_n$. Then E has measure zero.*

Proof. Fix $\epsilon > 0$. It suffices to exhibit a countable collection of intervals whose union is E and whose lengths sum to at most ϵ.

For each n select a sequence of intervals $\{I_{n,k}\}$ so that

$$E_n \subseteq \cup_{k=1}^\infty I_{n,k} \quad \text{and} \quad \sum_{k=1}^\infty \ell(I_{n,k}) \le \epsilon 2^{-n}.$$

(We can do this since each E_n has measure zero.) Then

$$E = \cup_{n=1}^\infty E_n \subseteq \cup_{n=1}^\infty \cup_{k=1}^\infty I_{n,k}$$

and

$$\sum_{n=1}^\infty \sum_{k=1}^\infty \ell(I_{n,k}) \le \sum_{n=1}^\infty \epsilon 2^{-n} = \epsilon,$$

so that the countable family $\{I_{n,k}\}$ has the requisite properties.

1.2.14. Theorem. *Let f be a bounded function defined on the interval $[a, b]$. Then f is Riemann integrable if and only if f is continuous almost everywhere.*

This theorem characterizes precisely Riemann-integrable functions. Notice that the theorem says that the set $\{x \in [a, b] : \omega(f; x) > 0\}$ must have measure zero; this is *not* the same as saying that f agrees almost everywhere with a continuous function (see Problem 1.21).

Proof of 1.2.14. We first show sufficiency.

Fix $\epsilon > 0$ and choose N so large that $N \geq 4(b-a)/\epsilon$; set

$$E_N = \left\{ x \in [a,b] : \omega(f;x) \geq \frac{1}{N} \right\},$$

so that E_N has measure zero.

Choose m and M so that $m \leq f(x) \leq M$ for all $x \in [a,b]$ and select a system $\{I_k\}$ of open intervals (applying Problem 1.20) so that $E_N \subseteq \cup_1^\infty I_k$ and

$$\sum_1^\infty \ell(I_k) \leq \frac{\epsilon}{2(M-m)}.$$

Finally, for each $x \in [a,b] \backslash E_N$ we may select an open interval J_x about x so that $\omega(f;J_x) \leq 2/N$.

Now the intervals $\{I_k\} \cup \{J_x\}$ are an open cover of $[a,b]$ and so admit a finite subcover

$$J_{x_1}, \ldots, J_{x_p}, I_1, \ldots, I_q.$$

By shrinking the foregoing intervals if necessary, we may assume that all of the intervals are disjoint and are contained in $[a,b]$.

Next, for $k = 1, \ldots, p$, set

$$M_k = \sup\{f(x) : x \in J_{x_k}\} \quad \text{and}$$
$$m_k = \inf\{f(x) : x \in J_{x_k}\}$$

so that $(M_k - m_k) \leq \omega(f;J_k) \leq 2/N$. Finally, set

$$s_1 = \sum_{k=1}^p m_k \chi_{J_{x_k}} + \sum_{j=1}^q m \chi_{I_j} \quad \text{and}$$

$$s_2 = \sum_{k=1}^p M_k \chi_{J_{x_k}} + \sum_{j=1}^q M \chi_{I_j}.$$

Clearly, $s_1 \leq f \leq s_2$. Moreover,

$$\int_a^b s_2(x) - s_1(x)\, dx = \sum_{k=1}^p (M_k - m_k)\ell(J_{x_k}) + \sum_{j=1}^q (M-m)\ell(I_j)$$

$$\leq \frac{1}{2N}(b-a) + (M-m)\sum_{j=1}^q \ell(I_j)$$

$$\leq \epsilon.$$

Since ϵ was arbitrary, this shows that f is Riemann integrable.

The proof of the converse is slightly more involved. Set

$$E = \{x \in [a,b] : f \text{ is not continuous at } x\}$$

and for $n = 1, 2, \ldots$ set

$$E_n = \left\{ x \in [a, b] : \omega(f; x) \geq \frac{1}{n} \right\}.$$

Since $E = \cup E_n$, it suffices to show that each E_n has measure zero.

Fix $\epsilon > 0$ and n and select continuous functions g and h so that $g \leq f \leq h$ and

$$\int_a^b h(x) - g(x) \, dx \leq \frac{\epsilon}{4n}.$$

Since both g and h are uniformly continuous on $[a, b]$, we may select $\delta > 0$ so that if $|s - t| \leq \delta$ and $s, t \in [a, b]$, then

$$|h(s) - h(t)| \leq \frac{1}{8n} \quad \text{and} \quad |g(s) - g(t)| \leq \frac{1}{8n}.$$

Next, for each $x \in E_n$, select an open interval I_x about x with $\ell(I_x) \leq \delta$ and with $\omega(f; I_x) \geq 1/2n$. Notice that if $t \in I_x$, then

$$\frac{1}{2n} \leq \omega(f; I_x) \leq \sup\{h(s) : s \in I_x\} - \inf\{g(s) : s \in I_x\}$$

$$\leq h(t) - g(t) + \frac{1}{4n}.$$

Since E_n is compact (Proposition 1.1.3), we may select intervals $I_{x_1}, \ldots, I_{x_p}$ which cover E_n. By shrinking some of the intervals if necessary, we may further assume that the intervals are disjoint and that each is contained in $[a, b]$.

Finally, we compute

$$\sum_{j=1}^p \ell(I_{x_j}) = \sum_{j=1}^p \int_{I_{x_j}} 1 \, dt$$

$$\leq \sum_{j=1}^p 2n \left(\int_{I_{x_j}} h(t) - g(t) + \frac{1}{4n} \, dt \right)$$

$$= 2n \sum_{j=1}^p \int_{I_{x_j}} h(t) - g(t) \, dt + \frac{1}{2} \sum_{j=1}^p \ell(I_{x_j}),$$

implying that

$$\frac{1}{2} \sum_{j=1}^p \ell(I_{x_j}) \leq 2n \sum_{j=1}^p \int_{I_{x_j}} h(t) - g(t) \, dt.$$

Since the intervals $\{I_{x_j}\}$ are disjoint and contained in $[a, b]$, this implies that

$$\frac{1}{2} \sum_{j=1}^p \ell(I_{x_j}) \leq 2n \int_a^b h(t) - g(t) \, dt \leq \frac{\epsilon}{2}$$

by choice of h and g. Thus $E_n \subseteq \bigcup_{j=1}^{p} I_{x_j}$ and $\sum_1^p \ell(I_{x_j}) \leq \epsilon$, showing that E_n has measure zero, as desired.

There are a number of deficiencies of the Riemann integral. First, such functions as the characteristic function of the rationals which agree almost everywhere with Riemann-integrable functions can fail to be Riemann integrable. Second, the integral is defined only for bounded functions, so that a function such as

$$f(x) = \begin{cases} x^{-\frac{1}{2}} & 0 < x \leq 1 \\ 0 & \text{elsewhere} \end{cases}$$

fails to be Riemann integrable on $[0, 1]$. In calculus, the latter problem is handled by the Cauchy Principal Value, using the monotone sequence

$$f_n(x) = \begin{cases} x^{-\frac{1}{2}} & \frac{1}{n} < x \leq 1 \\ 0 & \text{elsewhere} \end{cases}$$

to approximate the integral of f.

This last example is indicative of a more general problem dealing with limits; it is possible to have a monotone sequence $\{f_n\}$ of Riemann-integrable functions which converge everywhere to a bounded function f with the property that

$$\lim \int_a^b f_n(x)\, dx$$

exists and is finite but *the limiting function f is not Riemann integrable* [see Problem 1.19]. We would like to be able formally to define

$$\int_a^b f(x)\, dx = \lim \int_a^b f_n(x)\, dx.$$

However, this definition *depends upon the sequence* $\{f_n\}$. In particular, it *might* happen that *another* monotone sequence $\{g_n\}$ could be found with

$$\lim g_n = f \quad \text{everywhere}$$

$$\lim \int_a^b g_n(x)\, dx \quad \text{exists and is finite}$$

$$but$$

$$\lim \int_a^b g_n(x)\, dx \neq \lim \int_a^b f_n(x)\, dx.$$

There would then be no basis for deciding, *a priori*, which sequence should be used to define $\int_a^b f(x)\, dx$. Circumventing this problem is the first major task of the next chapter.

PROBLEMS

1.12.   Prove Proposition 1.2.2(i), (iii), and (iv).

1.13. Prove Proposition 1.2.5(i), (iii), and (iv).

1.14. Let g be a Riemann-integrable function defined on the interval $[a,b]$. If $\alpha \neq 0$, then verify

$$\int_{\alpha a + \beta}^{\alpha b + \beta} g(x)\, dx = \alpha \int_a^b g(\alpha x + \beta)\, dx.$$

1.15. Let $S \subseteq [a,b]$ and suppose that for every $\epsilon > 0$, there is a continuous function g with $g \geq \chi_S$ and

$$\int_a^b g(x)\, dx \leq \epsilon.$$

Show that S has measure zero. Is the converse true?

1.16. *The Cantor set.* Recall that every number $x \in [a,b]$ has a *ternary expansion* of the form

$$x = \sum_{k=1}^{\infty} x_k 3^{-k},$$

where $x_k \in \{0,1,2\}$ (see Problem A.8.).

(a) Show that the ternary expansion for x is unique unless $x = a3^{-m}, 0 < a < 3^m$ and 3 does not divide a. If $x = a3^{-m}, 0 < a < 3^m$, and 3 does not divide a, show that x has both a finite and an infinite expansion.

(b) Suppose that $x = a3^{-m}, 0 < a < 3^m$, and that 3 does not divide a. Show that the finite expansion for x is of the form

$$x = \sum_{k=1}^{m} x_k 3^{-k}, \tag{1}$$

where $x_m = 1$ if $a = 1 \bmod 3$ and $x_m = 2$ if $a = 2 \bmod 3$. If $x_m = 2$, show that the infinite for expansion for x is in the form

$$x = \sum_{k=1}^{m-1} x_k 3^{-k} + 0 \cdot 3^{-m} + \sum_{k=m+1}^{\infty} 2 \cdot 3^{-k}. \tag{2}$$

(c) Assign to each $x \in [0,1]$ a *unique* ternary expansion by using (1) when x has two expansions and $x_m = 1$ and by using (2) when x has two expansions and $x_m = 2$. Set

$$P_n = \left\{ x \in [0,1] : \{x_1, \ldots, x_n\} \subseteq \{0,2\} \quad \text{where} \quad x = \sum_{k=1}^{\infty} x_k 3^{-k} \right\}.$$

Show that

$$P_1 = [0, \tfrac{1}{3}] \cup [\tfrac{2}{3}, 1] \quad \text{and}$$

$$P_2 = [0, \tfrac{1}{9}] \cup [\tfrac{2}{9}, \tfrac{1}{3}] \cup [\tfrac{2}{3}, \tfrac{7}{9}] \cup [\tfrac{8}{9}, 1].$$

(Note that $\tfrac{1}{3} = \sum_{k=2}^{\infty} 2 \cdot 3^{-k}$.) In particular, if $C = \bigcap P_n$, then C is the set described in 1.2.10(iii).

(d) Show that C is uncountable.

(e) Show that

$$\chi_{P_n}(t) = \begin{cases} \chi_{P_{n-1}}(3t) & 0 \le t \le \frac{1}{3} \\ 0 & \frac{1}{3} < t < \frac{2}{3} \\ \chi_{P_{n-1}}(3t-2) & \frac{2}{3} \le t \le 1. \end{cases}$$

(f) Show that

$$\int_0^1 \chi_{P_n}(t)\, dt = \left(\frac{2}{3}\right)^n.$$

(g) Show that the Cantor set C has measure zero.

(h) Show that χ_C is Riemann integrable and that $\int_0^1 \chi_C(t)\, dt = 0$.

1.17. *Lebesgue's Singular Function.*

(a) Show that every x in the Cantor set C can be uniquely represented as

$$x = \sum_{k=1}^{\infty} 2x_k 3^{-k},$$

where $x_k \in \{0, 1\}$.

(b) Let x in the Cantor set be represented as in (a); Define a function $\hat{\psi}$ on the Cantor set by

$$\hat{\psi}\left(\sum_{k=1}^{\infty} 2x_k 3^{-k}\right) = \sum_{k=1}^{\infty} x_k 2^{-k}.$$

Show that $\hat{\psi}(C) = [0, 1]$ and that $\hat{\psi}$ is nondecreasing.

(c) Extend the function $\hat{\psi}$ defined in (b) to all of $[0, 1]$ by setting

$$\psi(t) = \sup\left\{\hat{\psi}(x) : x \in C \text{ and } x < t\right\}.$$

Show that the resulting function ψ is continuous and nondecreasing.

1.18. If p is an odd positive integer, we can construct the *Cantor middle p^{th} set*, C_p, by writing

$$x = \sum_{n=1}^{\infty} x_n p^{-n},$$

where $x_n \in \{0, \ldots, p-1\}$ and following the construction in 1.16. Show that C_p has measure zero if and only if $p = 3$.

1.19. Find a bounded sequence $\{f_n\}$ of Riemann-integrable functions with the following properties:

$\{f_n\}$ converges everywhere on $[0, 1]$ to a function f;

$\lim \int_0^1 f_n(t)\, dt$ exists and is finite;

the limiting function f fails to be Riemann integrable.

1.20. Let $E \subseteq \Re$ be a set of measure zero. Show that for each $\epsilon > 0$ there is a family $\{I_n\}$ of *disjoint, open* intervals with

$$E \subseteq \cup_{n=1}^{\infty} I_n \quad \text{and}$$

$$\sum_{n=1}^{\infty} \ell(I_n) \le \epsilon.$$

1.21. Find a function f which is *not* Riemann integrable but which agrees almost everywhere with a continuous function.

Chapter 2

The Lebesgue Integral

2.1. BASIC PROPERTIES. Shortly after Riemann introduced his theory of integration, it became evident that this theory did not adequately describe the behavior of integrals and limits. As we have already seen, it is possible to have a sequence $\{g_n\}$ of Riemann integrable functions such that

$$\lim g_n(x) \; = \; g(x) \quad \text{everywhere}$$

$$\lim \int_a^b g_n(x)\,dx \quad \text{exists and is finite}$$

but the limiting function is not Riemann-integrable. This is a highly undesirable pathology; much effort during the latter half of the nineteenth century was devoted to extending the Riemann integral to a new integral which would be more regular with regard to the limiting process. There were a number of approaches, notably by Cantor, Caratheodory, Borel, and Baire, but that of Lebesgue turned out to be the most successful. It is Lebesgue's theory which is now the most widely used by mathematicians and to which we now turn. In all of the theorems, the idea is to increase the number of "integrable" functions in order to avoid pathologies of the type noted above.

Lebesgue first defined the "measure of a set" — an abstraction of "length of an interval" — and then defined the integral based on measurable dissections. Subsequently, F. Riesz observed that much of the theory could be developed without reference to measure theory. The latter approach is both more elementary and more direct, so we are following Riesz's approach. A slightly more abstract version of the theory we present, referred to as the "Daniell integral," is discussed in Section 7.5. Since measure theory is an important topic on its own, a major portion of this chapter is devoted to a discussion of the properties of measurable sets.

2.1.1. Definition. A continuous function f defined on the real numbers is said to have *compact support* if there is an interval $[a, b]$ such that $f(x) = 0$ for all $x \notin [a, b]$. The collection of all real-valued continuous functions having compact support is denoted by $\mathcal{C}_C(\mathfrak{R})$.

2.1.2. Definition. Denote by $\mathcal{L}^+(\mathfrak{R})$ the class of all nonnegative extended real-valued functions f for which there is a nondecreasing sequence $\{f_n\} \subset \mathcal{C}_C(\mathfrak{R})$ satisfying

$$\lim_{n \to \infty} f_n(x) \; = \; f(x) \quad \text{almost everywhere.}$$

We will call $\{f_n\}$ an *approximating sequence* for f.

It is evident that if $f \in \mathcal{L}^+(\Re)$ and $\{f_n\}$ is an approximating sequence for f, then the Riemann Integral $\int_\Re f_n(x)\,dx$ exists for each n and $\{\int_\Re f_n(x)\,dx\}$ is a nondecreasing sequence of real numbers. Following the analogy of the Cauchy Principal Value for improper integrals, we would like to define the integral of f to be $\lim_{n\to\infty} \int_\Re f_n(x)\,dx$. However, such a definition could possibly depend upon the selection of the approximating sequence $\{f_n\} \subset \mathcal{C}_C(\Re)$. Our first goal is to show that this limit is in fact independent of the particular sequence selected. To this end, we begin with the following fundamental lemma.

2.1.3. Lemma. *Let $\{\varphi_n\}$ be a nonincreasing sequence of nonnegative functions in $\mathcal{C}_C(\Re)$ and suppose that $\lim \varphi_n(x) = 0$ almost everywhere. Then $\lim \int_\Re \varphi_n(x)\,dx = 0$.*

Proof. If $\lim \varphi_n(x) = 0$ for *all* x, then $\{\varphi_n\}$ converges *uniformly* to zero by Dini's Theorem, and the conclusion of the lemma is immediate. The idea of the proof is to apply Dini's Theorem to a sequence $\{\psi_n\}$ which "almost" agrees with $\{\varphi_n\}$.

Select an interval $[a, b]$ so that $\varphi_1(x) = 0$ for all $x \notin [a, b]$; set

$$M = \sup\{\varphi_1(x) : a \le x \le b\}.$$

Since the sequence $\{\varphi_n\}$ is nonincreasing, it follows that

$$\varphi_n(x) \le M \quad \text{for all} \quad n, x;$$

$$\varphi_n(x) = 0 \quad \text{if} \quad x \notin [a, b];$$

$$\lim \varphi_n(x) \quad \text{exists} \quad \text{for all} \quad x; \quad \text{and}$$

$$\lim \int_\Re \varphi_n(x)\,dx \quad \text{exists}.$$

Set $E = \{x : \lim \varphi_n(x) \ne 0\}$, and fix $\epsilon > 0$. Select a sequence $\{(a_i, b_i)\}$ of disjoint open intervals satisfying:
  (1) $E \subset \cup_1^\infty (a_i, b_i)$; and
  (2) $\sum_1^\infty (b_i - a_i) \le \frac{\epsilon}{2M}$
(see Problem 1.20). Set $G_n = \cup_1^n (a_i, b_i)$ and $G_\infty = \cup_1^\infty (a_i, b_i)$. Next we construct a nondecreasing sequence $\{\psi_n\}$ of continuous functions satisfying:
  (3) $0 \le \psi_n(x) \le \chi_{G_n}(x)$ for all x;
  (4) $\lim \psi_n(x) = \chi_{G_\infty}(x)$ for all x; and
  (5) $\int_\Re \psi_n(x)\,dx \le \frac{\epsilon}{2M}$.
for each n.

To construct ψ_n, we use piecewise linear approximations of the characteristic functions of the intervals (a_i, b_i). For fixed integers n and i, define $g_{n,i}$ as follows:

$$g_{n,i} \equiv 0 \quad \text{if} \quad b_i - a_i < 2/n; \quad \text{otherwise}$$

$$g_{n,i} = \begin{cases} 0 & x \le a_i \\ n(x - a_i) & a_i < x \le a_i + 1/n \\ 1 & a_i + 1/n < x \le b_i - 1/n \\ n(b_i - x) & b_i - 1/n < x \le b_i \\ 0 & b_i < x. \end{cases}$$

(Draw a sketch of $g_{n,i}$; constructions of this type are often very useful.) Now if

$$\psi_n(x) = \sum_{i=1}^{n} g_{n,i}(x),$$

then it is routine to check that the sequence $\{\psi_n\}$ has the requisite properties (3)—(5).

The next step of the proof is the following

Claim: *If n is sufficiently large, then*

$$\sup\{\varphi_n(x) - M\psi_n(x) : a \le x \le b\} \le \frac{\epsilon}{2(b - a)}.$$

[The point of the claim is that for n large enough, $\varphi_n(x) - M\psi_n(x)$ can be made *uniformly* small. Certainly, this claim is true for each fixed x; that it is true uniformly in x follows from Dini's Theorem.]

To see the Claim, set

$$h_n(x) = \max\{\varphi_n(x) - M\psi_n(x); 0\}.$$

Now if $x \in [a, b] \backslash \cup_1^\infty (a_i, b_i)$, then $\psi_n(x) = 0$ for all n [by (3)], and so $\lim h_n(x) = \lim \varphi_n(x) = 0$. On the other hand, if $x \in \cup_1^\infty (a_i, b_i)$, then for n sufficiently large, $\psi_n(x) = 1$ and so

$$\varphi_n(x) - M\psi_n(x) = \varphi_n(x) - M \le 0,$$

implying that $h_n(x) = 0$ if n sufficiently large. In either case

$$\lim h_n(x) = 0$$

for all x in $[a, b]$. Since $\{h_n\}$ is nonincreasing, Dini's Theorem implies that $\{h_n\}$ converges to zero uniformly. Since $\varphi_n - M\psi_n \le h_n$, this establishes the Claim.

To complete the proof, note that by the Claim

$$\lim \int_{\Re} \varphi_n(x) - M\psi_n(x)\, dx \le \frac{\epsilon}{2}$$

and so

$$0 \leq \lim \int_{\Re} \varphi_n(x)\, dx = \lim \int_{\Re} \varphi_n(x) - M\psi_n(x)\, dx + \lim \int_{\Re} M\psi_n(x)\, dx$$

$$\leq \frac{\epsilon}{2} + \frac{\epsilon}{2}$$

$$= \epsilon.$$

Since ϵ was arbitrary, this shows that $\lim \int_{\Re} \varphi_n(x)\, dx = 0$, as desired. (This lemma can also be proved by using step functions—see Problem 2.7. The proof with step functions is messier, but a little more intuitive.)

We are now in a position to prove that the value of

$$\lim_{n \to \infty} \int_{\Re} f_n(x)\, dx$$

does not depend upon the particular approximating sequence selected.

2.1.4. Lemma. *Let f and g be elements of $\mathcal{L}^+(\Re)$ and let $\{f_n\}$ and $\{g_n\}$ be approximating sequences in $\mathcal{C}_C(\Re)$ for f and g, respectively. If $f(x) \leq g(x)$ almost everywhere, then*

$$\lim_{n \to \infty} \int_{\Re} f_n(x)\, dx \leq \lim_{n \to \infty} \int_{\Re} g_n(x)\, dx.$$

Proof. As we have already noted, each of the foregoing limits exists, since the sequences are nondecreasing (the limits may be infinite). Fix n and for each k, define

$$h_k(x) = \max\{0;\, f_n(x) - g_k(x)\},$$

so $\{h_k\} \subset \mathcal{C}_C(\Re)$ is a nonincreasing sequence of nonnegative functions. Also, since $f(x) \leq g(x)$ almost everywhere and $\lim_{k \to \infty} g_k(x) = g(x)$ almost everywhere, it follows that $\lim_{k \to \infty} h_k(x) = 0$ almost everywhere. Lemma 3.3 then implies that $\lim_{k \to \infty} \int_{\Re} h_k(x)\, dx = 0$. Plainly, $f_n(x) - g_k(x) \leq h_k(x)$ for all k and x, and thus

$$\lim_{k \to \infty} \int_{\Re} f_n(x) - g_k(x)\, dx \;\leq\; \lim_{k \to \infty} \int_{\Re} h_k(x)\, dx \;=\; 0.$$

Thus $\int_{\Re} f_n(x)\, dx \leq \lim_{k \to \infty} \int_{\Re} g_k(x)\, dx$; letting n tend to infinity completes the proof.

2.1.5. Corollary. *If $f \in \mathcal{L}^+(\Re)$ and $\{f_n\}$ and $\{g_n\}$ are any two approximating* *sequences for f, then*

$$\lim_{n\to\infty} \int_{\Re} f_n(x)\,dx = \lim_{n\to\infty} \int_{\Re} g_n(x)\,dx.$$

Proof. Taking $f = g$ in Lemma 2.1.4, we obtain

$$\lim_{n\to\infty} \int_{\Re} f_n(x)\,dx \le \lim_{n\to\infty} \int_{\Re} g_n(x)\,dx;$$

a second application of the lemma yields the reverse inequality.

2.1.6. Definition. If $f \in \mathcal{L}^+(\Re)$, we define the *Lebesgue integral of f* to be

$$\int_{\Re} f\,d\lambda = \lim_{n\to\infty} \int_{\Re} f_n(x)\,dx,$$

where $\{f_n\} \subset \mathcal{C}_C(\Re)$ is any nondecreasing sequence converging to f almost everywhere.

By Corollary 2.1.5, $\int_{\Re} f\,d\lambda$ is well-defined. We may also rephrase Lemma 2.1.4 in terms of the integral.

2.1.7. Proposition. *Let f and g be elements of $\mathcal{L}^+(\Re)$ and suppose that $f(x) \le g(x)$ almost everywhere. Then $\int_{\Re} f\,d\lambda \le \int_{\Re} g\,d\lambda$.*

Some elementary properties of the integral are now readily derived.

2.1.8. Proposition. *Let f and g be elements of $\mathcal{L}^+(\Re)$ and let $\alpha \in [0, \infty)$. Then each of the functions $(f + g)$, αf, $\max(f, g)$, $\min(f, g)$, and $f \cdot g$ is an element of $\mathcal{L}^+(\Re)$. Moreover,*
  (a) $\int_{\Re} (f + g)\,d\lambda = \int_{\Re} f\,d\lambda + \int_{\Re} g\,d\lambda$
  (b) $\int_{\Re} (\alpha f)\,d\lambda = \alpha \int_{\Re} f\,d\lambda.$

Proof. We will prove only the assertions regarding $(f+g)$; the remaining assertions can be derived in a similar fashion and are left to the problems.

Let $\{f_n\} \subset \mathcal{C}_C(\Re)$ and $\{g_n\} \subset \mathcal{C}_C(\Re)$ be approximating sequences for f and g, respectively. Then $\{(f_n + g_n)\} \subset \mathcal{C}_C(\Re)$ is a nondecreasing sequence of continuous functions, and

$$\lim_{n\to\infty} f_n(x) + g_n(x) = f(x) + g(x) \quad \text{almost everywhere.}$$

Thus

$$\int_{\Re} (f + g)\,d\lambda = \lim_{n\to\infty} \int_{\Re} f_n(x) + g_n(x)\,dx = \lim_{n\to\infty} \int_{\Re} f_n(x)\,dx + \int_{\Re} g_n(x)\,dx.$$

Now, $f_n \ge f_1$ and $g_n \ge g_1$, so

$$\lim_{n\to\infty} \int_{\Re} f_n(x)\,dx \ne -\infty$$

and

$$\lim_{n \to \infty} \int_{\Re} g_n(x)\, dx \neq -\infty.$$

Thus, the limit on the right above is the sum of the limits, and

$$\int_{\Re} (f + g)\, d\lambda \;=\; \int_{\Re} f\, d\lambda + \int_{\Re} g\, d\lambda.$$

Certainly, every nonnegative continuous function is in $\mathcal{L}^+(\Re)$ [indeed, every nonnegative Riemann-integrable function is in $\mathcal{L}^+(\Re)$ —see the problems at the end of this section]. For our immediate purposes, it will suffice to show that characteristic functions of intervals are in $\mathcal{L}^+(\Re)$.

2.1.9 Proposition. *If $-\infty < a < b < \infty$, then each of the following functions is an element of $\mathcal{L}^+(\Re)$:*

$$\chi_{(-\infty,a]}, \quad \chi_{[a,b]}, \quad \chi_{(b,\infty)}, \quad \chi_{(a,b)}, \quad \chi_{[a,b)}, \quad \chi_{(a,b]}, \quad \chi_{(-\infty,a)}, \quad \chi_{[b,\infty)}.$$

Moreover, $\int_{\Re} \chi_{(a,b)}\, d\lambda \;=\; b - a$.

Proof. We will show only that $\chi_{(a,b)} \in \mathcal{L}^+(\Re)$, the proof for the other functions being essentially the same.

Set $\delta = (b-a)/2$. [For each $n = 1, 2, \ldots$, notice that $a + (\delta/n) \le b - (\delta/n)$.]
Set

$$g_n(x) = \begin{cases} 0 & x \le a \\ (n/\delta)(x - a) & a < x \le a + (\delta/n) \\ 1 & a + (\delta/n) < x \le b - (\delta/n) \\ (n/\delta)(b - x) & b - (\delta/n) < x \le b \\ 0 & b < x. \end{cases}$$

It is evident that $\lim g_n(x) = \chi_{(a,b)}(x)$ everywhere [and that

$$\lim_{n \to \infty} \int_{\Re} g_n(x)\, dx \;=\; (b - a)].$$

2.1.10. Definition. The class $L(\Re)$ is the collection of all functions f for which there are functions g and h in $\mathcal{L}^+(\Re)$ such that $f(x) = g(x) - h(x)$ almost everywhere.

For functions $f = g - h$ in $L(\Re)$ we would like to define $\int_{\Re} f\, d\lambda$ to be $\int_{\Re} g\, d\lambda - \int_{\Re} h\, d\lambda$; however, this difference could be of the form "$\infty - \infty$" and hence undefined. Thus we need to restrict our attention to a smaller class.

2.1.11. Definition. We denote by $\mathcal{L}_1(\Re)$ the collection of all functions $f \in L(\Re)$ corresponding to which there are functions g and h in $\mathcal{L}^+(\Re)$ satisfying

(a) $f(x) = g(x) - h(x)$ almost everywhere;

(b) both $\int_{\Re} g\, d\lambda < \infty$ and $\int_{\Re} h\, d\lambda < \infty$.

Now if f is an element of $\mathcal{L}(\Re)$, the difference $\int_{\Re} g\, d\lambda - \int_{\Re} h\, d\lambda$ makes sense; it remains to argue that the value of this difference does not depend upon the particular decomposition of $f = g - h$ selected.

2.1.12. Lemma. *Suppose that f_1, f_2, g_1, and g_2 are elements of $\mathcal{L}^+(\Re)$ and that all four functions have a finite integral. If $f_1(x) - g_1(x) = f_2(x) - g_2(x)$ almost everywhere, then*

$$\int_{\Re} f_1\, d\lambda - \int_{\Re} g_1\, d\lambda = \int_{\Re} f_2\, d\lambda - \int_{\Re} g_2\, d\lambda.$$

Proof. By assumption,

$$f_1(x) + g_2(x) = f_2(x) + g_1(x) \quad \text{almost everywhere}$$

and so, by 2.1.7,

$$\int_{\Re} (f_1 + g_2)\, d\lambda = \int_{\Re} (f_2 + g_1)\, d\lambda.$$

This implies, by 2.1.8(a), that

$$\int_{\Re} f_1\, d\lambda + \int_{\Re} g_2\, d\lambda = \int_{\Re} f_2\, d\lambda + \int_{\Re} g_1\, d\lambda.$$

Rearranging gives the result.

2.1.13. Definition. If $f \in \mathcal{L}_1(\Re)$, then we define the *Lebesgue integral of f* to be

$$\int_{\Re} f\, d\lambda = \int_{\Re} g\, d\lambda - \int_{\Re} h\, d\lambda,$$

where g, h are functions in $\mathcal{L}^+(\Re)$ satisfying $f(x) = g(x) - h(x)$ almost everywhere and g and h have finite integral.

Lemma 2.1.12 assures that the definition is independent of the decomposition of f into the difference of $\mathcal{L}^+(\Re)$ functions. Note that if $f \in \mathcal{L}_1(\Re)$, then $|\int_{\Re} f\, d\lambda| < \infty$; other properties which follow readily are summarized below.

2.1.14 Proposition. *Let f, $g \in L(\Re)$ and let $\alpha \in \Re$. Then $f + g$ and $\alpha f \in L(\Re)$. Moreover, if f, $g \in \mathcal{L}_1(\Re)$, then:*

(a) *if $f \le g$ almost everywhere, $\int_{\Re} f\, d\lambda \le \int_{\Re} g\, d\lambda$;*

(b) *$f + g \in \mathcal{L}_1(\Re)$ and $\int_{\Re} (f + g)\, d\lambda = \int_{\Re} f\, d\lambda + \int_{\Re} g\, d\lambda$;*

(c) *$\alpha f \in \mathcal{L}_1(\Re)$ and $\int_{\Re} (\alpha f)\, d\lambda = \alpha \int_{\Re} f\, d\lambda.$*

Proof. We establish (a) and leave (b) and (c) to the exercises. Choose functions f_1, f_2, g_1, and $g_2 \in \mathcal{L}^+(\Re)$ such that all four have finite integral and $f(x) = f_1(x) - f_2(x)$ and $g(x) = g_1(x) - g_2(x)$ almost everywhere. Then

$$f_1(x) + g_2(x) \leq g_1(x) + f_2(x) \quad \text{almost everywhere}$$

and so, as in the proof of 2.1.12,

$$\int_\Re f_1 \, d\lambda + \int_\Re g_2 \, d\lambda \leq \int_\Re g_1 \, d\lambda + \int_\Re f_2 \, d\lambda,$$

from which (a) is evident.

Some of the more useful attributes of continuous functions are the "lattice properties," namely that both $f \wedge g = \min(f, g)$ and $f \vee g = \max(f, g)$ are continuous functions when f and g are continuous. The corresponding fact for functions f, $g \in \mathcal{L}_1(\Re)$ is somewhat more difficult to prove. We begin with some preliminary observations.

2.1.15. Lemma. *If $f \in L(\Re)$, then $|f| \in L(\Re)$.*

Proof. Select g, $h \in \mathcal{L}^+(\Re)$ such that $f(x) = g(x) - h(x)$ almost everywhere. Since $g(x) \geq 0$ and $h(x) \geq 0$,

$$|g(x) - h(x)| = g(x) + h(x) - 2\min\{g(x), h(x)\}.$$

Since $(g + h) \in \mathcal{L}^+(\Re)$ and $\min\{g, h\} \in \mathcal{L}^+(\Re)$, it follows that $|f| \in L(\Re)$.

The same proof, *mutatis mutandis*, shows the following.

2.1.16. Corollary. *If $f \in \mathcal{L}_1(\Re)$, then $|f| \in \mathcal{L}_1(\Re)$.*

More interesting is the following, which is more or less immediate from the foregoing and a simple algebraic identity.

2.1.17. Corollary. *If $f \in L(\Re)$, and if $f^+(x) = \max\{0, f(x)\}$ and $f^-(x) = -\min\{0, f(x)\}$, then $f^+ \in L(\Re)$, $f^- \in L(\Re)$. If, in addition, $f \in \mathcal{L}_1(\Re)$, then f^+ and f^- are in $\mathcal{L}_1(\Re)$ and*
  (a) $\int_\Re f \, d\lambda = \int_\Re f^+ \, d\lambda - \int_\Re f^- \, d\lambda$;
  (b) $\int_\Re |f| \, d\lambda = \int_\Re f^+ \, d\lambda + \int_\Re f^- \, d\lambda$.

Proof. It is evident that

$$f^+ = \tfrac{1}{2}(|f| + f) \quad \text{and} \quad f^- = \tfrac{1}{2}(|f| - f).$$

By 2.1.14, f^+ and $f^- \in L(\Re)$. If $f \in \mathcal{L}_1(\Re)$, then by 2.1.16 both $f^+ \in \mathcal{L}_1(\Re)$ and $f^- \in \mathcal{L}_1(\Re)$. Since $f = f^+ - f^-$ and $|f| = f^+ + f^-$, (a) and (b) now follow readily from 2.1.14(b) and (c).

2.1.18. Corollary. *If $f \in \mathcal{L}_1(\Re)$, then $|\int_\Re f \, d\lambda| \le \int_\Re |f| \, d\lambda$.*

The inequality in Corollary 2.1.18 is more or less obvious for the Riemann integral; it is rather surprising that it is so much more difficult to prove for the Lebesgue integral.

Proof. We apply 2.1.17:

$$
\begin{aligned}
\left| \int_\Re f \, d\lambda \right| &= \left| \int_\Re f^+ \, d\lambda - \int_\Re f^- \, d\lambda \right| \\
&\le \left| \int_\Re f^+ \, d\lambda \right| + \left| \int_\Re f^- \, d\lambda \right| \\
&= \int_\Re f^+ \, d\lambda + \int_\Re f^- \, d\lambda \\
&= \int_\Re |f| \, d\lambda.
\end{aligned}
$$

Combining 2.1.17 with another simple algebraic identity gives the desired lattice properties for $\mathcal{L}_1(\Re)$.

2.1.19. Proposition. *If f, $g \in L(\Re)$, then both $\max(f, g) \in L(\Re)$ and $\min(f, g) \in L(\Re)$. If f, $g \in \mathcal{L}_1(\Re)$, then both $\max(f, g) \in \mathcal{L}_1(\Re)$ and $\min(f, g) \in \mathcal{L}_1(\Re)$.*

Proof. The conclusions are immediate from the identities

$$
\begin{aligned}
\max\{f(x), g(x)\} &= (f - g)^+(x) + g(x) \quad \text{and} \\
-\min\{f(x), g(x)\} &= (f - g)^+(x) - f(x).
\end{aligned}
$$

2.1.20. Proposition. *If f, $g \in L(\Re)$, then $(fg) \in L(\Re)$.*

Proof. First suppose that $f = g$. Select f_1, $f_2 \in \mathcal{L}^+(\Re)$ such that $f(x) = f_1(x) - f_2(x)$ almost everywhere. Then

$$
(f(x))^2 = f_1^2(x) + f_2^2(x) - 2f_1(x)f_2(x).
$$

By 2.1.8, each of the functions f_1^2, f_2^2, and $f_1 f_2$ are in $L^+(\Re)$, and hence $f^2 \in L(\Re)$. Now for the general case, observe that

$$
fg = \tfrac{1}{4}\left((f + g)^2 - (f - g)^2\right).
$$

Since $(f + g)^2$ and $(f - g)^2 \in L(\Re)$, it follows that $fg \in L(\Re)$.

It need *not* be the case that the product of two $\mathcal{L}_1$ functions is again an

$\mathcal{L}_1$ function. For example, if

$$f(x) = \begin{cases} x^{-\frac{1}{2}} & 0 < x \le 1 \\ 0 & \text{elsewhere,} \end{cases}$$

then $f \in \mathcal{L}_1(\Re)$ and $f^2 \notin \mathcal{L}_1(\Re)$.

We conclude this section by observing that the results above can be restricted to certain subsets of the reals (which include, for example, closed intervals).

2.1.21. Definition. A set $E \subset \Re$ is (Lebesgue) *measurable* if for all $f \in \mathcal{L}_1(\Re)$, the function $f\chi_E$ is an element of $\mathcal{L}_1(\Re)$. The collection of all Lebesgue measurable sets is denoted by $\mathcal{M}$.

2.1.22. Definition. Given $E \in \mathcal{M}$, the collection $\mathcal{L}_1(E)$ is the class of all functions $f \in L(\Re)$ for which $f\chi_E \in \mathcal{L}_1(\Re)$. If $f \in \mathcal{L}_1(E)$, then we define the Lebesgue integral of f to be

$$\int_E f \, d\lambda = \int_\Re f\chi_E \, d\lambda.$$

In Section 2.3 we will study the measurable sets in greater detail. For now we summarize the basic properties of $\int_E f \, d\lambda$; most of these are immediate from the definition and corresponding properties of $\int_\Re f \, d\lambda$.

2.1.23. Proposition. *Let f, $g \in \mathcal{L}_1(E)$ and let $\alpha \in \Re$. Then the functions $(f+g)$, (αf), $\min(f, g)$, and $\max(f, g)$ are all in $\mathcal{L}_1(E)$. Moreover,*
  (a) *if $f \le g$ almost everywhere, then $\int_E f \, d\lambda \le \int_E g \, d\lambda$;*
  (b) *$\int_E (f + g) \, d\lambda = \int_E f \, d\lambda + \int_E g \, d\lambda$;*
  (c) *$\int_E (\alpha f) \, d\lambda = \alpha \int_E f \, d\lambda$.*

Proof. Problem 2.3.

2.1.24. Proposition. *Suppose that A and B are measurable sets and $f \in \mathcal{L}_1(A) \cup \mathcal{L}_1(B)$. If $A \cap B$ has measure 0, then $f \in \mathcal{L}_1(A \cup B)$ and*

$$\int_{A\cup B} f \, d\lambda = \int_A f \, d\lambda + \int_B f \, d\lambda.$$

Proof. This follows at once from the definition and the fact that $\chi_{A \cup B} = \chi_A + \chi_B$ almost everywhere.

PROBLEMS

2.1. Complete the proof of Proposition 2.1.8.

2.2.   Prove 2.1.14(b) and (c).

2.3.   Prove Proposition 2.1.23.

2.4.   Let C_p denote the Cantor "middle p^{th}" set (see Problem 1.18). Show that $\chi_{C_p} \in \mathcal{L}_1(\Re)$ for any odd integer $p > 1$, and compute $\int_{\Re} \chi_{C_p} \, d\lambda$.

2.5.   Let f be a Riemann-integrable function defined on an interval $[a, b]$. Show that $f \in \mathcal{L}_1[a, b]$.

2.6.   **Definition.** A set G is said to be a "G_δ" if there is a sequence $\{U_n\}$ of open sets such that $G = \cap U_n$. A set F is said to be an "F_σ" if $\Re \backslash F$ is a G_δ.

Show that all G_δ and F_σ sets are measurable.

2.7.   *Another proof of Lemma 2.1.3.* Let $\{\varphi_n\}$ be as in Lemma 2.1.3. Choose M, $[a, b]$, E and $\{(a_i, b_i)\}$ as in the proof of Lemma 2.1.3.
   (a)   For each $x \in [a, b] \backslash E$, find an open interval $I(x)$ and an integer $N(x)$ so that $n \geq N(x)$ and $y \in I(x)$ imply that

$$0 \leq \varphi_n(y) \leq \frac{\epsilon}{2(b-a)}.$$

   (b)   Find a collection

$$\{I(x_1), \cdots, I(x_n), (a_1, b_1), \cdots (a_p, b_p)\}$$

which covers $[a, b]$. By taking appropriate combinations of the intervals $\{I(x_k)\}$, construct disjoint intervals $J_k \subset [a, b]$ such that $\cup_1^n J_k = \cup_1^n I(x_k)$.
   (c)   Define

$$s = \sum_1^n \frac{\epsilon}{2(b-a)} \chi_{J_k} + \sum_1^p M \chi_{(a_i, b_i)}.$$

Show that $\int_{\Re} S(x) \, dx \leq \epsilon$ while, for n sufficiently large, $S(x) = \varphi_n(x)$.

2.8.   Let $f \in \mathcal{L}_1(\Re)$ and let $g(x) = f(\alpha x + \beta)$, where $\alpha, \beta \in \Re$. Show that $g \in \mathcal{L}_1(\Re)$ and that

$$\int_{\Re} f \, d\lambda = \alpha \int_{\Re} g \, d\lambda.$$

2.9.   Let

$$g(x) = \begin{cases} x^{-\frac{1}{2}} & 0 < x \leq 1 \\ 0 & \text{elsewhere.} \end{cases}$$

Show that $g \in \mathcal{L}_1(\Re)$ and find $\int_{\Re} g \, d\lambda$.

2.10.

   (a)   Suppose that $f \in L(\Re)$ and that $|f| \in \mathcal{L}_1(\Re)$; show that $f \in \mathcal{L}_1(\Re)$. (This is harder than it looks! Later, we will be able to give an easy proof.)
   (b)   Show that E is measurable if and only if $\chi_E \in L(\Re)$.

2.2. CONVERGENCE THEOREMS. We are now in a position to prove the basic convergence theorems regarding the Lebesgue integral. In general, all these theorems deal with the same basic type of question: When can integral signs and limits be interchanged? Since integrals are themselves realized as limits, the theorems deal with the interchangeability of different limiting processes. We begin with a lemma about series in $\mathcal{L}^+(\Re)$.

2.2.1. Lemma. *Let* $\{g_n\}$ *be a sequence in* $\mathcal{L}^+(\Re)$, *and let* g *be the extended real-valued function* $g(x) = \sum_1^\infty g_n(x)$. *Then* $g \in \mathcal{L}^+(\Re)$ *and* $\int_\Re g \, d\lambda = \sum_1^\infty \int_\Re g_n \, d\lambda$, *i.e.,* $\int_\Re \left(\sum_1^\infty g_n\right) d\lambda = \sum_1^\infty \int_\Re g_n \, d\lambda$.

Proof. For each n, select a nondecreasing sequence $\{\psi_{n,m}\} \subset \mathcal{C}_C(\Re)$ such that

$$\lim_{m \to \infty} \psi_{n,m}(x) = g_n(x) \quad \text{almost everywhere.}$$

Next set $\Phi_m(x) = \sum_{n=1}^m \psi_{n,m}(x)$, so that $\{\Phi_m\}$ is a nondecreasing sequence in $\mathcal{C}_C(\Re)$. Moreover, for each m and almost all x

$$\Phi_m(x) = \sum_{n=1}^m \psi_{n,m}(x) \le \sum_{n=1}^m g_n(x) \le g(x). \tag{1}$$

Now define Φ to be the extended real-valued function $\Phi(x) = \lim_{m \to \infty} \Phi_m(x)$. Then $\Phi \in \mathcal{L}^+(\Re)$ and, by (1),

$$\int_\Re \Phi \, d\lambda = \lim_{m \to \infty} \int_\Re \Phi_m(x) \, dx \le \lim_{m \to \infty} \sum_{n=1}^m \int_\Re g_n \, d\lambda = \sum_1^\infty \int_\Re g_n \, d\lambda. \tag{2}$$

For the reverse inequalities, fix k and note that for $m \ge k$,

$$\Phi_m(x) = \sum_{n=1}^m \psi_{n,m}(x) \ge \sum_{n=1}^k \psi_{n,m}(x);$$

letting m tend to infinity gives for almost all x,

$$\Phi(x) \ge \sum_{n=1}^k g_n(x).$$

Since k is arbitrary, this implies that $\Phi(x) \ge \sum_1^\infty g_n(x) \equiv g(x)$ almost everywhere, and so, by (1), $\Phi(x) = g(x)$ almost everywhere. Also,

$$\int_\Re \Phi \, d\lambda \ge \sum_{n=1}^k \int_\Re g_n \, d\lambda;$$

since k was arbitrary, this implies that $\int_\Re \Phi \, d\lambda \ge \sum \int_\Re g_n \, d\lambda$. This inequality, when combined with (2), gives the result.

A similar result holds for the class $\mathcal{L}_1(\Re)$; we shall need the following lemmas. In general, nonnegative functions in $\mathcal{L}_1(\Re)$ need *not* be in $\mathcal{L}^+(\Re)$; however, they are "almost" in $\mathcal{L}^+$, as the following lemma shows.

2.2.2. Lemma. *Let $f \in \mathcal{L}_1(\Re)$ and let $\epsilon > 0$. If $f \geq 0$ almost everywhere, then there are functions g, $h \in \mathcal{L}^+(\Re)$ such that $f = g - h$ almost everywhere and $\int_{\Re} h \, d\lambda < \epsilon$.*

Proof. By definition, there are functions φ, $\psi \in \mathcal{L}^+(\Re)$ such that $f = \varphi - \psi$ almost everywhere and both $\int_{\Re} \varphi \, d\lambda < \infty$ and $\int_{\Re} \psi \, d\lambda < \infty$. Select a nondecreasing sequence $\{\psi_n\} \subset \mathcal{C}_C(\Re)$ such that $\lim \psi_n = \psi$ almost everywhere and $\int_{\Re} \psi \, d\lambda = \lim \int_{\Re} \psi_n(x) \, dx$. Thus, for n sufficiently large, $0 \leq \int_{\Re} (\psi - \psi_n) \, d\lambda < \epsilon$.

Now set $g = \max\{\varphi - \psi_n, 0\}$ and $h = \max\{\psi - \psi_n, 0\}$. Since $\psi_n \leq \psi$ almost everywhere, $h = \psi - \psi_n$ almost everywhere. Since $0 \leq f = \varphi - \psi$ almost everywhere, it follows that $\psi_n \leq \psi \leq \varphi$ almost everywhere, and hence that $g = \varphi - \psi_n$ almost everywhere. Since $\psi_n \in \mathcal{C}_C(\Re)$, it is evident that both g and h are elements of $\mathcal{L}^+(\Re)$ and for almost all x,

$$f(x) = \varphi(x) - \psi(x) = g(x) - h(x).$$

If the function $g(x) = \sum_1^\infty g_n(x)$ of Lemma 2.2.1 is also in $\mathcal{L}_1$, then the sum cannot diverge "too often"; this will be important to know when we consider the difference of two such sums (as we must do if the functions g_n are in $\mathcal{L}_1$ rather than $\mathcal{L}^+$). The next lemma addresses this issue.

2.2.3 Lemma. *Let $f \in \mathcal{L}^+(\Re)$ and suppose that $\int_{\Re} f \, d\lambda < \infty$. Then the set $\{x : f(x) = +\infty\}$ has measure zero.*

Proof. Select a nondecreasing sequence $\{\varphi_n\} \subset \mathcal{C}_c(\Re)$ such that $\lim \varphi_n = f$ a.e.; fix $\epsilon > 0$ and set $\alpha = \int_{\Re} f \, d\lambda$. Without loss of generality we may and do assume that α is not zero. Define

$$E_n = \{x \in \Re : \varphi_n(x) > \alpha \epsilon^{-1}\} \quad \text{and}$$
$$E = \cup_n E_n.$$

Since each E_n is open, E is open, so there is a family of disjoint intervals $\{(a_i, b_i)\}$ such that

$$E = \cup_1^\infty (a_i, b_i)$$

(possibly $a_i = -\infty$ or $b_i = +\infty$ for some i). By 2.1.9, $\chi_{(a_i, b_i)} \in \mathcal{L}^+(\Re)$ for each i and so, by 2.2.1,

$$\chi_E = \sum \chi_{(a_i, b_i)} \in \mathcal{L}^+(\Re).$$

Next observe that $\alpha \epsilon^{-1} \chi_E \leq f$ a.e., for if $x \in E$, then there is an n for which

$$\alpha \epsilon^{-1} \leq \varphi_n(x) \leq f(x).$$

Thus, by 2.1.7,

$$\int_{\Re} \alpha \epsilon^{-1} \chi_E \, d\lambda \le \int_{\Re} f \, d\lambda = \alpha,$$

or

$$\int_{\Re} \chi_E \, d\lambda \le \epsilon.$$

Since $\int_{\Re} \chi_E \, d\lambda < \infty$, it now follows that each interval (a_i, b_i) is finite and, by 2.2.1 and 2.1.9,

$$\epsilon \ge \int_{\Re} \chi_E \, d\lambda = \sum \int_{\Re} \chi_{(a_i, b_i)} \, d\lambda = \sum (b_i - a_i).$$

Now set $A = \{x : \lim \varphi_n(x) = f(x)\}$ and let $B = \Re \backslash A$, so that B has measure zero. Note that if $x \in A$ and $f(x) = +\infty$, then there is an n for which $\varphi_n(x) \ge \alpha \epsilon^{-1}$, so $x \in E$. Thus

$$\{x \in A : f(x) = +\infty\} \subset E = \cup_1^\infty (a_i, b_i)$$

and $\sum (b_i - a_i) \le \epsilon$. Since ϵ was arbitrary, this shows that $\{x \in A : f(x) = +\infty\}$ has measure zero. Since $\{x \in \Re : f(x) = +\infty\} \subset \{x \in A : f(x) = +\infty\} \cup B$, and each of the latter sets has measure zero, the former does as well, completing the proof.

We can now extend the basic convergence result embodied in 2.2.1 to the class $\mathcal{L}_1(\Re)$; most of the work to prove the next theorem has been done in the preceding two lemmas.

2.2.4. Theorem. *Let $\{f_n\}$ be a sequence of nonnegative functions in $\mathcal{L}_1(\Re)$ and suppose that $\sum_1^\infty \int_{\Re} f_n \, d\lambda < \infty$. Then $\sum_1^\infty f_n \in \mathcal{L}_1(\Re)$ and*

$$\int_{\Re} \left(\sum_1^\infty f_n \right) d\lambda = \sum_1^\infty \int_{\Re} f_n \, d\lambda.$$

Proof. Fix $\epsilon > 0$ and select $g_n, h_n \in \mathcal{L}^+(\Re)$ such that $f_n = g_n - h_n$ almost everywhere and $\int_{\Re} h_n \, d\lambda \le \epsilon/2^n$. Set $M = \sum_1^\infty \int_{\Re} f_n \, d\lambda$, and observe that

$$\sum_{n=1}^N \int_{\Re} g_n \, d\lambda = \sum_1^N \int_{\Re} (f_n + h_n) \, d\lambda$$

and thus

$$\sum_{n=1}^\infty \int_{\Re} g_n \, d\lambda \le M + \epsilon.$$

By 2.2.1, $\sum_1^\infty g_n \in \mathcal{L}^+(\Re)$, and by the above and 2.2.1, $\sum_1^\infty g_n \in \mathcal{L}^+(\Re) \cap \mathcal{L}_1(\Re)$. Similarly, $\sum_1^\infty h_n \in \mathcal{L}^+(\Re) \cap \mathcal{L}_1(\Re)$. From this, both

$$\sum_1^\infty g_n(x) < \infty \quad \text{and} \quad \sum_1^\infty h_n(x) < \infty$$

for almost all x, and hence

$$\sum_1^\infty f_n(x) = \sum_1^\infty g_n(x) - \sum_1^\infty h_n(x) \, dx$$

almost everywhere, establishing $f \in \mathcal{L}_1(\Re)$. Another application of 2.2.1 yields

$$\int_\Re \sum_1^\infty f_n \, d\lambda = \sum_1^\infty \int_\Re f_n \, d\lambda.$$

Theorem 2.2.4 has manifold consequences, and numerous very powerful convergence results can be deduced from it. We begin with the following.

2.2.5. Monotone Convergence Theorem. *Let $\{f_n\}$ be a nondecreasing sequence in $\mathcal{L}_1(\Re)$ and suppose that $\{\int_\Re f_n \, d\lambda\}$ is bounded. Then $(\lim f_n) \in \mathcal{L}_1(\Re)$ and*

$$\int_\Re (\lim f_n) \, d\lambda = \lim \int_\Re f_n \, d\lambda.$$

Proof. Set $g_0 = f_0$ and for $n \geq 1$, $g_n = f_n - f_{n-1}$. Then each $g_n \geq 0$ and $\sum_0^N g_n = f_N$, so that all the hypotheses of 2.2.4 are fulfilled.

The class of Lebesgue integrable functions $\mathcal{L}_1(\Re)$ was essentially constructed by forming all almost everywhere pointwise limits of continuous functions. The resulting class was somewhat larger than the original class; it is natural to ask if repeating the process, this time with pointwise almost everywhere limits of functions in $\mathcal{L}_1(\Re)$, might result in a yet larger class. The next result shows that this is not the case: We have, in some sense, identified the "largest" class that can be obtained in this way. This result also extends the lattice properties in $\mathcal{L}_1(\Re)$ from finite to infinite operations.

2.2.6. Corollary. *If $\{f_n\}$ is a sequence in $\mathcal{L}_1(\Re)$ and if $\{|\int_\Re f_n \, d\lambda|\}$ is bounded, then each of the following functions is an element of $\mathcal{L}_1(\Re)$:*

$$\sup\{f_n\}; \ \inf\{f_n\}; \ \overline{\lim} f_n; \ \underline{\lim} f_n.$$

If, in addition, $\lim f_n(x)$ exists almost everywhere, then $(\lim f_n)$ is in $\mathcal{L}_1(\Re)$.

Proof. We will show that if $g(x) = \inf\{f_n(x)\}$, then $g \in \mathcal{L}_1(\Re)$; the proof for the other cases is similar.

Set $h_n = \min\{f_1, \ldots, f_n\}$ and set $g_n = f_1 - h_n$. Then $\{g_n\}$ is a nondecreasing sequence of nonnegative functions in $\mathcal{L}_1(\Re)$ and $|\int_\Re g_n \, d\lambda| \leq \int_\Re |f_1| \, d\lambda + |\int_\Re h_n \, d\lambda| \leq 2\max\{\int_\Re f_j \, d\lambda : j = 1, \ldots, n\}$, so that $\{\int_\Re g_n \, d\lambda\}$ is bounded. Consequently, $g_\infty = \lim g_n$ is in $\mathcal{L}_1(\Re)$ by 2.2.4. But now it follows at once that $f_1 - g_\infty = \inf\{f_n\}$.

2.2.7. Fatou's Lemma. *Let $\{g_n\}$ be a sequence of nonnegative functions in $\mathcal{L}_1(\Re)$; suppose that $\{\int_\Re g_n \, d\lambda\}$ is bounded, and $g = \lim g_n$ almost everywhere. Then*

$$\int_\Re g \, d\lambda \leq \varliminf \int_\Re g_n \, d\lambda.$$

Proof. Set $h_n = \inf\{g_n, g_{n+1}, \ldots\}$, so $\{h_n\}$ is a nondecreasing sequence and $\lim h_n = g$ almost everywhere. Then, by 2.2.5,

$$\int_\Re g \, d\lambda = \lim \int_\Re h_m \, d\lambda \leq \varliminf \int_\Re g_n \, d\lambda.$$

We remark that the inequality above may be strict even if $g_n \to g$ *uniformly* (see Problem 2.11).

The final result in the section is probably the most useful and widely applied convergence theorem in the theory. The problems at the end of this section contain several applications. The proof is deceptively brief—notice that we use, indirectly, virtually everything in this and the preceding section.

2.2.8. Lebesgue's Dominated Convergence Theorem. *Let $\{f_n\}$ be a sequence in $\mathcal{L}_1(\Re)$ and suppose that $\lim_n f_n = f$ almost everywhere. If there is a $g \in \mathcal{L}_1(\Re)$ such that $|f_n| \leq g$ almost everywhere, then $f \in \mathcal{L}_1(\Re)$ and*

$$\int_\Re f \, d\lambda = \lim \int_\Re f_n \, d\lambda.$$

Proof. The sequence $\{g - f_n\}$ is nonnegative, so Fatou's Lemma implies that

$$\int_\Re (g - f) \, d\lambda \leq \varliminf \int_\Re (g - f_n) \, d\lambda,$$

from which $\varlimsup \int_\Re f_n \, d\lambda \leq \int_\Re f \, d\lambda$. Similarly, the sequence $\{g + f_n\}$ is nonnegative, so Fatou's Lemma again implies that

$$\int_\Re (g + f) \, d\lambda \leq \varliminf \int_\Re (g + f_n) \, d\lambda$$

and thus

$$\int_{\Re} f \, d\lambda \ \le \ \underline{\lim} \int_{\Re} f_n.$$

which completes the proof.

PROBLEMS

2.11. Find a sequence $\{g_n\}$ of functions in $C_c(\Re)$ satisfying:
  (a) $\{g_n\}$ converges uniformly to a continuous function;
  (b) $\{\int_{\Re} g_n \, d\lambda\}$ is convergent; but
  (c) $\lim \int_{\Re} g_n \, d\lambda \neq \int_{\Re} g \, d\lambda$.

  This shows that the inequality in 2.2.7 may be strict.

2.12. Let $\{f_n\}$ be a sequence of nonnegative functions in $\mathcal{L}_1(\Re)$ for which $\{\int_{\Re} f_n \, d\lambda\}$ is bounded. Show that

$$\int_{\Re} \underline{\lim} f_n \, d\lambda \ \le \ \underline{\lim} \int_{\Re} f_n \, d\lambda.$$

2.13. Let $f \in \mathcal{L}_1(\Re)$ and show that

$$\lim \int_{\Re} \cos(nt) f(t) \, dt \ = \ 0 \quad \text{and}$$

$$\lim \int_{\Re} \sin(nt) f(t) \, dt \ = \ 0.$$

(This famous problem is known as the **Riemann-Lebesgue Lemma**.)

2.14. Show that the Monotone Convergence Theorem need not hold for decreasing sequences.

2.15. Suppose that $g \in \mathcal{L}_1[0,1]$, $F \in C[0,1]$ and that for each n,

$$\int_0^1 e^{-x/n} g(x) \, dx \ = \ n \int_0^1 e^{-nx} F(x) \, dx.$$

Show that

$$\int_0^1 g \, d\lambda \ = \ F(0).$$

2.16. Let $\{f_n\} \subset \mathcal{L}_1(\Re)$ and suppose that for some $f \in \mathcal{L}_1(\Re)$,

$$\lim f_n = f \quad \text{a.e.} \quad \text{and} \quad \lim \int_{\Re} f_n \, d\lambda = \int_{\Re} f \, d\lambda.$$

Show that for all measurable sets E,

$$\lim \int_E f_n \, d\lambda \ = \ \int_E f \, d\lambda.$$

2.17. *Beppo Levi's Theorem.* Let $\{f_n\} \subseteq \mathcal{L}_1(\Re)$ be a nondecreasing sequence and suppose that $\int_\Re f_k^- \, d\lambda < \infty$ for some k. Show that

$$\lim \int_\Re f_n \, d\lambda = \int_\Re \lim f_n \, d\lambda.$$

2.18. Show that

$$\lim \int_{-n}^{n} \left(1 + \frac{x}{n}\right)^n e^{-\frac{x^2}{2}} \, dx = \sqrt{\frac{\pi}{2e}}.$$

2.3. MEASURE THEORY. As we already observed, Lebesgue originally developed his theory from the concept of "measurability." This notion continues to play a fundamental role in integration theory, so we will discuss it in some detail in the next section. The basic idea is to describe those sets whose "lengths" (measures) can be computed as Lebesgue integrals. (Recall that 2.1.9 already tells us that intervals fall into this category.) We begin by recalling Definition 2.1.21.

2.3.1. Definition. A set $E \subset \Re$ is (Lebesgue) measurable if for each $f \in \mathcal{L}_1(\Re)$, the function $f\chi_E \in \mathcal{L}_1(\Re)$. The class of all measurable sets is denoted by $\mathcal{M}$.

One of the very first things which we can deduce is that $\mathcal{M}$ is closed under countable unions and under complements, i.e., that $\mathcal{M}$ satisfies the following definition.

2.3.2. Definition. A collection $\mathcal{A}$ of subsets of $\Re$ is a *σ-algebra* if
  (a) $\emptyset \in \mathcal{A}$ and $\Re \in \mathcal{A}$;
  (b) $\{A_n\} \subset \mathcal{A}$ implies that $\cup_1^\infty A_n \in \mathcal{A}$; and
  (c) $A \in \mathcal{A}$ implies that $A^c = (\Re \backslash A) \in \mathcal{A}$.

2.3.3. Proposition. $\mathcal{M}$ *is a* σ-algebra.

Proof. It is evident that both $\emptyset \in \mathcal{M}$ and $\Re \in \mathcal{M}$. Now suppose that $E_1, E_2 \in \mathcal{M}$ and fix $f \in \mathcal{L}_1(\Re)$. Each of the functions $f^+\chi_{E_1}, f^+\chi_{E_2}, f^-\chi_{E_1}$, and $f^-\chi_{E_2}$ are in $\mathcal{L}_1(\Re)$ and thus so are the functions

$$f^+\chi_{E_1 \cup E_2} = \max\{f^+\chi_{E_1}, f^+\chi_{E_2}\} \quad \text{and}$$

$$f^-\chi_{E_1 \cup E_2} = \max\{f^-\chi_{E_1}, f^-\chi_{E_2}.\}$$

Thus $f\chi_{E_1 \cup E_2} = (f^+ - f^-)\chi_{E_1 \cup E_2} \in \mathcal{L}_1(\Re)$, showing that $E_1 \cup E_2 \in \mathcal{M}$. An obvious induction shows that if $\{E_1, \ldots, E_n\} \subset \mathcal{M}$, then $\cup_1^n E_j \in \mathcal{M}$.

Next let $\{E_n\}_{n=1}^{\infty} \subset \mathcal{M}$, set $F_n = \cup_1^n E_j$, $E = \cup_1^{\infty} E_n$, and fix $f \in \mathcal{L}_1(\Re)$. Each of the functions $f\chi_{F_n}$ is in $\mathcal{L}_1(\Re)$ and $|f\chi_{F_n}| \leq |f|$. Thus by the Dominated Convergence Theorem, $f \cdot \chi_E = \lim f\chi_{F_n}$ is in $\mathcal{L}_1(\Re)$. Since $f \in \mathcal{L}_1(\Re)$ was arbitrary, this shows that $E \in \mathcal{M}$.

For (c), again fix $E \in \mathcal{M}$ and $f \in \mathcal{L}_1(\Re)$. Then $(f - f \cdot \chi_E) \in \mathcal{L}_1(\Re)$, and since $(f - f\chi_E) = f(1 - \chi_E) = f\chi_{E^c}$, $E^c \in \mathcal{M}$.

Of course, it follows easily that if $\mathcal{M}$ is closed under both complements and countable unions, then it must be closed under countable intersections as well.

2.3.4. Corollary. *If $\{E_n\} \subset \mathcal{M}$, then $\cap_1^{\infty} E_n \in \mathcal{M}$.*

Proof. If $A_n = E_n^c$, then $\cup_1^{\infty} A_n \in \mathcal{M}$. But, via DeMorgan's Laws, $\cup_1^{\infty} A_n = (\cap_1^{\infty} E_n)^c$. Thus $(\cap_1^{\infty} E_n)^c \in \mathcal{M}$, whence, via 2.3.2(c), $\cap_1^{\infty} E_n \in \mathcal{M}$.

2.3.5. Definition. The *(Lebesgue) measure* of a set $E \in \mathcal{M}$ is the extended real number

$$\lambda(E) = \sup \int_{-n}^{n} \chi_E \, d\lambda = \lim_{n \to \infty} \int_{-n}^{n} \chi_E \, d\lambda.$$

[Since $\chi_{[-n,n]} \in \mathcal{L}_1(\Re)$ for all n, it follows that the above is well-defined. Notice also that if $\lambda(E) < \infty$, then $\chi_E \in \mathcal{L}_1(\Re)$, by the Monotone Convergence Theorem.]

The following property of Lebesgue measure is fundamental. Students familiar with probability will recognize this as the formula for computing the probability of the union of two events— in fact, there is a very strong connection between measure theory and probability theory. This connection, discussed in the exercises, is one of the factors making measure theory such an important topic.

2.3.6. Theorem. *If E and F are elements of $\mathcal{M}$, then*

$$\lambda(E \cup F) = \lambda(E) + \lambda(F) - \lambda(E \cap F),$$

provided that the right-hand side is not of the form "$\infty - \infty$."

Proof. This is immediate from the identity

$$\chi_{E \cup F} = \chi_E + \chi_F - \chi_{E \cap F}.$$

2.3.7. Proposition. *If $E, F \in \mathcal{M}$ and $F \subset E$, then $\lambda(F) \leq \lambda(E)$.*

Proof. Since $F \subset E$, $\chi_F \leq \chi_E$, so

$$\lambda(F) = \lim_{n \to \infty} \int_{-n}^{n} \chi_F \, d\lambda \leq \lim_{n \to \infty} \int_{-n}^{n} \chi_E \, d\lambda = \lambda(E) \quad \text{by 2.1.14(a).}$$

As a consequence of 2.3.6, if E and F are disjoint measurable sets and $E \cap F = \emptyset$, then $\lambda(E \cup F) = \lambda(E) + \lambda(F)$. Because of the Monotone Convergence Theorem, this extends to countable unions of pairwise disjoint sets.

2.3.8. Theorem. *Let $\{E_n\} \subset M$ and suppose, for $n \neq m$, that $E_n \cap E_m = \emptyset$. Then $\lambda(\cup_1^\infty E_n) = \sum_1^\infty \lambda(E_n)$.*

Proof. First observe that 2.3.6 implies, for each n, that

$$\lambda(\cup_1^n E_j) = \sum_1^n \lambda(E_j).$$

By 2.3.7,

$$\lambda(\cup_1^\infty E_j) \geq \lambda(\cup_1^n E_j)$$

and hence

$$\sum_1^\infty \lambda(E_j) \leq \lambda(\cup_1^\infty E_j). \tag{1}$$

Now if $\sum_1^\infty \lambda(E_j) = \infty$, this completes the proof. On the other hand, if $\sum_1^\infty \lambda(E_j) < \infty$, then $\lambda(\cup_1^n E_j) \leq \sum_1^\infty \lambda(E_j) < \infty$. Thus, if $f_n = \chi_{(E_1 \cup \ldots \cup E_n)}$, then each $f_n \in \mathcal{L}_1(\Re)$ and $f_n \leq f_{n+1}$. Then the Monotone Convergence Theorem implies that

$$\chi_{\cup E_n} = \lim f_n \in \mathcal{L}_1(\Re)$$

and

$$\lambda(\cup E_n) = \int_\Re \chi_{\cup E_n} \, d\lambda$$

$$= \lim \int_\Re f_n \, d\lambda \leq \sum_1^\infty \lambda(E_j).$$

Since this is the reverse of (1), the result follows.

If the sets fail to be pairwise disjoint, the best we can do is get an upper bound on the measure of the union.

2.3.9. Corollary. *Let $\{E_n\} \subset M$. Then*

$$\lambda(\cup_1^\infty E_n) \leq \sum_1^\infty \lambda(E_n).$$

Proof. Set $F_1 = E_1$ and, for $n > 1$, set $F_n = E_n \backslash (\cup_1^{n-1} E_j)$. Then each $F_n \in M$, $F_n \subset E_n$ for each n and, if $n \neq m$, $F_n \cap F_m = \emptyset$. Since $\cup_1^\infty E_n = \cup_1^\infty F_n$, it follows that

$$\lambda(\cup_1^\infty E_n) = \lambda(\cup_1^\infty F_n) = \sum_1^\infty \lambda(F_n) \le \sum_1^\infty \lambda(E_n),$$

applying 2.3.8 and 2.3.7.

Next, we show that Lebesgue measure is "continuous" in the algebra $\mathcal{M}$.

2.3.10. Theorem. *Let* $\{E_n\} \subset \mathcal{M}$ *and suppose, for each* n, *that* $E_{n+1} \subset E_n$. If $\lambda(E_1) < \infty$, *then*

$$\lambda(\cap_1^\infty E_n) = \lim \lambda(E_n).$$

Proof. Set $E = \cap_1^\infty E_n$ and $f_n = \chi_{E_n}$. Then f_n converges to χ_E and $|f_n| \le \chi_{E_1}$. Since $\chi_{E_1} \in \mathcal{L}_1(\Re)$ by hypothesis, the conclusion follows at once from Lebesgue's Dominated Convergence Theorem.

The foregoing three results are central features of Lebesgue measure, and indeed, measures in general.

We have yet to describe which sets are measurable. An important class, which almost corresponds to $\mathcal{M}$, is the following.

2.3.11. Definition. The smallest σ-algebra which contains all the open sets is referred to as the class of *Borel sets*, denoted by $\mathcal{B}$.

2.3.12. Proposition. *The Borel sets are a subset of* $\mathcal{M}$, *i.e.,* $\mathcal{B} \subset \mathcal{M}$.

Proof. It suffices to show that if $a < b$, then $(a, b) \in \mathcal{M}$. We have seen that $\chi_{(a,b)} \in \mathcal{L}^+(\Re)$ and $\int_\Re \chi_{(a,b)} d\lambda = (b - a)$. Thus, if $f \in \mathcal{L}_1(\Re)$, 2.1.20 implies that $f\chi_{(a,b)} \in L(\Re)$.

Now if $f \in \mathcal{L}_1(\Re)$, there are functions $g, h \in \mathcal{L}^+(\Re)$ such that $f = g - h$ almost everywhere and both g and h have finite integrals. Since $\int_\Re g\chi_{(a,b)} d\lambda \le \int_\Re g d\lambda$ and $\int_\Re h\chi_{(a,b)} d\lambda \le \int_\Re h d\lambda$, it follows that $f\chi_{(a,b)} \in \mathcal{L}_1(\Re)$. Since f was arbitrary, this shows that $(a, b) \in \mathcal{M}$.

The final result in this section will show that every measurable set can be written as the union of a Borel set and a set of measure zero (so measurable sets are "almost" Borel sets). We will actually show something a little stronger, namely that every measurable set can be written as the union of a closed set and a set of arbitrarily small measure (so measurable sets are "almost" closed sets). It will be easier to work with the complementary statement, namely that every measurable set "misses" being open by a set of arbitrarily small measure. The next lemma verifies this for bounded measurable sets.

2.3.13. Lemma. *Let $E \in \mathcal{M}$ and suppose that $E \subset (a, b)$. Then for each $\epsilon > 0$ there is an open set U such that $E \subset U$ and $\lambda(U \setminus E) < \epsilon$.*

Proof. Since $E \subset (a, b)$, $\chi_E \in \mathcal{L}_1(\Re)$, and so there are functions g and h in $\mathcal{L}^+(\Re)$ each having finite integral such that $\chi_E = g - h$ almost everywhere. By Lemma 4.2, we may assume that $\int_\Re h \, d\lambda < \frac{\epsilon}{4}$. By replacing g and h by $g \cdot \chi_{(a,b)}$ and $h \cdot \chi_{(a,b)}$, respectively, we may further assume that both $g(x) = 0$ and $h(x) = 0$ if $x \notin (a, b)$.

Now select a nondecreasing sequence $\{g_n\} \subset \mathcal{C}_C(\Re)$ such that g_n converges to g almost everywhere. Set

$$A_1 = \{x \in (a, b) : g_n(x) \text{ does not converge to } g(x)\} \text{ and}$$
$$A_2 = \{x \in (a, b) : \chi_E(x) \neq g(x) - h(x)\}.$$

Then both A_1 and A_2 have measure zero, so there are open sets $U_1 \supset A_1$ and $U_2 \supset A_2$ such that $U_1 \subset (a, b)$, $U_2 \subset (a, b)$, and

$$\lambda(U_1) < \frac{\epsilon}{4} \quad \text{and} \quad \lambda(U_2) < \frac{\epsilon}{4}.$$

Now take U_3 to be the open set

$$U_3 = \cup_1^\infty \{x \in (a, b) : g_n(x) > 1 - \gamma\},$$

where $\gamma = \frac{\epsilon}{4(b-a)}$, and set $U = U_1 \cup U_2 \cup U_3$.

To see that $E \subset U$, suppose that $x \in E$ and $x \notin (U_1 \cup U_2)$. Then $x \notin A_2$, so $\chi_E(x) \leq g(x)$, and thus $g(x) \geq 1$. Since $x \notin A_1$, there is an n such that $g_n(x) > 1 - \gamma$, establishing $E \subset U$. Observe further that if $x \in U_3 \setminus U_1$, then $g(x) > 1 - \gamma$, i.e., $1 - g(x) < \gamma$.

Thus

$$\lambda(U \setminus E) = \int_a^b \chi_U - \chi_E \, d\lambda$$

$$= \int_U 1 \, d\lambda - \int_a^b g \, d\lambda + \int_a^b h \, d\lambda$$

$$\leq \lambda(U_1 \cup U_2 \cup U_3) - \int_{U_3 \setminus U_1} g \, d\lambda + \frac{\epsilon}{4}$$

$$\leq \lambda(U_3 \setminus U_1) + \lambda(U_2) + \lambda(U_1) - \int_{U_3 \setminus U_1} g \, d\lambda + \frac{\epsilon}{4}$$

$$\leq \int_{U_3 \setminus U_1} (1 - g) \, d\lambda + \frac{3\epsilon}{4}$$

$$\leq \int_a^b \gamma \, d\lambda + \frac{3\epsilon}{4} \leq \epsilon,$$

establishing the result.

Now, by writing an arbitrary measurable set E as $E = \cup(E \cap (n, n+1])$, we can extend the above to all members of $\mathcal{M}$.

2.3.14. Theorem. *Let $E \in \mathcal{M}$. Then for each $\epsilon > 0$ there is an open set $U \supset E$ such that $\lambda(U \backslash E) < \epsilon$.*

Proof. For each n, set $E_n = E \cap (n, n+1)$. By Lemma 2.3.13 there is an open set $U_n \supset E_n$ such that

$$\lambda(U_n \backslash E_n) \leq \epsilon 2^{-|n|+4}.$$

Set $V_n = (n - \epsilon 2^{-|n|+5}, n + \epsilon 2^{-|n|+5})$, and set $U = \cup_{-\infty}^{\infty}(U_n \cup V_n)$. Then

$$\lambda(U \backslash E) = \lambda\left(\left[\cup_{-\infty}^{\infty}(U_n \backslash E_n)\right] \cup \left[\cup_{-\infty}^{\infty}(V_n \backslash E_n)\right]\right)$$

$$\leq \lambda\left(\cup_{-\infty}^{\infty}(U_n \backslash E_n)\right) + \lambda\left(\cup_{-\infty}^{\infty}V_n\right)$$

$$\leq \sum_{-\infty}^{\infty}\lambda(U_n \backslash E_n) + \sum_{-\infty}^{\infty}\lambda(V_n)$$

$$\leq \sum_{-\infty}^{\infty}\epsilon 2^{-|n|-4} + \sum_{-\infty}^{\infty}\epsilon 2^{-|n|-4}$$

$$= \frac{3\epsilon}{4},$$

proving the result.

The desired conclusions about members of $\mathcal{M}$ now follow as an easy exercise.

2.3.15. Theorem. *For a set $E \subset \Re$, the following are equivalent:*
(a) *$E \in \mathcal{M}$;*
(b) *There is a collection $\{U_n\}$ of open sets such that $E \subset \cap_1^{\infty}U_n$ and $\lambda((\cap_1^{\infty}U_n) \backslash E) = 0$;*
(c) *There is a collection $\{F_n\}$ of closed sets such that $E \supset \cup_1^{\infty}F_n$ and $\lambda(E \backslash (\cup_1^{\infty}F_n)) = 0$.*
(d) *For each $\epsilon > 0$ there is a closed set $F \subset E$ and an open set $U \supset E$ such that $\lambda(U \backslash F) \leq \epsilon$.*

PROBLEMS

2.19. Prove Theorem 2.3.15.

2.20. Either prove the following assertion or find a counterexample:

Let f be a continuous bijection (i.e., one to one and onto) from $[0, 1]$ onto $[0, 1]$; then f sends sets of measure zero to sets of measure zero.

2.21. *Outer measure.* **Definition.** Let $A \subset \Re$; define the *outer measure* $\lambda^*(A)$ of A to be

$$\lambda^*(A) = \inf \left\{ \sum (b_i - a_i) : \{(a_i, b_i)\} \text{ is a sequence of} \right.$$

$$\left. \text{disjoint intervals covering } A \right\}.$$

(a) Show that, if $A \in M$ then $\lambda^*(A) = \lambda(A)$.

(b) Show that $E \in M$ if and only if for all $A \subset \Re$

$$\lambda^*(A) = \lambda^*(A \cap E) + \lambda^*(A \cap E^C).$$

2.22. Let $A \in M$ and define $-A = \{x : -x \in A\}$ and for $t \in \Re$, $A + t = \{x : x - t \in A\}$. Show that both $-A \in M$ and $A + t \in M$ and that

$$\lambda(-A) = \lambda(A + t) = \lambda(A)$$

(i.e., that λ is *translation invariant*).

2.23. Let $E \in M$; show that

$$\lambda(E) = \sup\{\lambda(F) : F \subset E \text{ and } F \text{ is closed}\}.$$

2.24. *A nonmeasurable set.* Define a binary operation $\oplus$ on $[0, 1)$ as follows:

$$x \oplus y = \begin{cases} x + y & \text{if } x + y < 1 \\ 1 - (x + y) & \text{if } x + y \geq 1. \end{cases}$$

(a) Let M be a measurable subset of $[0, 1)$ and let $y \in [0, 1)$. Show that $(M \oplus y) = \{z \oplus y : z \in M\}$ is measurable and that $\lambda(M \oplus y) = \lambda(M)$.

Now define an equivalence relation "$\sim$" on $[0, 1)$ by saying that $x \sim y$ if and only if $x - y$ is rational. For each $x \in [0, 1)$, define $[x] = \{y \in [0, 1) : x \sim y\}$.

(b) Show that "$\sim$" is an equivalence relation and hence either $[x] = [y]$ or $[x] \cap [y] = \emptyset$.

Next select one representative element from each equivalence class, and denote by E the collection of all such representatives. (The Axiom of Choice guarantees that we may make such a selection—see Appendix B.)

(c) Let $\{r_n\}$ be an enumeration of the rational numbers in $[0, 1)$ and let $E_n = \{x \in [0, 1) : x \oplus r_n \in E\}$. Show that $[0, 1) = \cup E_n$ and that if $n \neq m$, then $E_n \cap E_m = \emptyset$.

(d) Conclude that E is **not** measurable.

2.25. Let f be a nonnegative function in $\mathcal{L}_1(\Re)$ and suppose that

$$\int_\Re f \, d\lambda = 1.$$

(Such a function is called a *probability density function.*) For a measurable set E define the set function μ_f by $\mu_f(E) = \int_E f \, d\lambda$. [Notice that $\mu_f(\Re) = 1$; μ_f is an example of a *probability measure.*]

(a) If $\{E_n\}$ is a sequence of disjoint measurable sets, show that

$$\mu_f\left(\bigcup E_n\right) = \sum \mu_f(E_n).$$

(b) Show that $\mu_f(A \cup B) = \mu_f(A) + \mu_f(B) - \mu_f(A \cap B)$ for all measurable sets A, B.

If A is measurable and if $\mu_f(A) \neq 0$, then we can define the set function $\mu_{f|A}$ as follows:

$$\mu_{f|A} = \frac{\mu_f(A \cap B)}{\mu_f(A)}.$$

(c) Show that $\mu_{f|A}(\Re) = 1$ and that if $\{E_n\}$ is a family of disjoint measurable sets, then

$$\mu_{f|A}\left(\bigcup E_n\right) = \sum \mu_{f|A}(E_n).$$

The measurable sets A and B are *independent* if

$$\mu_f(A \cap B) = \mu_f(A)\mu_f(B).$$

(d) Let A and B be measurable sets with $\mu_f(A) \neq 0$; show that A and B are independent if and only if $\mu_{f|A}(B) = \mu_f(B)$.

(e) Prove *Bayes' Rule*: if A and B are measurable sets and if $\mu_f(A) \neq 0 \neq \mu_f(B)$, then

$$\mu_{f|A}(B) = \frac{\mu_{f|B}(A)\mu_f(B)}{\mu_f(A)}.$$

(f) Suppose that $xf(x)$ and $x^2 f(x)$ are members of $\mathcal{L}_1(\Re)$ and set

$$\mu = \int_{\Re} xf(x)\,dx \quad \text{and}$$

$$\sigma^2 = \int_{\Re} x^2 f(x)\,dx.$$

Verify *Cebysev's inequality*:

$$\int_{\mu-\epsilon}^{\mu+\epsilon} f(x)\,dx \leq \frac{\sigma^2}{\epsilon^2}$$

for all $\epsilon > 0$.

2.26. Prove *Steinhaus' Theorem*: Let A be a closed bounded set of real numbers and suppose that $\lambda(A) > 0$. Then there is an $\epsilon > 0$ such that if $|x| < \epsilon$, then $x = u - v$ for some pair $u, v \in A$.

2.27. Find an example of a sequence $\{E_n\}$ of measurable sets with $E_{n+1} \subseteq E_n$ for which $\lambda(\cup E_n) \neq \lim \lambda(E_n)$.

2.28. Show that there is a measurable set E which is *not* a Borel set.

2.29. Show that for each $\epsilon > 0$ there is an open set $E \subseteq [0, 1]$ such that $\lambda(E) = \epsilon$ and the closure of E is $[0, 1]$.

2.30. Suppose that $A, B \subseteq [0, 1]$ and $\lambda(A) = 1$. Show that $\lambda(A \cap B) = \lambda(B)$.

2.31. If $\{E_n\}$ is a sequence of measurable sets, show that both

$$LI(E_n) \equiv \bigcup_{i=1}^{\infty} \bigcap_{j=i}^{\infty} E_j \quad \text{and} \quad LS(E_n) \equiv \bigcap_{i=1}^{\infty} \bigcup_{j=i}^{\infty} E_j$$

are measurable. [Note that $x \in LS(E_n)$ if and only if x is in infinitely many of the sets E_n, while $x \in LI(E_n)$ if and only if x is in all but finitely many of the sets E_n. These sets are sometimes called the *limit superior* and *limit inferior* of the sequence $\{E_n\}$. We will use these sets in proving Egorov's Theorem in Section 2.6.]

2.4. MEASURABLE FUNCTIONS.

We are now in a position to give an alternative description of the class $L(\Re)$. In Section 2.3 we showed that measurable sets are "almost" open sets. In the same way, we shall see that integrable functions are "almost continuous"—this is scarcely surprising since a function in $L(\Re)$ is almost everywhere a limit of continuous functions. The appropriate class of functions to consider is given by the following definition; note the similarity with the characterization of "continuous functions" given in 1.1.5(iii).

2.4.1. Definition. Let f be an extended real-valued function defined on $\Re$; f is said to be *measurable* if whenever U is an open set, then

$$\{x : f(x) \in U\} \in M.$$

We will show that $f \in L(\Re)$ if and only if f is a measurable function. We begin with a sequence of preliminary results. The first of these gives a characterization of measurable functions in terms of inverse images of intervals; this characterization is especially useful in establishing measurability of particular functions.

2.4.2. Proposition. *An extended real-valued function g is measurable if and only if for each $r \in \Re$, any of the following sets are measurable:*
   (a) $\{x : g(x) \leq r\}$
   (b) $\{x : g(x) < r\}$
   (c) $\{x : g(x) > r\}$
   (d) $\{x : g(x) \geq r\}$

Proof. If g is measurable, then the sets in both (b) and (c) are measurable. Since (a) and (d) are the complements to (c) and (b) and M is closed under complements, (a) and (d) are also measurable.

For the converse, suppose that U is an open set. Then $U = \cup_1^{\infty}(a_n, b_n)$ where the intervals $\{(a_n, b_n)\}$ are disjoint. Since $(a_n, b_n) = (-\infty, b_n) \cap (a_n, \infty)$, each of the sets

$$E_n = \{x : f(x) \in (a_n, b_n)\}$$

is measurable, and thus

$$\cup E_n = \{x : f(x) \in U\}$$

is also measurable.

Although it turns out that the class of all measurable functions has the same vector space and lattice properties as the class $L(\mathfrak{R})$, even the most basic of these properties are not immediately obvious from the definition. However, axioms of σ-algebras and the preceding proposition place many of these properties in easy reach.

2.4.3. Proposition. *If f and g are measurable functions and $\alpha \in \mathfrak{R}$, then the functions $f + g$ and αf are measurable.*

Proof. It is apparent from 2.4.2 that αf is measurable if f is. To see that $f + g$ is measurable, fix $r \in \mathfrak{R}$ and observe that if $f(x) + g(x) < r$, then it is possible to "wedge" a rational number q between r and $f(x) + g(x)$, i.e.,

$$\{x : f(x) + g(x) < r\} = \cup_{q \in Q} \left(\{x : f(x) + q \le r\} \cap \{x : g(x) \le q\} \right).$$

The latter set is measurable since $\mathcal{M}$ is closed under countable unions and intersections, and hence $f + g$ is measurable.

It turns out to be even easier to show that the measurable functions are closed under the lattice and limit operations.

2.4.4. Proposition. *Let $\{f_n\}$ be a family of measurable functions. Then the following functions are measurable.*
   (a) $\sup(f_n)$
   (b) $\inf(f_n)$
   (c) $\overline{\lim}(f_n)$
   (d) $\underline{\lim}(f_n)$.
   If, in addition, $\lim f_n$ exists almost everywhere, then $\lim f_n$ is also measurable.

Proof. All the conclusions follow readily from (a) (Problem 2.32.). To verify (a), fix $r \in \mathfrak{R}$ and note that

$$\{x : \sup\{f_n(x)\} \le r\} = \cap_n \{x : f_n(x) \le r\}$$

and hence $\sup f_n$ is measurable.

2.4.5. Corollary. *Let f_n be a family of nonnegative measurable functions. Then $\sum f_n$ is measurable.*

A first approximation to showing f is measurable if and only if f is in $L(\Re)$ will be to show that functions in $\mathcal{L}^+(\Re)$ are measurable; this is essentially the content of the next lemma. The proof combines the ideas used in the preceding two propositions.

2.4.6. Lemma. *Let $g \in \mathcal{L}^+(\Re)$, and let $r \in \Re$. Then*

$$\{x : g(x) > r\} \in \mathcal{M}.$$

Proof. Select a nondecreasing sequence $\{g_n\} \subset \mathcal{C}(\Re)$ such that g_n converges to g on the set E. Since E^C has measure 0, $E \in \mathcal{M}$. Now

$$\{x \in E : g(x) > r\} = \cup_1^\infty \{x \in E : g_n(x) > r\}$$

since $\{g_n\}$ is nondecreasing. Since each g_n is continuous,

$$E_n = \{x \in \Re : g_n(x) > r\}$$

is open, and hence in $\mathcal{M}$. Thus $E_n \cap E \in \mathcal{M}$ and $\cup_1^\infty E_n \cap E \in \mathcal{M}$. This shows that $\{x \in E : g(x) > r\}$ is in $\mathcal{M}$. Since $\lambda(E^C) = 0$, $\{x \in E^C : g(x) > r\}$ also has measure zero, and thus is also in $\mathcal{M}$. The union of these two sets must also be in $\mathcal{M}$, which establishes the result.

The next corollary now follows in exactly the same manner as 2.4.2.

2.4.7. Corollary. *Let $g \in \mathcal{L}^+(\Re)$ and let $r \in \Re$. Then each of the following sets is in $\mathcal{M}$:*

    (a) $\{x : g(x) \leq r\}$
    (b) $\{x : g(x) \geq r\}$
    (c) $\{x : g(x) < r\}$
    (d) $\{x : g(x) = r\}$.

In particular, g is a measurable function.

The next step is to show that Lemma 2.4.6 holds for functions in $L(\Re)$; once this is done, it will follow at once that every function in $L(\Re)$ is measurable.

2.4.8. Lemma. *Let $f \in L(\Re)$ and let $r \in \Re$. Then*

$$\{x : f(x) < r\} \in \mathcal{M}.$$

Proof. Select $g, h \in \mathcal{L}^+(\Re)$ such that $f = g - h$ almost everywhere. It suffices to show that

$$\{x : g(x) - h(x) < r\} \in M.$$

But observe that

$$\{x : g(x) - h(x) < r\} = \cup_{q \in Q}\Big(\{x : g(x) < r + q\} \cap \{x : q < h(x)\}\Big).$$

By 2.4.7, each of the sets on the right-hand side of the expression above are in M. Hence, since M is closed under countable unions, the set $\{x : f(x) < r\}$ is in M also.

The following theorem is the first half of our desired equivalence.

2.4.9. Theorem. *If $f \in L(\Re)$ and $r \in \Re$, then each of the following sets is in M:*

(a) $\{x : f(x) \leq r\}$
(b) $\{x : f(x) < r\}$
(c) $\{x : f(x) \geq r\}$
(d) $\{x : f(x) > r\}$
(e) $\{x : f(x) = r\}$. *In particular, f is measurable.*

Proof. This follows exactly as with 2.4.7.

The converse to 2.4.9 is much harder; we need to show that measurable functions can, in some way, be "built up" from continuous functions. Once again, we will first establish an intermediate result which relates measurable functions and measurable sets in a very concrete way. Once this is in place, we can use the results about measurable sets from Section 2.3 to deduce our result.

2.4.10. Definition. Let $\{\alpha_1, \dots, \alpha_n\} \subset \Re$ and let $\{E_1, \dots, E_n\}$ be a family of disjoint measurable sets, each having finite measure. A *simple function α* is a function of the form $\alpha = \sum_1^n \alpha_j \chi_{E_j}$; the class of all simple functions will be denoted by $\sum$.

Certainly, every simple function is a measurable function. If we could approximate measurable functions with simple functions (much as we approximated continuous functions with step functions in studying the Riemann integral), we would take a big step in showing that measurable functions are in $L(\Re)$. In order to simplify convergence arguments it will be simpler to deal with nonnegative measurable functions first.

2.4.11. Theorem. *Let f be a nonnegative measurable function. Then there is a nondecreasing sequence $\{\sigma_n\} \subset \sum$ such that $\lim \sigma_n = f$.*

Proof. For each pair of natural numbers (n, k) define

$$E(n, k) = \{x : k2^{-n} \leq f(x) < (k+1)2^{-n}\} \cap [-n, n]$$

and set

$$\varphi_n = 2^{-n} \sum_{k=1}^{2^{2n}} k \chi_{E(n,k)}.$$

Since, for fixed n, $f(x)$ lies in at most one of the intervals $\{[k2^{-n}, (k+1)2^{-n})\}$, the sets $\{E(n, k)\}_{k=0}^{2^{2n}}$ are pairwise disjoint. Since each is contained in the interval $[-n, n]$, each has finite measure, and so $\varphi_n \in \sum$.

We will next show that $\varphi_{n+1} \geq \varphi_n$; to see this, fix x and suppose that $\varphi_n(x) \neq 0$ since if $\varphi_n(x) = 0$, the conclusion is obvious. Since $\varphi_n(x) \neq 0$, $x \in E(n, k)$ for some k between 1 and 2^{2n}, so

$$k2^{-n} \leq f(x) < (k+1)2^{-n} \quad \text{and} \quad -n \leq x \leq n.$$

Now we distinguish two cases. [Note that $-(n+1) \leq x \leq n+1$.]

Case 1. $k2^{-n} \leq f(x) < 1/2(k2^{-n} + (k+1)2^{-n})$.

In this case,

$$2k2^{-(n+1)} \leq f(x) < (2k+1)2^{-(n+1)},$$

and so $x \in E(n+1, 2k)$. Then

$$\begin{aligned} \varphi_{n+1}(x) &= 2^{-(n+1)}(2k) \\ &= 2^{-n}k \\ &= \varphi_n(x). \end{aligned}$$

Case 2. $1/2(k2^{-n} + (k+1)2^{-n}) \leq f(x) < (k+1)2^{-n}$.

In this case

$$(2k+1)2^{-(n+1)} \leq f(x) < (2k+2)2^{-(n+1)},$$

and so $x \in E(n+1, 2k+1)$. Then

$$\begin{aligned} \varphi_{n+1}(x) &= 2^{-(n+1)}(2k+1) \\ &> 2^{-(n+1)}2k \\ &= \varphi_n(x), \end{aligned}$$

so in either case $\varphi_n(x) \leq \varphi_{n+1}(x)$.

Now suppose that $f(x) < \infty$; we will show that $\lim \varphi_n(x) = f(x)$. If $f(x) = 0$, then each $\varphi_n(x) = 0$, so we suppose that $f(x) > 0$. Fix $\epsilon > 0$ and choose N so large that $n \geq N$ implies that $2^n > f(x)$, $|x| \leq n$, and $2^{-n} \leq \epsilon$. For such n, there is exactly one interval $I = [k2^{-n}, (k+1)2^{-n})$ such that $f(x) \in I$

as k ranges from 0 to 2^{2n} (this is why the sum runs to 2^{2n}), and thus $x \in E(n, k)$. Thus

$$0 = k2^{-n} - k2^{-n} \leq f(x) - \varphi_n(x) \leq (k+1)2^{-n} - k2^{-n} = 2^{-n} \leq \epsilon$$

and so $\lim \varphi_n(x) = f(x)$ if $f(x) < \infty$.

Next set

$$A_n = \{x : f(x) = +\infty \text{ and } |x| \leq n\} = \cap_k \{x : f(x) \geq k \text{ and } |x| \leq n\}$$

so that each A_n is measurable, $A_n \cap E(n, k) = \emptyset$, and $\lambda(A_n) < \infty$. Set

$$\sigma_n = \varphi_n + n\chi_{A_n}$$

so that $\{\sigma_n\}$ is a nondecreasing sequence of simple functions and $\lim \sigma_n = f$.

By decomposing f into its positive and negative parts, Theorem 2.4.11 easily extends to all of $L(\Re)$.

2.4.12. Corollary. *If $f \in L(\Re)$, then there is a sequence $\{\sigma_n\}$ of simple functions such that $\lim \sigma_n = f$ and $|\sigma_n| \leq |f|$.*

Proof. By 2.4.11, there are nondecreasing sequences of simple functions $\{\varphi_n\}$ and $\{\psi_n\}$ such that

$$\lim \varphi_n = f^+ \quad \text{and} \quad \lim \psi_n = f^-.$$

Then $\sigma_n = \varphi_n - \psi_n$ is a simple function, $\lim \sigma_n = f$ and

$$-|f| = -f^+ - f^- \leq \varphi_n - \psi_n \leq f^+ + f^- = |f|,$$

so $|\sigma_n| \leq |f|$.

In order to apply the results above, we need to know when the limit of the sequence $\{\sigma_n\}$ is a member of $L(\Re)$; this is analogous to 2.2.5 (the Monotone Convergence Theorem).

2.4.13. Lemma. *Let $\{f_n\}$ be a sequence of nonnegative functions in $\mathcal{L}_1(\Re)$ and suppose that $f(x) = \sum_1^{\infty} f_n(x)$. Then $f \in L(\Re)$.*

Proof. By 2.2.2, for each n there are functions $\varphi_n, \psi_n \in \mathcal{L}^+(\Re)$ for which $f_n = \varphi_n - \psi_n$ a.e. and $\int_{\Re} \psi_n \, d\lambda \leq 2^{-n}$. By 2.2.1, $\varphi = \sum_1^{\infty} \varphi_n$ and $\psi = \sum_1^{\infty} \psi_n$ are in $\mathcal{L}^+(\Re)$ and $\int_{\Re} \varphi \, d\lambda = \sum_1^{\infty} \int_{\Re} \varphi_n \, d\lambda$, $\int_{\Re} \psi \, d\lambda = \sum_1^{\infty} \psi_n \, d\lambda \leq 1$.

Thus by 2.2.3, $\psi(x) < \infty$ a.e. and so the function $\varphi - \psi$ is defined a.e. and $f = \sum f_n = \sum (\varphi_n - \psi_n) = \varphi - \psi$ a.e.

2.4.14. Theorem. *If f is measurable, then $f \in L(\Re)$.*

Proof. It suffices to prove this in the case that f is nonnegative, for then in the general case $g = g^+ - g^-$ is the difference of two nonnegative functions in $L(\Re)$ and hence in $L(\Re)$.

Thus suppose that $f \geq 0$ and select a nondecreasing sequence $\{\sigma_n\}$ of simple functions such that $\lim \sigma_n = f$. Set $f_1 = \sigma_1$ and for $n > 1$, set $f_n = \sigma_n - \sigma_{n-1}$, so each f_n is a nonnegative function. Since each $\sigma_n \in \mathcal{L}_1(\Re)$, so is each f_n. Moreover, $\sum_{j=1}^{n} f_j = \sigma_n$ and so $f = \sum_1^{\infty} f_n$. By 2.4.13, $\sum_1^{\infty} f_n \in L(\Re)$, completing the proof.

PROBLEMS

2.32. Complete the proof of Proposition 2.4.4.

2.33. Let f be a measurable function and let g be a continuous function; show that $h = g \circ f$ is measurable.

2.34. Show that the composition of measurable functions need not be measurable.

2.35. If $f \in L(\Re)$, show that there is a sequence $\{g_n\}$ of continuous functions such that $|g_n| \leq |f|$ a.e. and $\lim g_n = f$ a.e.

2.36. If $f \in L(\Re)$, show that there is a sequence of step functions $\{s_n\}$ such that $|s_n| \leq |f|$ a.e. and $\lim s_n = f$ a.e.

2.37. Let $f \in L(\Re)$ and suppose that $|f| \in \mathcal{L}_1(\Re)$. Show that $f \in \mathcal{L}_1(\Re)$.

2.38. Let f and g be nonnegative functions in $L(\Re)$ and suppose that $f \geq g$ and that $f \in \mathcal{L}_1(\Re)$. Show that $g \in \mathcal{L}_1(\Re)$.

2.39. Show that E is measurable if and only if $\chi_E \in L(\Re)$.

2.40. Let $f \in L(\Re)$ and let $0 \leq p < \infty$. Show that $|f|^p \in L(\Re)$.

2.41. Show that every upper [respectively, lower] semicontinuous function is measurable.

2.42. Show that every nondecreasing [respectively, nonincreasing] function is measurable.

2.43. Show that Lebesgue's singular function (see Problem 1.17) is measurable.

2.44. Let $f \in L(\Re)$; define g by

$$g(x) = \begin{cases} 0 & \text{if } f(x) \text{ is rational} \\ 1 & \text{otherwise.} \end{cases}$$

Show that g is measurable.

2.45. Let $f \in \mathcal{L}_1(\Re)$ and let $\epsilon > 0$. Show that there is an upper semicontinuous function φ bounded from above and a lower semicontinuous function ψ bounded from below satisfying $\varphi \leq f \leq \psi$ and $\int_\Re \psi - \varphi \, d\lambda < \epsilon$.

2.5. LITTLEWOOD'S THREE PRINCIPLES. The famous English mathematician J. E. Littlewood has described three general principles which provide an intuitive basis for many applications of Lebesgue's integration theory. These

principles, although heuristic in character, often suggest how to approach partic-ular problems, and an application of one or more of the principles almost always simplifies seemingly complex problems. The principles are:

    (1) Every measurable set is "nearly" an open set;

    (2) Every convergent sequence of measurable functions "nearly" converges uniformly; and

    (3) Every measurable function is "nearly" continuous.

      We saw various forms of the first principle in Section 2.3 in 2.3.14 and 2.3.15. The last principle was exploited in the Section 2.4 (although it was not formally stated). The rigorous formulation of the latter two principles is the main topic of this section; the second principle is a consequence of Egorov's Theorem.

2.5.1. Egorov's Theorem. *Let E be a measurable set having finite measure and let $\{f_n\}$ be a sequence of measurable functions defined on E; suppose that there is a measurable function f defined on E satisfying $\lim f_n = f$ a.e. on E. Then for each $\epsilon > 0$ there is a set $A \subset E$ such that*

    (i) $\lambda(E \backslash A) \leq \epsilon$; and

    (ii) $\{f_n\}$ converges to f uniformly on A.

Proof. For each pair of natural numbers (n, k) set

$$A(n, k) = \left\{ x \in E : |f(x) - f_k(x)| \geq \frac{1}{n} \right\};$$

fix $\epsilon > 0$.

      Keeping n fixed, set

$$B(n, j) = \cup_{k=j}^{\infty} A(n, k);$$

note that $\{B(n, j)\}_{j=1}^{\infty}$ is a nonincreasing sequence of measurable sets. Moreover,

$$\cap_{j=1}^{\infty} B(n; j) \subset \{x \in E : \{f_n(x)\} \text{ does not converge to } f(x)\}$$

and hence $\lambda\left(\cap_{j=1}^{\infty} B(n; j)\right) = 0$. Thus by Theorem 2.3.10, $\lim_{j} \lambda(B(n; j)) = 0$ for each n, and consequently we may select $j(n)$ so large that $\lambda(B(n; j(n))) \leq \epsilon 2^{-n}$.

      Now set $F = \cup_{n=1}^{\infty} B(n; j(n))$ so that $\lambda(F) \leq \sum \lambda(B(n; j(n))) \leq \epsilon$. Take $A = E \backslash F$; fix $\delta > 0$ and select n so large that $\frac{1}{n} \leq \delta$. If $m \geq j(n)$ and $x \in A$, then $x \notin B(n; j(n))$ and so

$$|f_m(x) - f(x)| \leq \frac{1}{n} < \delta.$$

Since this holds for all $x \in A$ and for $j(n)$ fixed, this establishes the result.

      The proof above is deceptively short and encompasses some important techniques. The sets

$$\cap_{j=1}^{\infty} B(n; j) = \cap_{j=1}^{\infty} \cup_{k=j}^{\infty} A(n, k)$$

can be thought of as a "limit superior" of the family $\{A(n, k)\}$. Note that $x \in \cap_{j=1}^{\infty} B(n; j)$ if and only if $x \in A(n, k)$ for infinitely many values of k. Thus this intersection is precisely the collection of all x for which $\overline{\lim}|f_k(x) - f(x)| \geq \frac{1}{n}$. The continuity property of measures in 2.3.10 combines with these observations to yield the result.

The third of Littlewood's principles, Lusin's Theorem, is deduced from the second. The latter is especially useful where integration techniques from calculus can simplify a problem.

2.5.2 Lusin's Theorem. *Let f be a measurable function defined on an interval $[a, b]$. Then for each $\epsilon > 0$ there is a closed set $A \subset [a, b]$ such that f is continuous on A and $\lambda([a, b]\backslash A) < \epsilon$.*

Proof. Since $f \cdot \chi_{[a,b]} \in L(\Re)$, there are functions φ and ψ in $\mathcal{L}^+(\Re)$ such that $f \cdot \chi_{[a,b]} = \varphi - \psi$ a.e. Select nondecreasing sequences $\{\varphi_n\}$ and $\{\psi_n\}$ in $\mathcal{C}_C(\Re)$ such that $\lim \varphi_n = \varphi$ and $\lim \psi_n = \psi$ a.e. Fix $\epsilon > 0$ and select $B \subset [a, b]$ so that $\lambda([a, b]\backslash B) < \frac{\epsilon}{2}$ and $\{\varphi_n - \psi_n\}$ converges uniformly to $\varphi - \psi$ on B. Set $E = \{x : \varphi - \psi \neq f\chi_{[a,b]}\}$, so $\lambda(E) = 0$. Now we may select an open set U so that $U \supset ([a, b]\backslash B) \cup E$ and $\lambda(U\backslash(([a, b]\backslash B) \cup E)) < \frac{\epsilon}{2}$. Then

$$\lambda(U) \leq \lambda(U\backslash(([a, b]\backslash B) \cup E))$$
$$+ \lambda([a, b]\backslash B) + \lambda(E)$$
$$\leq \epsilon.$$

Thus if $A = [a, b]\backslash U$, then $\lambda([a, b]\backslash A) \leq \lambda(U) \leq \epsilon$, A is closed, and $\{\varphi_n - \psi_n\}$ converges to f uniformly on A. Since $\varphi_n - \psi_n$ is continuous for each n and A is compact, this implies that f is continuous on A.

Somewhat more can be said: There is a continuous function g defined on all of $[a, b]$ such that $g = f$ on A; i.e., $f\chi_A$ can be extended to a continuous function defined on $[a, b]$. To prove this, we first verify the following lemma.

2.5.3. Lemma. *Let $[a, b]$ be a closed bounded interval and let A be a closed subset of $[a, b]$. If f is a continuous function defined on A, then for each $\epsilon > 0$ there is a continuous function g defined on $[a, b]$ such that*

$$\sup\{|f(t) - g(t)| : t \in A\} \leq \epsilon.$$

Proof. Fix $\epsilon > 0$; since A is compact, f is uniformly continuous on A and so there is a $\delta > 0$ such that $|s - t| < \delta$ implies that $|f(s) - f(t)| < \frac{\epsilon}{2}$. Choose n so large that if $\sigma = (b - a)/n$, then $\sigma < (\delta/2)$ and for $k = 0, \ldots, n$ set $x_k = a + k\sigma$. Let

$$k_0 = \min\{k \geq 0 : [a, x_{k+1}] \cap A \neq \emptyset\}.$$

For $k = 0, \dots, k_0 - 1$, set $t_k = x_k$; for $k \geq k_0$, set

$$t_k = \sup([a, x_{k+1}] \cap A).$$

Next define

$$g(x_k) = \begin{cases} 0 & k = 0, \dots, k_0 - 1 \\ f(t_k) & k \geq k_0. \end{cases}$$

For $x \in [x_k, x_{k+1}]$, set

$$g(x) = \sigma^{-1}(g(x_{k+1}) - g(x_k))(x - x_k) + g(x_k)$$

so that the graph of g is the line segment joining $(x_k, g(x_k))$ and $(x_{k+1}, g(x_{k+1}))$, and hence g is continuous.

Now if $t \in A$, then $t \in [x_k, x_{k+1}]$ for some k, and so

$$\begin{aligned}
|g(t) - f(t)| &\leq |g(t) - g(x_k)| + |g(x_k) - f(t)| \\
&= \sigma^{-1}|g(x_{k+1}) - g(x_k)|\,|t - x_k| + |f(t_k) - f(t)| \\
&\leq |f(t_{k+1}) - f(t_k)| + |f(t_k) - f(t)|.
\end{aligned}$$

Now $t \in [x_k, x_{k+1}] \cap A$, so $t_k \in [x_k, x_{k+1}]$, and thus $|t - t_k| \leq \frac{\delta}{2}$. Moreover, $x_k \leq t_k \leq t_{k+1} \leq x_{k+2}$, and so $|t_k - t_{k+1}| \leq |x_{k+2} - x_k| \leq 2\sigma \leq \delta$. Thus $|f(t_k) - f(t_{k+1})| \leq \frac{\epsilon}{2}$ and $|f(t_k) - f(t)| \leq \frac{\epsilon}{2}$, so $|g(t) - f(t)| \leq \epsilon$.

The approximation embodied in g in 2.5.3 is frequently used in results of this type; the student should draw a sketch to help understand how the approximation works.

2.5.4. Theorem. *Let A be a closed subset of a closed, bounded interval $[a, b]$ and let f be a continuous function defined on A. Then there is a continuous function g defined on all of $[a, b]$ satisfying*

(i) $\sup\{|g(t)| : a \leq t \leq b\} \leq \sup\{|f(t)| : t \in A\}$; and

(ii) $f(t) = g(t)$ if $t \in A$.

Proof. Set $\alpha = \sup\{|f(t)| : t \in A\}$; if $\alpha = 0$, the result is trivial, so we suppose that $\alpha > 0$. Select $\varphi_n \in C[a, b]$ such that

$$\sup\{|\varphi_n(t) - f(t)| : t \in A\} \leq \alpha 2^{-n-2}.$$

Set $\psi_0 = \varphi_0$ and for $n \geq 1$, set $\psi_n = \varphi_n - \varphi_{n-1}$. Then $\varphi_n = \sum_{j=0}^{n} \psi_j$ and if $t \in A$, $|\psi_n(t)| \leq \alpha 2^{-n-1}$. Now set

$$g_n = (\psi_n \wedge \alpha 2^{-n-1}) \vee (-\alpha 2^{-n-1})$$

so, if $t \in A$, $g_n = \psi_n$, while for all t, $|g_n(t)| \leq \alpha 2^{-n-1}$. Thus the sequence $\{\sum_0^n g_j\}$ converges uniformly on $[a, b]$ to a continuous function g. If $t \in A$,

$$g(t) = \lim \sum_0^n g_j(t) = \lim \varphi_n(t) = f(t),$$

showing that g satisfies (ii). If $t \in [a, b]$ is arbitrary,

$$|g(t)| \leq |\sum_{j=0}^{\infty} g_j(t)| \leq \alpha,$$

showing that g satisfies (i) as well.

2.5.5. Corollary. *Let f be a measurable function defined on $[a, b]$ and let $\epsilon > 0$. Then there is a continuous function g defined on all of $[a, b]$ and a set $A \subset [a, b]$ such that:*

(i) $\lambda([a, b] \backslash A) < \epsilon$;
(ii) $g(t) = f(t)$ for all $t \in A$; and
(iii) $\sup\{|g(t)| : t \in [a, b]\} \leq \sup\{|f(t)| : t \in [a, b]\}$.

PROBLEMS

2.46. Only one of Egoroff's Theorem and Lusin's Theorem remains true when the functions are defined on all of $\Re$. Prove the one that is true and find a counterexample to the one that fails.

2.47. Let E be a bounded measurable set and let $x > 0$; define

$$xE = \left\{ t \in \Re : \frac{1}{x} t \in E \right\}.$$

Show that xE is measurable and that $\lambda(xE) = x\lambda(E)$.

2.48. Let $f \in \mathcal{L}_1(\Re)$ and suppose that there is a constant M such that $|f(t)| \leq M$ a.e. Show that

$$\lim_{x \to \infty} \int_a^b f(xt)\, dt = 0$$

for all real numbers $a < b$.

2.49. *Convergence in Measure.* **Definition.** A sequence $\{f_n\}$ of measurable functions is said to *converge in measure* to the measurable function f if for each $\epsilon > 0$

$$\lim_{n \to \infty} \lambda\{x : |f(x) - f_n(x)| \geq \epsilon\} = 0.$$

(a) Let $\{f_n\}$ be a sequence of measurable functions which converges in measure to f. Show there is a subsequence $\{f_{n_k}\}$ such that $\{f_{n_k}\}$ converges to f a.e. [Hint: Set $A_{n,k} = \{x : |f_n(x) - f(x)| \geq 2^{-k}\}$.]
(b) Find a sequence $\{f_n\} \subset \mathcal{L}_1[0, 1]$ such that $\{f_n\}$ converges to zero in measure but $\{f_n(x)\}$ converges for no x.

2.50. Let $f \in \mathcal{L}_1(\Re)$; for $h \in \Re$ define $f_h(x) = f(x + h)$. Show that

$$\int_{\Re} |f_h - f|\, d\lambda \to 0 \quad \text{as} \quad h \to 0.$$

2.51. Let $f \in \mathcal{L}_1(\Re)$; show that for almost all u

$$\lim_{\epsilon \to 0} \frac{1}{2\epsilon} \int_{u-\epsilon}^{u+\epsilon} |f(x) - f(u)| \, d\lambda = f(u).$$

2.6. MORE CONVERGENCE THEOREMS.

The results of the preceding sections enable us to extend slightly the definition of the integral given in Section 2.1 and to somewhat sharpen the Monotone Convergence Theorem and Fatou's Lemma.

2.6.1. Lemma. *Let f be a nonnegative, measurable function and set*

$$I(f) = \sup\{\int_\Re \sigma \, d\lambda : \sigma \text{ is a simple function}$$

$$\text{and } \sigma \le f\}.$$

If $f \in \mathcal{L}_1(\Re)$, then $I(f) = \int_\Re f \, d\lambda$.

Proof. Select a nondecreasing sequence of simple functions $\{\sigma_n\}$ such that $\lim \sigma_n = f$. Select a sequence $\{\beta_n\}$ of simple functions such that $\beta_n \le f$ and $\lim \int \beta_n \, d\lambda = I(f)$; set $\alpha_n = \beta_n \vee \sigma_n$, so α_n is a simple function, $\alpha_n \le f$.
Thus

$$f \ge \overline{\lim} \alpha_n \ge \underline{\lim} \sigma_n = f,$$

and so $f = \lim \alpha_n$. Moreover,

$$I(f) = \lim \int \beta_n \, d\lambda$$

$$\le \underline{\lim} \int \alpha_n \, d\lambda \le \overline{\lim} \int \alpha_n \, d\lambda \le I(f),$$

so $\lim \int \alpha_n \, d\lambda = I(f)$. Now if $f \in \mathcal{L}_1(\Re)$, then the Dominated Convergence Theorem implies that

$$I(f) = \lim \int \alpha_n \, d\lambda = \int f \, d\lambda.$$

2.6.2. Definition. If f is a nonnegative measurable function, define

$$\int_\Re f \, d\lambda = \sup\{\int_\Re \sigma \, d\lambda : \sigma \text{ is a simple function and } \sigma \le f\}.$$

In view of 2.6.1, this definition reduces to that given in Section 2.1 if $f \in \mathcal{L}_1(\Re)$, and thus extends the Lebesgue integral to a slightly larger class. Although it remains the case that $\int f + g = \int f + \int g$, this is no longer obvious from the definitions—see the problems at the end of this section.

Fatou's Lemma and the Monotone Convergence Theorem continue to hold using this extended notion of integration.

2.6.3. Theorem. *Let $\{f_n\}$ be a sequence of nonnegative measurable functions such that $\lim f_n(x) = f(x)$ a.e. Then*

$$\int_{\Re} f \, d\lambda \leq \underline{\lim} \int_{\Re} f_n \, d\lambda.$$

Proof. If $f \in \mathcal{L}_1(\Re)$, this may be proved readily by appealing to 2.2.7, and so we establish only the case when $\int_{\Re} f \, d\lambda = +\infty$.

Fix $M > 0$ and select a simple function σ such that $0 \leq \sigma \leq f$ and $\int \sigma \, d\lambda \geq 2M$. By the Monotone Convergence Theorem

$$\lim \int_{-n}^{n} \sigma \, d\lambda = \int \sigma \, d\lambda$$

and so for $n \geq N$ sufficiently large,

$$\int_{-n}^{n} \sigma \, d\lambda \geq M.$$

Let $K = \max\{\sigma(x)\}$.

Now fix $\epsilon > 0$. By Lusin's Theorem, there is a closed set $A_1 \subset [-N, N]$ and a continuous function g defined on $[-N, N]$ such that

$$\lambda([-N, N]\backslash A_1) \leq \frac{\epsilon}{2K}$$

$$\sigma(t) = g(t) \quad \text{for} \quad t \in A_1.$$

Set $g_n = f_n \wedge \sigma$, so that $\lim g_n = \sigma$ a.e. By Egorov's Theorem, there is a closed set $A_2 \subset [-N, N]$ such that $\{g_n\}$ converges to σ uniformly on A_2, and $\lambda([-N, N]\backslash A_2) \leq \frac{\epsilon}{2K}$. Set $A = A_1 \cap A_2$, so $\lambda([-N, N]\backslash A) \leq \epsilon/K$.

Now choose n so large that $j \geq n$ implies that $|g_j(t) - \sigma(t)| \leq \frac{\epsilon}{2N}$ for all

$t \in A$. Then if $j \geq n$,

$$\int_{\Re} f_j \, d\lambda \geq \int_{-N}^{N} f_j \, d\lambda$$

$$\geq \int_{-N}^{N} g_j \, d\lambda$$

$$\geq \int_{A} g_j \, d\lambda$$

$$\geq \int_{A} g - (\epsilon/2N) \, d\lambda$$

$$= \int_{A} g \, d\lambda - (\epsilon/2N)\lambda(A)$$

$$\geq \int_{-N}^{N} \sigma \, d\lambda - \int_{[-N,N]\backslash A} \sigma \, d\lambda - \epsilon$$

$$\geq M - K\lambda([-N, N]\backslash A) - \epsilon$$
$$\geq M - 2\epsilon.$$

Since ϵ was arbitrary, this shows that

$$\underline{\lim} \int f_n \, d\lambda \geq M;$$

since M was arbitrary, this gives the desired conclusion.

2.6.4. Theorem. *Let $\{f_n\}$ be a sequence of nonnegative measurable functions and suppose there is a nonnegative measurable function f such that*
   (i) $f = \lim f_n$ a.e.; and
   (ii) $f_n \leq f$ a.e. for each n.
Then $\int f \, d\lambda = \lim \int f_n \, d\lambda$.

Proof. Since each $f_n \leq f$ a.e., it is evident that

$$\overline{\lim} \int f_n \, d\lambda \leq \int f \, d\lambda;$$

2.6.3 gives $\underline{\lim} \int f_n \, d\lambda \geq \int f \, d\lambda$, completing the proof.

It is noteworthy that in 2.6.4, we no longer require $\{f_n\}$ to be nondecreasing as in 2.2.5, but need only the weaker condition (ii) above. The main advantage of these results, of course, is in the elimination of the assumption that $\{\int f_n \, d\lambda\}$ be bounded.

We conclude this section with an easy result which extends the continuity property of λ embodied in 2.3.10.

2.6.5. Proposition. *Let* $f \in \mathcal{L}_1(\Re)$. *For each* $\epsilon > 0$ *there is a* $\delta > 0$ *such that if* $\lambda(A) < \delta$, *then* $\int_A |f|\, d\lambda < \epsilon$.

Proof. Set $f_n = |f| \wedge n$, so $\{f_n\}$ is a sequence of nonnegative functions which converges to $|f|$ a.e. Fix $\epsilon > 0$ and apply 2.6.3 to select an n so that $\int |f|\, d\lambda \leq \int |f_n|\, d\lambda + \frac{\epsilon}{2}$. Now if $\delta = \frac{\epsilon}{2n}$ and $\lambda(A) < \delta$, then

$$\int_A |f|\, d\lambda = \int_A |f| - f_n\, d\lambda + \int_A f_n\, d\lambda$$

$$\leq \frac{\epsilon}{2} + n\lambda(A) = \epsilon.$$

PROBLEMS

2.52. Let $f \in \mathcal{L}_1(\Re)$ and define $F(x) = \int_0^x f\, d\lambda$.

(a) Show that F is continuous.

(b) Show that, if $\lambda(E) = 0$, then $\lambda\big(F(E)\big) = 0$.

Not every continuous function has property (b)—see Problem 2.20. Problems 2.52 and 2.20 show that the functions which are "indefinite integrals" are a proper subclass of the continuous functions.

2.53. *More on convergence in measure.* Show that "convergence a.e." can be replaced by "convergence in measure" in 2.6.3, 2.6.4 and 2.2.8.

2.54. Let f be a nonnegative measurable function. Show that there are functions φ, $\psi \in \mathcal{L}^+(\Re)$ such that $f = \varphi - \psi$ a.e. and $\int_{\Re} \psi\, d\lambda < \infty$. Are there nonnegative measurable functions which are *not* in $\mathcal{L}^+(\Re)$?

2.55. Compute:

(a) $\displaystyle \lim_{n\to\infty} n^2 \int_0^n \sum_{k=0}^n \frac{(-xn)^k}{k!}\, dx.$

(b) $\displaystyle \lim_{n\to\infty} n \int_0^{n^2} \sum_{k=0}^n \frac{(-xn)^k}{k!}\, dx.$

2.56. If f and g are nonnegative measurable functions, show that

$$\int_{\Re} f + g\, d\lambda = \int_{\Re} f\, d\lambda + \int_{\Re} g\, d\lambda.$$

Chapter 3

Banach Spaces

3.1. BANACH SPACES. In this and subsequent sections we will begin to apply the preceding theory in various specific settings. The appropriate abstract context for many such applications is that of a Banach space. In this introductory section we give the basic axioms for these spaces and also some preliminary examples.

3.1.1 Definition. By a *vector space* (over the reals) we shall mean an ordered triple $(V, +, \cdot)$ where V is a set, "$+$" is a function from $V \times V$ to V, and "$\cdot$" is a function from $\Re \times V$ to V which satisfies [writing "$x + y$" for "$+(x, y)$" and "αy" for "$\cdot(\alpha, y)$"]:

(i) There is an element $\ominus \in V$ such that $\ominus + v = v$ for all $v \in V$;

(ii) $0v = \ominus$ for all $v \in V$ and $1 \cdot v = v$ for all $v \in V$;

(iii) $(\alpha + \beta)v = \alpha v + \beta v$ for all $\alpha, \beta \in \Re$ and $v \in V$;

(iv) $u + v = v + u$ for all $u, v \in V$; and

(v) $\alpha(u + v) = \alpha u + \alpha v$ for all $\alpha \in \Re$ and $u, v \in V$.

The elements of V are *vectors* and the elements of $\Re$ are *scalars*. Where no confusion will result we will generally use the same symbol ("0") for the zero element of both the vector space V and the scalar field $\Re$.

3.1.2 Definition. If V is a vector space, a *norm* is a function $\|\cdot\|$ from V to $[0, \infty)$ satisfying

(i) $\|v\| = 0$ if and only if $v = \ominus$;

(ii) $\|u + v\| \leq \|u\| + \|v\|$ for all $u, v \in V$;

(iii) $\|\alpha v\| = |\alpha| \|v\|$ for all $\alpha \in \Re$ and $v \in V$.

A vector space on which a norm is given is a *normed linear space*.

The most familiar normed linear space is, of course, $\Re^n$ with

$$\|(x_1, \ldots, x_n)\| = (x_1^2 + \cdots + x_2^2)^{\frac{1}{2}}.$$

(Property 3.1.2(ii) is not trivial for this norm—see exercise 3.1.) Before proceeding, we give some other examples.

3.1.3 Examples

(i) If $V = \mathcal{C}[a, b]$ and for $f \in \mathcal{C}[a, b]$, $\|f\| = \max\{|f(t)| : a \leq t \leq b\}$, then

$(V; \|\cdot\|)$ is a normed linear space. For obvious reasons this is referred to as the "norm of uniform convergence"; unless we specify otherwise we will always mean this norm when we mention the space $C[a, b]$.

(ii)   A second norm can be imposed on V: For $f \in C[a, b]$ define $\|f\|_1 = \int_a^b |f(t)| \, dt$. Since $\|f\|_1 \leq (b - a)^{-1} \|f\|$, this norm is, in some sense, "weaker" than that of example (i).

(iii)   Denote by ℓ_1 the class of all sequences $\{x_n\}$ such that $\sum_1^\infty |x_n| < \infty$; ℓ_1 is a normed linear space with norm $\|(x_j)\| = \sum_1^\infty |x_j|$.

(iv)   Denote by ℓ_∞ the class of all bounded sequences with norm $\|(x_j)\| = \sup |x_j|$; ℓ_∞ is a normed linear space.

(v)   Our next examples require slightly more work. Define an equivalence relation $\sim$ on $\mathcal{L}_1(\Re)$ by saying that $f \sim g$ if $f = g$ a.e., and set $\big|[f]\big| = \{g \in \mathcal{L}_1(\Re) : g \sim f\}$. Note that if $f \sim g$, then $\int_\Re |f - g| \, d\lambda = 0$. Take $V = \{[f] : f \in \mathcal{L}_1(\Re)\}$ and define a norm on V by

$$\| [f] \|_1 \ = \ \int_\Re |g| \, d\lambda,$$

where g is any element in $[f]$. Then $(V; \|\cdot\|)$ is a normed linear space. This is generally referred to as $\mathcal{L}_1(\Re)$, and the technicality of passing through equivalence classes $[f]$ is usually dispensed with, i.e., we that say two vectors $f, g \in \mathcal{L}_1(\Re)$ are the same if they are equal almost everywhere.

(vi)   Denote by $\mathcal{L}_\infty(\Re)$ the class of all measurable functions f with the property that there is a number $M > 0$ for which

$$\lambda\{x \in \Re : |f(x)| \geq M\} = 0.$$

As in (v) we adopt the convention that two vectors $f, g \in \mathcal{L}_\infty(\Re)$ are the same if $f = g$ a.e.; for $f \in \mathcal{L}_\infty(\Re)$ the *essential supremum* of f is the number $\|f\|_\infty$ given by

$$\|f\|_\infty \ = \ \inf\{M > 0 : \lambda\{x \in \Re : |f(x)| \geq M\} = 0\}.$$

The space $\mathcal{L}_\infty(\Re)$ with norm $\|\cdot\|_\infty$ is a normed linear space; the reason for the "∞" subscript is suggested by Problem 3.7.

3.1.4 Definition. Let $(V; \|\cdot\|)$ be a normed linear space; a sequence $\{v_n\} \subset V$ is said to be *Cauchy* if for each $\epsilon > 0$ there is an $N > 0$ such that $n, m \geq N$ implies that

$$\|v_n - v_m\| \leq \epsilon.$$

A sequence $\{v_n\} \subset V$ is said to be *convergent* if there is a $v_\infty \in V$ for which the sequence of real numbers $\{\|v_n - v_\infty\|\}$ converges to zero. The space $(V; \|\cdot\|)$ is said to be *complete* if every Cauchy sequence is convergent; in this case the normed linear space $(V; \|\cdot\|)$ is called a *Banach space*. A set $A \subset V$ is *closed* if, whenever $\{x_n\} \subset A$ converges to x_∞, $x_\infty \in A$.

Example 3.1.3(i) is thus a Banach space by Theorem 1.1.11; with the exception of 3.1.3(ii), each of the spaces above is a Banach space — a fact that we shall soon deduce.

Note that $C_C(\Re)$ is a vector space and is a subset of both $\mathcal{L}_1(\Re)$ and $\mathcal{L}_\infty(\Re)$; in general, if V_1 and V_2 are vector spaces and $V_1 \subset V_2$, we will say that V_1 is a subspace of V_2. More generally:

3.1.5 Definition. Let V be a vector space and let $A \subset V$. The *linear span* of A is the set $sp(A)$ given by

$$sp(A) = \cap \{V' \subset V : V' \text{ is a subspace of } V \text{ and } V' \supset A\}.$$

(Since $V \supset A$, the intersection above is nonempty.) If, in addition, V is a normed linear space, the *closed linear span* of A, $\overline{sp}(A)$, is given by

$$\overline{sp}(A) = \cap \{V' \subset V : V' \text{ is a closed subspace of } V \text{ and } V' \supset A\}.$$

For example, the polynomials are a subspace, but not a closed subspace, of $C[a, b]$.

3.1.6 Definition. A set $D \subset V$ is *dense* in V if for each $x \in V$ and each $\epsilon > 0$, the set $\{y \in V : \|y - x\| \le \epsilon\} \cap D$ is nonempty. If V has a countable dense subset, then V is *separable*.

Ultimately, we shall show that $\mathcal{L}_1(\Re)$ is separable; it is a fairly routine consequence of Lusin's Theorem and 2.6.5 that $C_C(\Re)$ is dense in $\mathcal{L}_1(\Re)$ (see 3.2.8). Also, the Dominated Convergence Theorem and 2.4.12 readily combine to show that the simple functions, $\sum$, are dense in $\mathcal{L}_1(\Re)$.

PROBLEMS

3.1. Let $x = (x_1, \cdots, x_n)$ and $y = (y_1, \cdots, y_n)$ be nonzero vectors in $\Re^n$; for $t \in \Re$ define:

$$f(t) = \|x + ty\|^2,$$

where $\|x\|^2 = \sum_1^n x_k^2$.

(a) Show that f has a unique global minimum at $\bar{t} = \|y\|^{-2} \sum_1^n x_k y_k$.

(b) Show that $\left(\sum_1^n x_k y_k\right)^2 \le \|x\|^2 \|y\|^2$.

(c) Verify that $\|\cdot\|$ is a norm on $\Re^n$.

3.2. Verify that each of 3.1.3(i)—(iv) is a normed linear space.

3.3. Verify that $C[a, b]$ is complete with the norm of uniform convergence, but is not complete with the $\mathcal{L}_1$ norm of 3.1.3(ii).

3.4. Show that $C[a, b]$ is separable. (*Hint:* Uniformly approximate $f \in C[a, b]$ with a piecewise linear function having "elbows" only at rational points and having only rational slopes. Use 2.5.3 as a guide.)

3.5 **Definition.** A function f is said to *vanish at infinity* if for each $\epsilon > 0$ there is a compact set $A \subset \Re$ such that $\sup\limits_{t \notin A} |f(t)| \leq \epsilon$. The class of all continuous functions which vanish at infinity is denoted by $C_0(\Re)$.

(a) Show that

$$\|f\| = \sup_{t \in \Re} |f(t)|$$

defines a norm on $C_0(\Re)$.

(b) Show that $C_G(\Re)$ is dense in $C_0(\Re)$ under the norm in (a).

(c) Show that $C_0(\Re)$ is a Banach space under the norm in (a).

(d) Show that $C_0(\Re)$ is separable.

3.6. Show that the simple functions Σ are dense in $\mathcal{L}_1(\Re)$.

3.2. HÖLDER AND MINKOWSKI INEQUALITIES.

In this section we begin our study of the $\mathcal{L}_p$ spaces. These spaces turn out to be of critical importance in many different areas of application. The relevance of the spaces to such diverse areas as differential equations and probability was evident to researchers at the turn of the century, but no adequate analysis of their structure was available until Lebesgue introduced his integral. As we shall see, the convergence theorems of Chapter 2 will play a central role.

3.2.1. Definition. If E is a measurable set and $p \in (0, \infty)$ is fixed, $\mathcal{L}_p(E)$ is the class of all measurable functions f for which $\int_E |f|^p \, d\lambda < \infty$. Our goal in this section will be to show that these spaces are normed linear spaces when $1 \leq p$ with norm [*]

$$\|f\|_p = \left(\int_E |f|^p \, d\lambda \right)^{\frac{1}{p}}.$$

Where no confusion will result (most of the time), we will omit reference to the set E and simply speak of $\mathcal{L}_p$. Note we do *not* assume that $\lambda(E) < \infty$ in general. Following Examples 3.1.3(v) and (vi), we say f and g are the "same" if $f = g$ a.e.

3.2.2. Lemma. *Let $p > 1$ and let x and y be nonnegative real numbers. If $q = \frac{p}{p-1}$, then*

$$xy \leq \frac{x^p}{p} + \frac{y^q}{q}.$$

If $0 < p < 1$, the inequality is reversed.

---

[*] If $0 < p < 1$, then $(\int f^p \, d\lambda)^{1/p}$ does *not* give a norm on $\mathcal{L}_p$ in the sense of 3.1.2. However, it is fairly standard notation to write $\|f\|_p$ in this case also, a practice which we shall follow.

Throughout we will assume that p and q have the relationship described above; note that $1/p + 1/q = 1$; p and q are said to be *conjugate pairs*. The spaces $\mathcal{L}_p$ and $\mathcal{L}_q$ are intimately connected, which will be the topic of a later section.

If $p = q = 2$, Lemma 3.2.2 asserts that $2xy \leq x^2 + y^2$; the lemma generalizes this obvious fact— indeed, most of our results about $\mathcal{L}_p$ will be extensions of obvious relations when $p = 2$. Specializing first to the case $p = 2$ frequently enhances intuition and sometimes suggests a method of proof.

Proof. If either $x = 0$ or $y = 0$, the conclusion is obvious, and so we suppose that $0 < x$ and $0 < y$. Now fix y and define, for $p > 1$,

$$\Phi(x) = \frac{x^p}{p} + \frac{y^q}{q} - xy.$$

Then $\Phi' = \frac{px^{p-1}}{p} - y = x^{p-1} - y$. Thus if $x \in \left(0, y^{\frac{1}{p-1}}\right)$, $\Phi'(x) < 0$, and if $x \in \left(y^{\frac{1}{p-1}}, \infty\right)$, $\Phi'(x) > 0$, i.e., Φ has a global minimum at $x = y^{\frac{1}{p-1}}$. Note that

$$\Phi\left(y^{\frac{1}{p-1}}\right) = \frac{y^q}{p} + \frac{y^q}{q} - y^q = 0$$

since $1/p + 1/q = 1$. Thus $\Phi(x) \geq 0$ for all x, whence $xy \leq x^p/p + y^q/q$. (Note that we have equality if and only if $x = y^{p-1}$.) Also, if $0 < p < 1$, then $x \in \left(0, y^{\frac{1}{p-1}}\right)$ implies that $\Phi' > 0$ (since $p - 1 < 0$) and $x \in \left(y^{\frac{1}{p-1}}, \infty\right)$ implies $\Phi'(x) < 0$. Thus, in this case, $\Phi\left(y^{\frac{1}{p-1}}\right)$ is a global maximum and the inequality is reversed.

3.2.3. Theorem (Hölder's Inequality). *Suppose that $1/p + 1/q = 1$, $f \in \mathcal{L}_p$ and $g \in \mathcal{L}_q$. Then $fg \in \mathcal{L}_1$ and if $p > 1$, $\int |fg| \, d\lambda \leq \|f\|_p \|g\|_q$. Also, if $0 < p < 1$ and if f and g are nonnegative functions for which $\int f^p \, d\lambda < \infty$ and $0 < \int g^q \, d\lambda < \infty$, then*

$$\int fg \, d\lambda \geq \left(\int f^p \, d\lambda\right)^{\frac{1}{p}} \left(\int g^q \, d\lambda\right)^{\frac{1}{q}}.$$

Proof. If either f or g are zero a.e., the result is obvious; thus we may assume that $\|f\|_p \neq 0$ and $\|g\|_q \neq 0$. Applying the lemma in case $1 < p$ gives

$$\frac{|f(t)| \, |g(t)|}{\|f\|_p \, \|g\|_q} \leq \frac{|f(t)|^p}{p\|f\|_p^p} + \frac{|g(t)|^q}{q\|g\|_q^q}.$$

Integrating gives

$$\left(\|f\|_p \|g\|_q\right)^{-1} \int |fg| \, d\lambda \leq \frac{1}{p\|f\|_p^p} \int |f|^p \, d\lambda + \frac{1}{q\|g\|_q^q} \int |g|^q \, d\lambda$$

$$= 1/p + 1/q$$
$$= 1$$

from which the result follows upon multiplying through by $\|f\|_p\|g\|_q$. If $0 < p < 1$, the inequalities are reversed in the lemma and the rest of the proof is unchanged.

Note that we have equality in Hölder's Inequality if and only if

$$\frac{|f(t)|^p}{\|f\|_p^p} = \frac{|g(t)|^q}{\|g\|_q^q} \quad \text{a.e.}$$

With the inequalities above in hand, the triangle inequality for $\mathcal{L}_p$ now follows. The case $p = 2$ can be seen as an immediate consequence of 3.2.3—just expand $|f + g|^2$. Historically, many of the inequalities for general p were found by writing a Taylor's series expansion for $|f+g|^p$. and mimicking the case when $p = 2$. Fortunately, such complex arguments are usually no longer required.

3.2.4. Theorem (Minkowski Inequality). *If f and g are in $\mathcal{L}_p$ and $p > 1$, then $f + g \in \mathcal{L}_p$ and $\|f + g\|_p \leq \|f\|_p + \|g\|_p$. (Thus $\|\cdot\|_p$ is a norm.)*

Proof. We first verify that $f + g \in \mathcal{L}_p$. Note that

$$|f + g|^p \leq (|f| + |g|)^p \leq (2\max\{|f|, |g|\})^p$$

$$= 2^p \max\{|f|^p, |g|^p\}$$

$$\leq 2^p(|f|^p + |g|^p).$$

Thus $|f + g|^p$ is bounded above by the function $2^p(|f|^p + |g|^p)$, which has finite integral by assumption, so $|f + g| \in \mathcal{L}_p$. From this $|f + g|^{p-1} \in \mathcal{L}_q$ since $(|f + g|^{p-1})^q = |f + g|^p$, so we may apply Hölder's Inequality as follows:

$$\int |f + g|^p \, d\lambda = \int |f + g|^{p-1}|f + g| \, d\lambda$$

$$\leq \int |f + g|^{p-1}|f| \, d\lambda + \int |f + g|^{p-1}|g| \, d\lambda \qquad (1)$$

$$\leq \| |f + g|^{p-1}\|_q \|f\|_p + \| |f + g|^{p-1}\|_q \|g\|_p.$$

Now $\| |f + g|^{p-1}\|_q = \left(\int |f + g|^p\right)^{(1/p)(p/q)} = \|f + g\|_p^{p-1}$. Dividing through by $\|f + g\|_p^{p-1}$ gives the result.

3.2.5. Theorem. *If $0 < p < 1$ and f and g are nonnegative functions in $\mathcal{L}_p$, then $\|f + g\|_p \geq \|f\|_p + \|g\|_p$.*

Proof. In the proof above, inequality in (1) is actually equality since $f, g \geq 0$. The next inequality is reversed and the remainder of the proof is essentially unchanged. (Some care must be taken to account for the places where $f + g = 0$, since $p - 1 < 0$; this is left as an easy exercise.)

The space ℓ_p is the collection of all sequences $x = \{x_n\}$ such that $\sum |x_n|^p < \infty$, with norm $\|x\|_p = \left(\sum |x_n|^p \right)^{\frac{1}{p}}$. It is a routine exercise to verify that the inequalities above carry over to this class without change.

Besides serving as a model to assist with pointwise inequalities, the case $p = 2$ deserves special attention; the spaces $\mathcal{L}_2$ and ℓ_2 are examples of *Hilbert spaces*, which we will study in greater detail in the sequel. These spaces enjoy many nice properties, most of which follow from the next theorem.

3.2.6. Theorem (Parallelogram Law). *If $f, g \in \mathcal{L}_2$, then*

$$\|f + g\|_2^2 + \|f - g\|_2^2 = 2\|f\|_2^2 + 2\|g\|_2^2.$$

Proof. This is an easy exercise in algebra:

$$\|f + g\|_2^2 + \|f - g\|_2^2 = \int |f + g|^2 \, d\lambda + \int |f - g|^2 \, d\lambda$$

$$= \int |f|^2 + 2fg + g^2 + |f|^2 - 2fg + g^2 \, d\lambda$$

$$= 2\|f\|_2^2 + 2\|g\|_2^2.$$

The parallelogram law is a statement about the two-dimensional subspace spanned by the vectors f and g. In particular, the result says that the sum of the squared lengths of the sides of a parallelogram equals the sum of the squared lengths of the diagonals.

We saw in the Chapter 2 that it was useful to be able to approximate measurable functions by simpler objects; similarly, it is useful to describe what types of $\mathcal{L}_p$ approximations are possible. This section concludes with two results of this type.

3.2.7. Theorem. *The simple functions $\sum$ are dense in $\mathcal{L}_p$ if $1 \leq p < \infty$.*

Proof. Clearly, $\sigma \subset \mathcal{L}_p$ for all $p \in [1, \infty)$. By 2.4.2 there is a sequence $\{\sigma_n\} \subset \Sigma$ such that $\sigma_n \to f$ a.e. and $|\sigma_n| \leq |f|$. Since $|\sigma_n - f|^p \leq (|\sigma_n| + |f|)^p \leq 2^p |f|^p$, the

Dominated Convergence Theorem implies that

$$\lim \int |\sigma_n - f|^p \, d\lambda = \int \lim |\sigma_n - f|^p \, d\lambda = 0,$$

from which the result follows.

3.2.8. Theorem. *The set $\mathcal{C}_C(\Re)$ is dense in $\mathcal{L}_p(\Re)$ if $1 \le p < \infty$.*

Proof. Fix $\epsilon > 0$ and let $f \in \mathcal{L}_p(\Re)$. In view of 3.2.7, we may assume that $f \in \Sigma$; in particular, if we let $M = \max\{|f(t)| : t \in \Re\}$, then $M < \infty$.

Choose a closed interval I so that

$$\int_{\Re \setminus I} |f|^p \, d\lambda < \left(\frac{\epsilon}{3}\right)^p.$$

Applying 2.5.5, we may select a closed set $A \subseteq I$ and a continuous function $g \in \mathcal{C}(I)$ so that

$$\lambda(I \setminus A) \le \left(\frac{\epsilon}{6}\right)^p \frac{1}{M}$$
$$f(t) = g(t) \quad \text{if} \quad t \in A; \quad \text{and}$$
$$\|g\|_\infty \le M.$$

In the usual manner we can extend g to a function $\hat{g} \in \mathcal{C}_c(\Re)$ with the property that

$$\int_{\Re \setminus I} |g|^p \, d\lambda < \left(\frac{\epsilon}{3}\right)^p.$$

Now, setting $\hat{I} = \Re \setminus I$ and $\hat{A} = I \setminus A$,

$$
\begin{aligned}
\|f - \hat{g}\|_p &= \|\chi_{\hat{I}}(f - \hat{g}) + \chi_I(f - \hat{g})\|_p \\
&\le \|\chi_{\hat{I}}(f - \hat{g})\|_p + \|\chi_I(f - \hat{g})\|_p \\
&\le \|\chi_{\hat{I}} f\|_p + \|\chi_{\hat{I}} g\|_p + \|\chi_{\hat{A}}(f - g)\|_p \\
&\le \frac{\epsilon}{3} + \frac{\epsilon}{3} + \|\chi_{\hat{A}} f\|_p + \|\chi_{\hat{A}} g\|_p \\
&\le \frac{2\epsilon}{3} + \lambda(\hat{A})M + \lambda(\hat{A})M \\
&\le \epsilon.
\end{aligned}
$$

Since ϵ was arbitrary, this completes the proof.

Theorem 3.2.8 is frequently very useful in verifying facts about the spaces $\mathcal{L}_p(\Re)$. Note that while we have used Lusin's Theorem to prove the above, a direct proof using the results of Section 2.2 is also possible.

PROBLEMS

3.7. Suppose that $\lambda(E) < \infty$ and $f \in \mathcal{L}_\infty(E)$. Show that $f \in \mathcal{L}_p(E)$ for all $p \ge 1$ and that $\lim_{p \to \infty} \|f\|_p = \|f\|_\infty$.

3.8. Show that the inequality in Minkowski's inequality is strict unless $f = tg$ a.e. for some scalar t.

3.9. Suppose that $\lambda(E) < \infty$ and that $1 \leq p_1 \leq p_2 < \infty$. Then $\mathcal{L}_{p_2}(E) \subset \mathcal{L}_{p_1}(E)$ and for $f \in \mathcal{L}_{p_2}(E)$,

$$\|f\|_{p_1} \leq \|f\|_{p_2}(\lambda(E))^{(p_2-p_1)/p_1 p_2}.$$

3.10. If $1 \leq p_1 \leq p_2 < \infty$, then $\ell_{p_1} \subset \ell_{p_2}$.

3.11. Let $f \in \mathcal{L}_1$ and $g \in \mathcal{L}_\infty$. Show that $gf \in \mathcal{L}_1$ and

$$\left| \int fg \, d\lambda \right| \leq \|f\|_1 \|g\|_\infty.$$

(This is Hölder's Inequality in the case $p = 1$.)

3.12. Let $1 < p < \infty$ and let $f \in \mathcal{L}_p[0, \infty)$. Show that for $x > 0$

$$\left| \int_0^\infty e^{-tx} f(t) \, dt \right| \leq \|f\|_p (xq)^{-\frac{1}{q}}.$$

3.13. Show that $\mathcal{L}_p$ is separable if $1 \leq p < \infty$. (Apply Problem 3.5 and Theorem 3.2.8.)

3.14. Show that ℓ_p is separable if $1 \leq p < \infty$.

3.15. Suppose that $f \in \mathcal{L}_p[0, \infty)$, $\quad \forall p \geq 1$. For what values of p is the function $\frac{f(t)}{\sqrt{1+t}}$ in $\mathcal{L}_1[0, \infty)$?

3.3. RIESZ-FISCHER THEOREM. In this section we show that the spaces $\mathcal{L}_p$ are complete for $1 \leq p < \infty$. When an early version of this theorem was announced in 1906 it created a sensation; with the machinery we have at hand it turns out to be relatively straightforward.

3.3.1. Theorem (Riesz-Fischer Theorem). *For $1 \leq p < \infty$ the spaces $\mathcal{L}_p$ and ℓ_p are complete normed linear spaces.*

The spaces $\mathcal{L}_\infty$ and ℓ_∞ are also complete — see Problem 3.16. We will prove this theorem only for $\mathcal{L}_p$ and leave the entirely similar proof for ℓ_p to the problems.

As with our earlier convergence results in Section 2.2, it is more convenient to write the proof using series rather than sequences. In particular, we begin by showing that a normed linear space is complete if and only if every "absolutely convergent" series is convergent.

3.3.2. Lemma. *Let X be a normed linear space. Then the following are* *equivalent:*

(i) X *is complete;*

(ii) *whenever* $\{x_n\}$ *is a sequence in* X *with* $\sum_n \|x_n\| < \infty$, *there is an* $x \in X$ *with* $x = \lim_{n \to \infty} \sum_{k=1}^{n} x_k$.

Proof. (i) $\Rightarrow$ (ii). Let $\{x_n\}$ be a sequence in X and suppose that $\sum_n \|x_n\| < \infty$. For each n set

$$y_n = \sum_{k=1}^{n} x_k.$$

Then for $n < m$,

$$\|y_n - y_m\| = \left\| \sum_{k=n+1}^{m} x_k \right\| \le \sum_{k=n+1}^{m} \|x_k\|.$$

Since $\{\sum_{k=1}^{n} \|x_k\|\}$ is a Cauchy sequence of real numbers, it follows that $\{y_n\}$ is a Cauchy sequence in X, and thus convergent to some $x \in X$ by (i).

(ii) $\Rightarrow$ (i). Let $\{x_n\}$ be a Cauchy sequence in X. Then for every $\epsilon > 0$ there is an N such that $n, m \ge N$ implies that $\|x_n - x_m\| \le \epsilon$. Taking $\epsilon = 2^{-k}$, we can find a sequence $N_1 < N_2 < \cdots$ such that $n, m > N_k$ implies that

$$\|x_n - x_m\| \le 2^{-k};$$

in particular,

$$\|x_{N_{k+1}} - x_{N_k}\| \le 2^{-k}.$$

Now define a new sequence $\{y_n\}$ by taking

$$y_1 = x_{N_1} \quad \text{and}$$
$$y_j = x_{N_j} - x_{N_{j-1}}.$$

Then

$$x_{N_k} = \sum_{j=1}^{k} y_j \quad \text{and}$$

$$\sum_{j=1}^{\infty} \|y_j\| = \|x_{N_1}\| + \sum_{j=2}^{\infty} \|x_{N_j} - x_{N_{j-1}}\|$$

$$\le \|x_{N_1}\| + \sum_{j=2}^{\infty} 2^{-j+1} < \infty.$$

Thus, by (ii), there is an $x \in X$ such that

$$x = \lim_{n \to \infty} \sum_{j=1}^{n} y_j;$$

all that remains is to show that $\lim \|x_n - x\| = 0$. To see this, fix $\epsilon > 0$ and choose k so large that $2^{-k+1} < \epsilon/2$. Then for $n \geq N_k$,

$$\|x_n - x\| \leq \|x_n - x_{N_k}\| + \|x_{N_k} - x\|$$

$$\leq 2^{-k} + \|\sum_{j=1}^{k} y_j - \sum_{j=1}^{\infty} y_j\|$$

$$\leq 2^{-k+1} + \left\|\sum_{j=k+1}^{\infty} y_j\right\|$$

$$\leq \epsilon.$$

Since ϵ was arbitrary, this completes the proof.

3.3.3. Lemma. *Let $\{g_j\}_{j=1}^{\infty}$ be a sequence of functions in $\mathcal{L}_p$ for which $\sum_1^{\infty} \|g_j\| < \infty$. Then $\{\sum_1^k g_j\}$ converges a.e. to a function $f \in \mathcal{L}_p$.*

Proof. Set $h_k = \sum_{j=1}^{k} |g_j|$ and observe that

$$\left(\int |h_k|^p\right)^{\frac{1}{p}} = \|\sum_{j=1}^{k} |g_j|\| \leq \sum_{j=1}^{\infty} \|g_j\| < \infty.$$

Thus the Monotone Convergence Theorem implies that

$$\left(\int \left(\sum_{j=1}^{\infty} (|g_j|)^p\right)^{\frac{1}{p}}\right) = \lim_{k \to \infty} \left(\int \left(\sum_{j=1}^{k} (|g_j|)^p\right)^{\frac{1}{p}}\right) < \infty,$$

and so $\sum_{j=1}^{\infty} |g_j| \in \mathcal{L}_p$. Since $|\sum_{j=1}^{k} g_j|^p \leq (\sum_{j=1}^{\infty} |g_j|)^p$, the Dominated Convergence Theorem implies that

$$\left(\int \left|\sum_{j=1}^{\infty} g_j\right|^p d\lambda\right)^{\frac{1}{p}} = \lim_{k \to \infty} \left(\int \left|\sum_{j=1}^{k} g_j\right|^p d\lambda\right)^{\frac{1}{p}} < \infty$$

and thus $\sum_{j=1}^{\infty} g_j \in \mathcal{L}_p$, as claimed.

The Riesz-Fisher Theorem for $\mathcal{L}_p$ is an immediate consequence of these two lemmas. Notice that, in addition, we have proved the following.

3.3.4. Corollary. *If $\|f_n - f\|_p \to 0$, then there is a subsequence $\{f_{N_k}\}$ of $\{f_n\}$ such that $f_{N_k} \to f$ a.e.*

In order to establish the Riesz-Fisher Theorem in ℓ_p, you will need to prove an analogue of 3.3.3 for sequences. As the critical steps in the proof of 3.3.3

are the applications of the Monotone Convergence Theorem and the Dominated Convergence Theorem, it will be necessary also to deduce analogues of these reults for sequence spaces. It is worth noting in passing that sequences are just functions from $\mathcal{N}$ to $\mathfrak{R}$; i.e., if $\xi : \mathcal{N} \to \mathfrak{R}$, then we can identify $\{\xi(j)\}$ with the sequence $\{\xi_j\}$. "Integrals" for functions from $\mathcal{N}$ to $\mathfrak{R}$ are just summations. Keeping these ideas in mind should make it easier to prove Problem 3.16. We will pursue this in much greater detail in Section 7.1.

PROBLEMS

3.16. *Riesz-Fisher Theorem for the spaces ℓ_p*

(a) *Fatou's Lemma for sequences.* For each n let $\{\xi_j^{(n)}\}$ be a sequence of non-negative numbers. Suppose that for each j

$$\lim_{n \to \infty} \xi_j^{(n)} = \xi_j.$$

Show that

$$\sum_{j=1}^{\infty} \xi_j \leq \liminf_{n \to \infty} \sum_{j=1}^{\infty} \xi_j^{(n)}.$$

(b) *Monotone Convergence Theorem for sequences.* For each n let $\{\xi_j^{(n)}\}$ be a sequence of nonnegative numbers and for each j, suppose that

$$\lim_{n \to \infty} \xi_j^{(n)} = \xi_j$$

and that $\xi_j^{(n)} \leq \xi_j$ for each j. Show that

$$\sum_{j=1}^{\infty} \xi_j = \lim_{n \to \infty} \sum_{j=1}^{\infty} \xi_j^{(n)}.$$

(c) *Dominated Convergence Theorem for sequences.* For each n let $\{\xi_j^{(n)}\}$ be a sequence of real numbers and for each j, suppose that

$$\lim_{n \to \infty} \xi_j^{(n)} = \xi_j.$$

Suppose in addition that there is a sequence $\{\beta_j\}$ for which $\sum_{j=1}^{\infty} |\beta_j| < \infty$ and $|\xi_j^{(n)}| \leq \beta_j$ for all n and j. Show that

$$\sum_{j=1}^{\infty} \xi_j = \lim_{n \to \infty} \sum_{j=1}^{\infty} \xi_j^{(n)}.$$

(d) Prove the Riesz-Fisher Theorem for ℓ_p.

3.17. Show that ℓ_∞ and $\mathcal{L}_\infty$ are Banach spaces.

3.18. Compute $\displaystyle\lim_{x \to 0} \sum_{1}^{\infty} \frac{\sin(nx)}{n^2}$.

3.19. Compute

$$\lim_{n \to \infty} \sum_{k=0}^{\infty} \sum_{j=k}^{k+n} \frac{s^k t^{j-k}}{k!(j-k)!}.$$

3.4. ABSTRACT HILBERT SPACES. A *Hilbert space* is a Banach space $\mathcal{H}$ whose norm satisfies the parallelogram law:

$$\|x + y\|^2 + \|x - y\|^2 = 2\|x\|^2 + 2\|y\|^2$$

whenever $x, y \in \mathcal{H}$. For examples, we have seen that $\mathcal{L}_2(E)$ and ℓ_2 are Hilbert spaces. Of course, $\mathfrak{R}^n$ endowed with the Euclidean norm is also an example of a Hilbert space. Ultimately, we shall see in this chapter that every infinite-dimensional separable Hilbert space can be identified with the space ℓ_2.

Hilbert spaces have been studied in great detail, not only because of their important connections with differential equations but also because they enjoy many special and very powerful properties. The first of these which we will discuss is the existence of an inner product.

3.4.1. Theorem. *Let $\mathcal{H}$ be a Hilbert space and define, for $x, y \in \mathcal{H}$, the inner product of x and y to be the number*

$$\langle x, y \rangle = \tfrac{1}{4} \left(\|x + y\|^2 - \|x - y\|^2 \right). \tag{1}$$

Then for all $x, y \in \mathcal{H}$ and scalars $\alpha \in \mathfrak{R}$
    (i) $\langle x, y \rangle = \langle y, x \rangle$
    (ii) $\langle x + y, z \rangle = \langle x, z \rangle + \langle y, z \rangle$
    (iii) $\langle \alpha x, y \rangle = \alpha \langle x, y \rangle$
    (iv) $\langle x, x \rangle = \|x\|^2$
Conversely, if there is pairing $\langle \cdot, \cdot \rangle$ defined on a space $\mathcal{H}$ satisfying (i) through (iv), then the norm of $\mathcal{H}$ satisfies the parallelogram law.

Proof. If $\mathcal{H}$ is a Hilbert space and $\langle x, y \rangle$ is given by (1), then (i) clearly holds. For (ii), fix f, g, and $h \in \mathcal{X}$ and observe (applying the parallelogram law) that

$$4 \langle f + g, h \rangle + 4 \langle f - g, h \rangle$$
$$= \|f + g + h\|^2 - \|f + g - h\|^2 + \|f - g + h\|^2$$
$$- \|f - g - h\|^2$$
$$= 2\|f + h\|^2 + 2\|g\|^2 - 2\|f - h\|^2 - 2\|g\|^2$$
$$= 8\langle f, h \rangle$$

and so

$$\langle f + g, h \rangle + \langle f - g, h \rangle = 2\langle f, h \rangle.$$

Since $\langle 0, h \rangle = 0$, taking $g = f$ in the above gives $\langle 2f, h \rangle = 2\langle f, h \rangle$ for all $f, h \in \mathcal{X}$. Now fix $x, y \in \mathcal{X}$ and set $f = (1/2)(x + y)$ and $g = (1/2)(x - y)$. Applying the above gives

$$\langle x, h \rangle + \langle y, h \rangle = 2 \left\langle \frac{x + y}{2}, h \right\rangle = \langle x + y, h \rangle$$

and (ii) follows.

It is evident from the definition that $\langle -x, y \rangle = -\langle x, y \rangle$ for all $x, y \in \mathcal{X}$. Induction on (ii) readily establishes that for any integer p, $\langle px, y \rangle = p\langle x, y \rangle$. From this, for any integer q,

$$(1/q)\langle x, y \rangle = 1/q\langle x, (\tfrac{1}{q})qy \rangle = \langle x, y/q \rangle.$$

Combining these observations, it follows that for any rational number r, $\langle rx, y \rangle = r\langle x, y \rangle$. Since $(\|x + \lambda y\|^2 - \|x - \lambda y\|^2)$ is continuous in λ, this establishes (iii); (iv) is immediate from the definition.

The converse is equally elementary; if $\langle \cdot, \cdot \rangle$ satisfies (i) through (iv), then

$$\|x + y\|^2 + \|x - y\|^2 = \langle x + y, x + y \rangle + \langle x - y, x - y \rangle$$
$$= \|x\|^2 + 2\langle x, y \rangle + \|y\|^2 + \|x\|^2$$
$$- 2\langle x, y \rangle + \|y\|^2$$
$$= 2\|x\|^2 + 2\|y\|^2.$$

Direct computations show in ℓ_2 that $\langle x, y \rangle = \sum x_n y_n$ and in $\mathcal{L}_2$, $\langle f, g \rangle = \int fg \, d\lambda$. For these specific examples, verification of (i) through (iv) is immediate.

3.4.2. Theorem (Cauchy-Schwarz). *If $x, y \in \mathcal{X}$, then $|\langle x, y \rangle| \leq \|x\| \|y\|$. The inequality is strict unless $x = \lambda y$ for some $\lambda \in \Re$.*

Proof. If $y = 0$, or if $x = y$, the inequality is obvious, so we suppose that $x \neq y \neq 0$, and let $\lambda \in \Re$. Then $0 \leq \langle x - \lambda y, x - \lambda y \rangle = \|x\|^2 - 2\lambda\langle x, y \rangle + \lambda^2\|y\|^2$. Setting $\lambda = \langle x, y \rangle \|y\|^{-2}$ gives the result. Note that the first inequalities are strict if $x - \lambda y \neq 0$, which establishes the second assertion.

We now turn to the theory of Fourier series. Elements $x, y \in \mathcal{H}$ are said to be *orthogonal* if $\langle x, y \rangle = 0$. A set $B \subset \mathcal{H}$ is *orthogonal* if for each $x, y \in B$ with $x \neq y$, it is the case that $\langle x, y \rangle = 0$. If, in addition, each element in B has norm 1, then B is said to be *orthonormal*.

3.4.3. Examples

(i) In ℓ_2 denote by e_n the vector which has a "1" in the nth position and a zero elsewhere; then $\{e_n\}$ is an orthonormal set. Note that for any $x \in \ell_2$, $x = \sum_1^\infty \langle x, e_n \rangle e_n$. In any Hilbert space $\mathcal{H}$ there is an orthonormal set $\{e_n\}$ having the property that every $x \in \mathcal{H}$ has the representation

$$x = \sum_1^\infty \langle x, e_n \rangle e_n;$$

this representation is called the Fourier series representation for x.

(ii) In $\mathcal{L}_2[0, 2\pi]$, set $e_n = \big(\cos(nx) + \sin(nx) \big) (2\pi)^{-\frac{1}{2}}$. Then the family $\{e_n\}_{|n|=0}^\infty$ is an orthonormal set in $\mathcal{L}_2[0, 2\pi]$. (This is not obvious without some elementary calculus—see Problem 3.32.) We will ultimately show that if $f \in \mathcal{L}_2[0, 2\pi]$, then $\lim_{N \to \infty} \| f - \sum_{|n|=0}^N \langle f, e_n \rangle e_n \|_2 = 0$, obtaining a Fourier series representation of f. This is a very deep result which has had a profound impact on mathematics.

One of the most useful attributes of orthonormal sets is the approximation property suggested by the two examples above, i.e. if e_n is an orthonormal set in a Hilbert space $\mathcal{H}$ and if $x \in \mathcal{H}$, then x may be approximated (in the Hilbert space norm) by sums of the form

$$\sum_1^n \alpha_k e_k.$$

Our first goal is to ascertain which choice of α_k gives the "best" approximation to x; the simple quadratic expansion in the following lemma gives an immediate answer.

3.4.4. Lemma. *Let $\mathcal{H}$ be a Hilbert space, let $\{e_1, \ldots, e_n\}$ be an orthonormal set in $\mathcal{H}$, and let $\{\alpha_1, \ldots \alpha_n\}$ be a collection of scalars. Then for all $x \in \mathcal{H}$,*

$$\left\| x - \sum_1^n \alpha_k e_k \right\|^2 = \|x\|^2 + \sum_1^n \big(\alpha_k - \langle x, e_k \rangle \big)^2 - \sum_1^n \langle x, e_k \rangle^2.$$

Proof. As with many arguments in Hilbert space, expanding the norm through inner products simplifies the problem; often completing the square is useful, too (as in this proof). Finally, an important identity for orthogonal vectors

$$\sum_1^n (\|e_k\|^2) = \left\| \sum_1^n (e_k) \right\|^2$$

is also used. To see this last identity, expand the norm on the right with inner products; the "mixed terms" $\langle e_j, e_k \rangle = 0$ when j is different from k.

Applying these ideas yields

$$\left\| x - \sum_1^n \alpha_k e_k \right\|^2 = \|x\|^2 - 2\left\langle x, \sum_1^n (\alpha_k e_k) \right\rangle + \left\| \sum_1^n (\alpha_k e_k) \right\|^2$$

$$= \|x\|^2 - \sum_1^n (2\alpha_k \langle x, e_k \rangle) + \sum_1^n (\alpha_k^2 \langle e_k, e_k \rangle)$$

$$= \|x\|^2 + \sum_1^n (\alpha_k^2 - 2\alpha_k \langle x, e_k \rangle)$$

$$= \|x\|^2 + \sum_1^n (\alpha_k - \langle x, e_k \rangle)^2 - \sum_1^n (\langle x, e_k \rangle^2),$$

establishing the result.

3.4.5. Corollary. *For any orthonormal collection* $\{e_1, \ldots, e_n\}$ *and any* $x \in \mathcal{H}$,

$$\sum_1^n \langle x, e_k \rangle^2 \leq \|x\|^2.$$

Moreover, the quantity $\|x - \sum_1^n \alpha_k e_k\|^2$ *is minimized if and only if* $\alpha_k = \langle x, e_k \rangle$.

Proof. Both conclusions are immediate from 3.4.4.

3.4.6. Definition. Let $\mathcal{A}$ be a (possibly uncountable) index set and let $\{x_\alpha : \alpha \in \mathcal{A}\}$ be a collection of nonnegative real numbers indexed by $\mathcal{A}$. We define $\sum_{\alpha \in \mathcal{A}} x_\alpha$ to be

$$\sum_{\alpha \in \mathcal{A}} x_\alpha = \sup \left\{ \sum_1^n x_{\alpha_j} : n \in \mathcal{N}, \alpha_1, \ldots, \alpha_n \in \mathcal{A} \right\}.$$

In case $\mathcal{A} = \mathcal{N}$, this corresponds to $\sum_1^\infty x_n$ (see Problem 3.24); if $\sum_{\alpha \in \mathcal{A}} x_\alpha < \infty$, then at most countably many of the numbers x_α are not zero (Problem 3.24).

3.4.7. Corollary (Bessel's Inequality). *Let* $\mathcal{B}$ *be an orthonormal set in a Hilbert space* $\mathcal{H}$. *Then for each* $x \in \mathcal{H}$,

$$\sum_{e \in \mathcal{B}} \langle x, e \rangle^2 \leq \|x\|^2;$$

in particular, $\langle e, x \rangle \neq 0$ for at most countably many $e \in \mathcal{B}$.

Proof. This is simply a restatement of 3.4.5.

3.4.8. Lemma. *Let $\{e_n\}$ be an orthonormal sequence in $\mathcal{X}$ and let $x \in \mathcal{X}$. Then*
(i) $\{\sum_1^n \langle x, e_j \rangle e_j\}$ is a Cauchy sequence in $\mathcal{X}$;
(ii) if $y = \lim_n \sum_1^n \langle x, e_j \rangle e_j$, then $\langle (x - y), e_n \rangle = 0$ for all n.

Proof. Since $\{e_n\}$ is an orthonormal sequence, expanding in inner products yields
for $n < m$

$$\left\| \sum_1^n \langle x, e_j \rangle e_j - \sum_1^m \langle x, e_j \rangle e_j \right\|^2 = \left\| \sum_{n+1}^m \langle x, e_j \rangle e_j \right\|^2$$
$$= \sum_{n+1}^m \langle x, e_j \rangle^2.$$

By Bessel's inequality, $\sum_{n+1}^m \langle x, e_j \rangle^2 \to 0$ as $n, m \to \infty$; this establishes (i).
For (ii), set $y_n = \sum_1^n \langle x, e_j \rangle e_j$ and fix m. Then for $n \geq m$,

$$|\langle x - y, e_m \rangle| \leq |\langle x - y_n, e_m \rangle| + |\langle y_n - y, e_m \rangle|$$
$$\leq \left| \langle x, e_m \rangle - \langle \sum_1^n \langle x, e_j \rangle e_j, e_m \rangle \right| + \|y_n - y\| \, \|e_m\|$$
$$= |\langle x, e_m \rangle - \langle \langle x, e_m \rangle e_m, e_m \rangle| + \|y_n - y\|$$
$$= \|y_n - y\|,$$

using $\langle e_j, e_m \rangle = 0$ if $j \neq m$ and $\|e_m\| = 1$. Since $\|y_n - y\| \to 0$ as $n \to \infty$, it follows
that $\langle x - y, e_m \rangle = 0$.

In view of 3.4.4, the series $\sum_1^\infty \langle x, e_j \rangle e_j$ converges to the point nearest
x in the closed subspace spanned by $\{e_n\}$. Conclusion 3.4.8(ii) gives sufficient
conditions for the series actually to converge to x; in particular the set $\{e_n\}$ must
satisfy the condition of the following definition.

3.4.9. Definition. An orthonormal set $\mathcal{B}$ in a Hilbert space $\mathcal{X}$ is *complete* if
$\langle x, e \rangle = 0$ for all $e \in \mathcal{B}$ implies that $x = 0$.

In many applications, complete orthonormal sets are actually constructed
to meet certain requirements [the trigonometric functions of 3.4.3(ii) are important
because of their connection with partial differential equations, for example]. The
general question of existence of complete orthonormal sets in arbitrary Hilbert
spaces is answered by the following theorem. The procedure used in the proof is
similar to the technique used to show that an arbitrary vector space has a basis.

3.4.10. Theorem. *Let $\mathcal{H}$ be a Hilbert space and suppose that $\mathcal{H} \neq \{0\}$. Then there is a complete orthonormal set $\mathcal{B} \subset \mathcal{H}$.*

Proof. Since $\mathcal{H} \neq \{0\}$, $\mathcal{H}$ contains a nonempty orthonormal set S (for example, $S = \{\|x\|^{-1}x\}$ for some $x \neq 0$ in $\mathcal{H}$). Let P denote the class of all orthonormal subsets of $\mathcal{H}$ which contain S, and partially order P by inclusion. If $\{S_\alpha\}$ is any chain in P, then $\cup_\alpha S_\alpha$ is again in P: If $x, y \in \cup_\alpha S_\alpha$, then $x \in S_{\alpha_1}$ and $y \in S_{\alpha_2}$ for some α_1, α_2. Since $\{S_\alpha\}$ is a chain, either $S_{\alpha_1} \subset S_{\alpha_2}$ or $S_{\alpha_2} \subset S_{\alpha_1}$; suppose without loss of generality that the former occurs. Then x and y are in the orthonormal set S_{α_2}, implying that $\langle x, y \rangle = 0$, as desired.

Thus by Zorn's Lemma the class P contains a maximal element $\mathcal{B}$. Now if $\mathcal{B}$ is not complete, there is a $y \in \mathcal{H}$ for which $\langle y, e \rangle = 0$ for all $e \in \mathcal{B}$ but $y \neq 0$. Setting $\mathcal{B}' = \mathcal{B} \cup \{\|y\|^{-1}y\}$, we contradict the maximality of $\mathcal{B}$, and thus $\mathcal{B}$ is complete.

The next theorem summarizes the basic facts about complete orthonormal sets in Hilbert space; most of the work has been incorporated in the previous results.

3.4.11. Theorem. *Let $\mathcal{H}$ be a Hilbert space and let $\mathcal{B}$ be an orthonormal set in $\mathcal{H}$. Then the following are equivalent.*
   (i) $\mathcal{B}$ is complete;
   (ii) For each $x \in \mathcal{H}$, $x = \sum_{e \in \mathcal{B}} \langle x, e \rangle e$ [Fourier series];
   (iii) For each $x \in \mathcal{H}$, $\|x\|^2 = \sum_{e \in \mathcal{B}} \langle x, e \rangle^2$ [Parseval's identity];
   (iv) The linear span of $\mathcal{B}$, $sp(\mathcal{B})$, is dense in $\mathcal{H}$.

Proof. (i) $\Rightarrow$ (ii). Fix $x \in \mathcal{H}$; by 3.4.7, the set $\mathcal{B}' = \{e \in \mathcal{B} : \langle x, e \rangle \neq 0\}$ is countable, so, by 3.4.8,

$$y = \sum_{e \in \mathcal{B}} \langle x, e \rangle e = \sum_{e \in \mathcal{B}'} \langle x, e \rangle e$$

is well-defined and $\langle (x - y), e \rangle = 0$ for all $e \in \mathcal{B}$. Since $\mathcal{B}$ is complete, this implies that $x = y$.

(ii) $\Rightarrow$ (iii). Let $\{e_n\}$ be an enumeration of the set $\mathcal{B}'$ defined in the proof of (i) $\Rightarrow$ (ii). Set $x_n = \sum_1^n \langle x, e_j \rangle e_j$, so, by (ii), $\|x - x_n\| \to 0$ as $n \to \infty$. Also, by the triangle inequality,

$$|\langle x, x \rangle - \langle x_n, x_n \rangle| = |\,\|x\|^2 - \|x_n\|^2\,| \to 0 \quad \text{as} \quad n \to \infty.$$

Applying $\langle e_j, e_k \rangle = 0$ if $j \neq k$ then gives

$$
\begin{aligned}
\|x\|^2 &= \lim_{n \to \infty} \langle x_n, x_n \rangle \\
&= \lim_{n \to \infty} \left\langle \sum_{j=1}^{n} \langle x, e_j \rangle e_j, \sum_{k=1}^{n} \langle x, e_k \rangle e_k \right\rangle \\
&= \lim_{n \to \infty} \sum_{j=1}^{n} \langle x, e_j \rangle \sum_{k=1}^{n} \langle x, e_k \rangle \langle e_j, e_k \rangle \\
&= \lim_{n \to \infty} \sum_{j=1}^{n} \langle x, e_j \rangle \langle x, e_j \rangle \langle e_j, e_j \rangle \\
&= \sum_{1}^{\infty} \langle x, e_j \rangle^2,
\end{aligned}
$$

proving (ii) $\Rightarrow$ (iii).

(iii) $\Rightarrow$ (i). This is immediate.

(ii) $\Rightarrow$ (iv). By (ii), every element of $\mathcal{X}$ can be approximated by finite linear combinations of elements of $\mathcal{B}$, showing that $sp(\mathcal{B})$ is dense in $\mathcal{X}$.

(iv) $\Rightarrow$ (i). Let $x \in \mathcal{X}$ and suppose that $\langle x, e \rangle = 0$ for all $e \in \mathcal{B}$. Then for all $y \in sp(\mathcal{B})$, $\langle x, y \rangle = 0$. Now choose $\{y_n\} \subset sp(\mathcal{B})$ such that $\|y_n - x\| \to 0$. Then

$$
\|x\|^2 = \langle x, x \rangle = \lim \langle x, y_n \rangle = 0,
$$

implying that $x = 0$. This completes the proof of 3.4.10.

In certain cases, a complete orthonormal set can actually be explicitly constructed from given elements; this is in particular true when the space has a countable dense subset.

Because of the results above — especially because of 3.4.11(ii)—complete orthonormal sets are very useful objects to have in a Hilbert space. While 3.4.11(iv) gives a criterion for an orthonormal set to be complete, we so far have no way of explicitly constructing such sets (except in the obvious case ℓ_2). In the next result we not only show that a separable Hilbert space must have a countable complete orthonormal set, but also give an explicit algorithm for constructing such a set, namely the *Gram-Schmidt orthonormalization procedure*.

3.4.12. Theorem. *A Hilbert space $\mathcal{X}$ is separable if and only if $\mathcal{X}$ has a complete orthonormal set which is either countable or finite.*

Proof. If $\{e_n\}$ is a complete orthonormal set in $\mathcal{H}$, then by 3.4.11(iv) the set

$$\left\{\sum_1^k \lambda_n e_n : \lambda_n \in \mathcal{Q}, \ k \in \mathcal{N}\right\}$$

is a countable dense subset of $\mathcal{H}$.

For the converse we follow the Gram-Schmidt orthonormalization procedure. Let $\mathcal{D} = \{d_n\}$ be a countable dense subset in $\mathcal{H}$; without loss of generality, we may assume that $d_n \neq 0$ for all n.

Set $e_1 = \|d_1\|^{-1} d_1$, so $\{e_1\}$ is an orthonormal set and $sp\{e_1\} = sp\{d_1\}$. To define e_2, choose the first integer n_2 so that d_{n_2} is not on the ray $sp\{e_1\}$ and set

$$y = d_{n_2} - \langle e_1, d_{n_2}\rangle e_1.$$

(Note that $\langle e_1, d_{n_2}\rangle e_1$ is the orthogonal projection of d_{n_2} onto e_1. A picture in two dimensions will help in visualizing the construction; the same picture should convince you that y is orthogonal to e_1.) Next set $e_2 = \|y\|^{-1} y$, and observe that $\|e_2\| = 1$ and

$$\langle e_2, e_1\rangle = \|y\|^{-1}(\langle d_{n_2}, e_1\rangle - \langle d_{n_2}, e_1\rangle\langle e_1, e_1\rangle) = 0,$$

as desired.

With the cases $n = 1$ and $n = 2$ in mind, we can now continue with the general case, proceeding by induction. Suppose that we have defined n vectors $\{e_1, \ldots, e_n\}$ and n integers $1 = m_1 < m_2 \cdots < m_n$ such that

(a) $\{e_1, \ldots, e_n\}$ is an orthonormal set.

(b) $sp\{e_1, \ldots, e_k\} = sp\{d_1, \ldots, d_{m_k}\}$ for $k = 1, \ldots, n$.

Now if $sp\{e_1, \ldots, e_n\} = \mathcal{H}$, we are done; otherwise, since $sp\{e_1, \ldots, e_n\}$ is closed (see Problem 3.25) there is a $d_\ell \in \mathcal{D}$ such that $\ell > m_k$ and $d_\ell \notin sp\{e_1, \ldots, e_n\}$. Take m_{n+1} to be the least such ℓ, so $d_{m_{n+1}} \notin sp\{e_1, \ldots, e_n\}$ and $m_{n+1} > m_n$. Now set

$$f = d_{m_{n+1}} - \sum_1^n \langle e_j, d_{m_{n+1}}\rangle e_j.$$

Since $d_{m_{n+1}} \notin sp\{e_1, \ldots, e_n\}$, $f \neq 0$ and, for $k = 1, \ldots, n$

$$\langle f, e_k\rangle = \langle d_{m_{n+1}}, e_k\rangle - \left\langle \sum_{j=1}^n \langle e_j, d_{m_{n+1}}\rangle e_j, e_k \right\rangle$$

$$= 0$$

since $\langle e_j, e_k\rangle = 0$ if $j \neq k$ and $\langle e_j, e_k\rangle = 1$ if $j = k$. Thus if $e_{n+1} = \|f\|^{-1} f$, the set $\{e_1, \ldots, e_{n+1}\}$ is orthonormal.

It is evident from our construction that, if $1 \leq k \leq m_{n+1}$, then $d_k \in sp\{e_1, \ldots, e_{n+1}\}$, and hence that

$$sp\{d_1, \ldots, d_{m_{n+1}}\} \subset sp\{e_1, \ldots, e_{n+1}\}.$$

Alternatively, since e_{n+1} is a linear combination of $\{e_1, \ldots, e_n, d_{m_{n+1}}\}$ and since for $m_n \leq j \leq m_{n+1}$,

$$sp\{e_1, \ldots, e_n\} \subseteq sp\{d_1, \ldots, d_j\},$$

it follows that

$$sp\{e_1, \ldots, e_{n+1}\} \subset sp\{d_1, \ldots, d_{m_{n+1}}\}.$$

This completes the proof of the inductive step.

Now take $\mathcal{B} = \{e_n\}_{n=1}^{\infty}$, so $\mathcal{B}$ is an orthonormal set $\mathcal{H}$. By (b), $\mathcal{D} \subset sp(\mathcal{B})$, and thus, by 3.4.10(iv), $\mathcal{B}$ is also complete.

3.4.13. Definition. The Banach spaces $\mathcal{H}_1$ and $\mathcal{H}_2$ are said to be *isometrically isomorphic* if there is a linear map T from $\mathcal{H}_2$ onto $\mathcal{H}_1$ such that $\|Tx\| = \|x\|$ for all $x \in \mathcal{H}_2$. The mapping T is called an *isometric isomorphism*.

If two Hilbert spaces are isometrically isomorphic, then they are essentially indistinguishable algebraically, analytically and geometrically. The next theorem says that all separable infinite dimensional Hilbert spaces are the "same" as ℓ_2; this is especially nice to know since ℓ_2 is much more readily understood than, say, $\mathcal{L}_2(\mathfrak{R})$.

3.4.14. Theorem. *Let $\mathcal{H}$ be a separable Hilbert space. Then $\mathcal{H}$ is isometrically isomorphic to either ℓ_2 or $\mathfrak{R}^n$ for some n.*

Proof. By 3.4.12, $\mathcal{H}$ has a complete orthonormal set $\{e_n\}$ which is either countable or finite; we prove the result in case $\{e_n\}$ is countable, the finite case being similar. Of course, there is a natural algebraic isomorphism between $sp(e_n)$ and ℓ_2 obtained by identifying e_n with the nth unit vector in ℓ_2. The fact that this is also an isomorphism follows rather easily from 3.4.11.

We define $T(x)$ from $\mathcal{H}$ to ℓ_2 by setting

$$T(x) = \{\langle x, e_n \rangle\}.$$

By Parseval's identity, $\|T(x)\| = \|x\|$. To see that T is onto, we apply 3.4.4 in the case $x = 0$ to obtain for $(\lambda_n) \in \ell_2$ that if $m < n$,

$$\left\| \sum_{m}^{n} \lambda_k e_k \right\|^2 = \sum_{m}^{n} \lambda_k^2.$$

Thus

$$\left\| \sum_{1}^{n} \lambda_k e_k - \sum_{1}^{m} \lambda_k e_k \right\|^2 = \sum_{m}^{n} \lambda_k^2,$$

showing that $\sum_{1}^{\infty} \lambda_k e_k$ converges in $\mathcal{H}$. Clearly, $T(\sum_{1}^{\infty} \lambda_k e_k) = (\lambda_k)$, showing that T is onto. Since it is evident that T is linear, this completes the proof.

PROBLEMS

3.20. Let $\mathcal{H}$ be a Hilbert space and let B be an orthonormal set. Show that B is complete if and only if

$$\text{for each } x, y \in \mathcal{H}, \quad \langle x, y \rangle = \sum_{e \in B} \langle x, e \rangle \langle y, e \rangle.$$

3.21. Let $\mathcal{H}$ be a separable Hilbert space and let B be an orthonormal set in $\mathcal{H}$. Show that B is at most countable.

3.22. Let $\mathcal{H}$ be a Hilbert space and let $x, y \in \mathcal{H}$. Show that if $0 \le t \le 1$, then

$$\|(1-t)x + ty\|^2 + t(1-t)\|x - y\|^2 = (1-t)\|x\|^2 + t\|y\|^2.$$

(If $t = 1/2$, this reduces to the parallelogram law.)

3.23. Definition. A set C contained in a vector space V is *convex* if $(1-t)x + ty \in C$ for all $x, y \in C$ and $0 \le t \le 1$.

    (a) Let C be a closed, convex subset of a Hilbert space $\mathcal{H}$ and let $y \in \mathcal{H}$. Show that there is a unique vector $P(y) \in C$ such that $\|y - Py\| = \inf\{\|y - x\| : x \in C\}$.

    (b) Show that $\|Py - Pw\| \le \|y - w\|$ for all $y, w \in \mathcal{H}$.

3.24. Let A be an index set and let $\{x_\alpha\}$ be a collection of nonnegative numbers indexed by A. Show that

    (a) if $\sum_{\alpha \in A} x_\alpha < \infty$, then $\{\alpha : x_\alpha \ne 0\}$ is countable or finite;

    (b) if $A = \mathcal{N}$, $\sum_{\alpha \in A} x_\alpha = \sum_1^\infty x_j$.

3.25. Let E be a finite-dimensional subspace of a Hilbert space $\mathcal{H}$. Show that E is closed.

3.26. Definition. Let $\{x_n\}$ be a sequence in a Hilbert space $\mathcal{H}$ and let $x_\infty \in \mathcal{H}$. If for each $y \in \mathcal{H}$

$$\lim \langle x_n, y \rangle = \langle x_\infty, y \rangle,$$

then we say that $\{x_n\}$ *converges weakly* to x_∞; we then write "$x_n \rightharpoonup x_\infty$."

Let $\{e_n\}$ be a complete orthonormal set in a separable Hilbert space $\mathcal{H}$; show that $e_n \rightharpoonup 0$.

3.27. Show that the sequence $\{\cos(nt) + \sin(nt)\}$ converges weakly to zero in $\mathcal{L}_2(\Re)$. (Compare with 2.13.)

3.28. Show that if $\lim \|x_n - x_\infty\| = 0$, then $x_n \rightharpoonup x_\infty$ for sequences $\{x_n\}$ in a Hilbert space $\mathcal{H}$.

3.29. Let $\{x_n\}$ be a sequence in a Hilbert space $\mathcal{H}$ which converges weakly to x_∞. Show that $\{x_n\}$ converges strongly to x_∞ (i.e., $\lim \|x_n - x_\infty\| = 0$) if $\lim \|x_n\| = \|x_\infty\|$.

3.30. Let $\{x_n\}$ and $\{y_n\}$ be sequences in a Hilbert space $\mathcal{H}$ and suppose that $x_n \rightharpoonup x_\infty$ and $\{y_n\}$ converges strongly to y_∞. Show that

$$\lim \langle x_n, y_n \rangle = \langle x_\infty, y_\infty \rangle.$$

Does this conclusion still hold if we only assume that $y_n \rightharpoonup y_\infty$?

3.31.

(a) Let $T : E_1 \to E_2$ be an isometric isomorphism between the Hilbert spaces E_1 and E_2. Show that T sends weakly convergent sequences to weakly convergent sequences.

(b) Let $\{x_n\} = \{(x_j^{(n)})\}$ be a bounded sequence in ℓ_2. Show that $\{x_n\}$ has a weakly convergent subsequence. [*Hint:* It suffices to select $\{x_{n_k}\}$ so that each coordinate $x_j^{(n_k)}$ has a limit $x_j^{(\infty)}$ as $n_k \to \infty$.]

(c) Let $\{x_n\}$ be a bounded sequence in a Hilbert space $\mathcal{H}$. Show that $\{x_n\}$ has a weakly convergent subsequence. [*Hint:* Take $\mathcal{H}'$ to be the separable Hilbert space $\mathcal{H}' = \overline{\mathrm{sp}}\{x_n\}$. Apply Theorem 3.4.14 to reduce to Problem 3.31(b) and then return to $\mathcal{H}$ via part (a).]

3.32 Show that the sequence of Example 3.4.3(iii) is orthonormal.

3.5. DUAL SPACES.

Suppose that $1 < p < \infty$ is fixed and $1/p + 1/q = 1$. Fix an element $g \in \mathcal{L}_q$ and define a mapping T from $\mathcal{L}_p$ to $\Re$ by

$$T(f) = \int fg \, d\lambda. \tag{1}$$

Clearly, T is linear and, moreover, by Hölder's Inequality,

$$|T(f)| \leq \int |f| \, |g| \, d\lambda \leq \|f\|_p \|g\|_q.$$

It is evident from this inequality that T is continuous (in fact, for $\varphi, \psi \in \mathcal{L}_p$,

$$|T(\varphi) - T(\psi)| = |T(\varphi - \psi)| \leq \|\varphi - \psi\|_p \|g\|_q).$$

As it turns out, *every* continuous linear mapping from $\mathcal{L}_p$ to $\Re$ arises in exactly this way; i.e., if $T : \mathcal{L}_p \to \Re$ is continuous and linear, then there is a $g \in \mathcal{L}_q$ such that $T(f) = \int fg \, d\lambda$. This is a fairly deep fact which we will prove in Section 3.8; in this section we introduce the basic definitions and concepts surrounding mappings of the foregoing type.

3.5.1. Definitions. Let X be a normed linear space; a *linear functional* defined on X is a mapping T from X to $\Re$ satisfying

$$T(\alpha x + \beta y) = \alpha T(x) + \beta T(y)$$

for all scalars α and β and all vectors x and y in X. The linear functional T is *continuous* if whenever $\{x_n\} \subset X$ converges to $x_\infty \in X$, then

$$\lim T(x_n) = T(x_\infty).$$

The *dual space* X^* of X is the collection of all continuous linear functionals defined on X.

Our remarks above show that, under the embedding (1), $\mathcal{L}_q \subset (\mathcal{L}_p)^*$. Note that X^* is a vector space; in fact, X^* is a Banach space (even when X is not complete). Before giving the norm on X^*, we prove a lemma.

3.5.2. Proposition. *Let T be a linear functional defined on a normed linear space X. Then T is continuous if and only if*

$$\sup\{|T(x)| : |x| \leq 1\} < \infty. \qquad (2)$$

Proof. We first show sufficiency; set $M = \sup\{|T(x)| : |x| \leq 1\}$. Note that if $y \in X$ and $y \neq 0$, then

$$\left|T\left(\frac{y}{\|y\|}\right)\right| \leq M;$$

linearity of T then implies that $|T(y)| \leq M\|y\|$. Now let $\{x_n\}$ be a sequence in X and suppose that $\lim x_n = x_\infty$. Since

$$|T(x_n) - T(x_\infty)| = |T(x_n - x_\infty)| \leq M\|x_n - x_\infty\|,$$

it follows at once that $\lim Tx_n = Tx_\infty$, showing that T is continuous.

For necessity, suppose that T is continuous but (2) fails. Then for each n we may select x_n such that $\|x_n\| \leq 1$ and $T(x_n) \geq n$. Setting $y_n = \frac{1}{n}x_n$, we see that $\lim y_n = 0$ but

$$T(0) = 0 < 1 \leq \underline{\lim} T(y_n),$$

contradicting continuity of T at the origin.

3.5.3. Definition. Let X be a normed linear space; for $T \in X^*$ define

$$\|T\| = \sup\{|T(x)| : x \in X, \|x\| \leq 1\}.$$

It is easily checked that $(X^*, \|\cdot\|)$ is a normed linear space (Problem 3.33). In fact, X^* must always be complete:

3.5.4. Theorem. *Let X be a normed linear space. Then X^* is a Banach space.*

Proof. Let $\{T_n\}$ be a Cauchy sequence in X^*, and fix $x \in X$. Since

$$|T_n(x) - T_m(x)| = |(T_n - T_m)(x)| \leq \|T_n - T_m\|\,\|x\|,$$

it follows that $\{T_n(x)\}$ is a Cauchy sequence of real numbers and thus

$$T_\infty(x) \equiv \lim_n T_n(x)$$

exists. Since each T_n is linear,

$$T_\infty(\alpha x + \beta y) = \lim_n T_n(\alpha x + \beta y)$$

$$= \lim_n \alpha T_n(x) + \lim_n \beta T_n(y)$$

$$= \alpha T_\infty(x) + \beta T_\infty(y)$$

and thus T_∞ is linear. Since $\{T_n\}$ is Cauchy, the set $\{\|T_n\|\}$ must be bounded; set $M = \sup\{\|T_n\|\}$. Then

$$|T_\infty(x)| = \lim_n |T_n(x)|$$

$$\leq \underline{\lim}\|T_n\| \, \|x\|$$

$$\leq M\|x\|$$

and so T_∞ is continuous.

All that remains to show is that $\lim \|T_\infty - T_n\| = 0$. To see this, fix $\epsilon > 0$ and select N so that if $m, n \geq N$, then $\|T_m - T_n\| \leq \epsilon$. Then for each $x \in X$ with $\|x\| \leq 1$,

$$|T_m(x) - T_n(x)| \leq \epsilon.$$

Now letting $m \to \infty$ gives, for each x with $\|x\| \leq 1$ and each $n \geq N$

$$|T_\infty(x) - T_n(x)| \leq \epsilon.$$

Taking the supremum over all such x gives, if $n \geq N$,

$$\|T_\infty - T_n\| \leq \epsilon,$$

showing that $\lim \|T_\infty - T_n\| = 0$.

Linear functionals have many properties which at first may seem rather peculiar; for example, if a linear functional is continuous at one point, then it is continuous everywhere. This and other properties are discussed in the problems.

PROBLEMS

3.33. Verify that 3.5.3 gives a norm on X^*.

3.34. Let T be a nonzero linear functional defined on a normed linear space E. Show that exactly one of the following is true:
  (a) T is continuous; or else
  (b) the null space of T, $N(T) = \{x \in E : T(x) = 0\}$, is dense in E.

3.35. Let E be a normed linear space. Show that E is finite-dimensional if and only if every linear functional defined on E is continuous.

3.36. Let E be an infinite-dimensional normed linear space. Show that there are disjoint, convex, dense subsets A and B contained in E such that $E = A \cup B$.

3.37. Let $T \in (\ell_1)^*$. Show that there is a unique vector $x = (x_n) \in \ell_\infty$ such that for all $y = (y_n) \in \ell_1$,

$$T(y) = \sum_1^\infty x_n y_n \quad \text{and} \quad \|x\| = \|T\|.$$

[*Hint:* Determine the behavior of T on the unit vectors

$$e_n = (0, \ldots, 0, 1, 0, \ldots),$$

where the "1" occurs in the nth position.]

3.38. Definition. The vector space c_0 is the class of all sequences $\{x_n\}$ of real numbers for which $\lim x_n = 0$; the vector space c is the class of all sequences $\{x_n\}$ for which $\lim x_n$ exists. The norm on c (and on $c_0 \subset c$) is

$$\|(x_n)\| = \sup |x_n| + \lim |x_n|.$$

In problem 3.37 a representation was given showing that $(\ell_1)^*$ and ℓ_∞ are isometrically isomorphic; as a shorthand for this represenation we write "$(\ell_1)^* = \ell_\infty$." This problem deals with similar representations for c_0 and c.

(a) Show that $(c_0)^* = \ell_1$;

(b) Show that $(c)^* = \ell_1$. (*Caution:* The representations are different.)

3.39. For every normed linear space E, show that $E \subset E^{**}$ (E^{**} is the dual space of E^*). (Spaces for which $E = E^{**}$ are called *reflexive* and have many nice properties. We will consider reflexive spaces in Chapter 8.)

3.40. Show that none of the following spaces are reflexive:

(a) c_0;

(b) c;

(c) ℓ_1.

3.41. Let X and Y be normed linear spaces and let $T : X \to Y$ be a *linear function*, i.e., $T(0) = 0$ and for all $x, y \in X$ and all scalars $\alpha \in \Re$, $T(\alpha x + y) = \alpha T(x) + T(y)$. Show that T is continuous if and only if

$$\sup\{\|T(x)\| : \|x\| \leq 1, \ x \in X\} < \infty.$$

3.6. RIESZ REPRESENTATION THEOREM FOR HILBERT SPACE.

In this section we will characterize the bounded linear functionals on Hilbert space; the ingenious proof which we present is due to E. J. McShane [*] and extends readily to the $\mathcal{L}_p$ spaces.

---

[*] *Linear functionals on certain Banach spaces*, Proceedings of the American Mathematical Society, **1**, 402-408 (1950).

Our goal will be to realize the action of a linear functional in terms of the inner product. The outline of the proof is best understood in terms of two-dimensional space. If L is a nonzero linear functional defined on $\Re^2$, then surely there is a vector x of norm less than or equal to 1 such that $L(x) = \|L\|$ (apply compactness of the unit ball). The vector x must in fact be of norm exactly 1 since otherwise a "small" perturbation by a vector u for which $L(u) > 0$ would give $\|L\| < L(x+u) \leq \|L\|$. Next notice that if $\ell = \{y : L(y) = 0\}$, then $x + \ell$ must be tangent to the unit circle in $\Re^2$. If this were not so, then there would be a vector x' along the line $x + \ell$ of norm strictly *less* than 1 which satisfies $L(x') = \|L\|$. All of this shows that the subspaces spanned by x and ℓ are perpendicular and so an arbitrary vector v in $\Re^2$ can be uniquely written as $v = l + \alpha x$, where $< x, l > = 0$ and $L(l) = 0$. In particular,

$$L(v) = \alpha L(x) = \alpha\|L\|\langle x, x\rangle = \|L\|\langle l + \alpha x, x\rangle,$$

showing that $L(v) = \langle v, \|L\|x\rangle$.

Translating this simple two-dimensional argument to arbitrary Hilbert spaces is our task. The first difficulty is the existence of a vector at which L attains its norm; fortunately, the geometry implicit in the parallelogram law is sufficient to compensate for the lack of compactness.

3.6.1. Lemma. *Let $\mathcal{X}$ be a Hilbert space and let L be a nonzero bounded linear functional defined on $\mathcal{X}$. Then there is a $g \in \mathcal{X}$ for which $\|g\| = 1$ and $L(g) = \|L\|$.*

Proof. Since $\|L\| = \sup\{|L(x)| : \|x\| \leq 1\} = \sup\{|L(w)| : \|w\| = 1\}$, we may select a sequence $\{f_n\}$ in $\mathcal{X}$ such that $\|f_n\| = 1$ for each n and $\lim |L(f_n)| = \|L\|$. Set $g_n = \operatorname{sgn}(L(f_n))f_n$, so that $\lim L(g_n) = \|L\|$. It suffices to show that $\{g_n\}$ is a Cauchy sequence in $\mathcal{X}$. In order to accomplish this, we will use the parallelogram law to estimate $\|g_n - g_m\|$; the technique embodied in the rest of the proof is frequently central to arguments in Hilbert space.

Fix $\epsilon > 0$ and choose $\xi > 0$ so small that $\xi < 2\|L\|$ and

$$(4 - \|L\|^{-2}(2\|L\| - \xi)^2)^{\frac{1}{2}} < \epsilon.$$

Next select N so large that $n \geq N$ implies that

$$0 < L(g_n) \geq \|L\| - \xi/2.$$

In view of the parallelogram law, if $n, m \geq N$ then

$$\|g_n + g_m\|^2 = 4 - \|g_n - g_m\|^2$$

since $\|g_n\| = \|f_n\| = 1$. Also, $0 < L(g_n + g_m)$, so $g_n + g_m \neq 0$.

Now observe that

$$\|L\| \geq \left(\frac{L(g_n + g_m)}{\|g_n + g_m\|}\right) = (4 - \|g_n - g_m\|^2)^{-\frac{1}{2}}(L(g_n) + L(g_m))$$

$$\geq (4 - \|g_n - g_m\|^2)^{-\frac{1}{2}}(2\|L\| - \xi).$$

Rewriting this, we obtain for $n, m \geq N$

$$\|g_n - g_m\| \leq (4 - \|L\|^{-2}(2\|L\| - \xi)^2)^{\frac{1}{2}} < \epsilon,$$

and thus $\{g_n\}$ is a Cauchy sequence. By completeness of $\mathcal{X}$, $\{g_n\}$ converges to some $g \in \mathcal{X}$ and $L(g) = \lim L(g_n) = \|L\|$; also, $\|g\| = \lim \|g_n\| = 1$, completing the proof.

Notice what makes the proof above work is that we can combine estimates on $\|g_n\|$, $\|g_m\|$, and $\|g_m + g_n\|$ with the parallelogram law in order to obtain the needed estimate on $\|g_n - g_m\|$.

Our intuition tells us that, in some sense, L is "tangent" to the unit ball at the point g; one way of phrasing this precisely is that the action of L should be described by the derivative of the norm. This is precisely the content of the next lemma.

3.6.2. Lemma. *Let E be a normed linear space and let $L \in E^*$. Suppose that g and $f \in E$ satisfy:*

(a) $\lim\limits_{t \to 0} \frac{\|g + tf\| - \|g\|}{t}$ *exists; and*

(b) $\|g\| = 1$ *and* $L(g) = \|L\|$.

Then $L(f) = \|L\| \lim\limits_{t \to 0} \frac{\|g + tf\| - \|g\|}{t}$.

Proof. Observe that for all t, $|L(\frac{g+tf}{\|g+tf\|})| \leq \|L\|$, and thus $L(g+tf) \leq \|L\|\|g+tf\|$. Using $\|g\| = 1$, this implies that

$$\lim_{t \to 0^-} \|L\| \left(\frac{\|g + tf\| - \|g\|}{t} \right) = \lim_{t \to 0^-} \frac{\|L\| \|g + tf\| - \|L\|}{t}$$

$$\leq \lim_{t \to 0^-} \frac{L(g + tf) - L(g)}{t}$$

$$= L(f)$$

$$\leq \lim_{t \to 0^+} \frac{\|L\| \|g + tf\| - \|L\|}{t}$$

$$= \lim_{t \to 0^+} \|L\| \left(\frac{\|g + tf\| - \|g\|}{t} \right).$$

Since the first and last terms are equal by (a), the conclusion follows.

Now all that remains is to compute the derivative of the norm in Hilbert space.

3.6.3. Riesz Representation Theorem for Hilbert Space. *Let $\mathcal{H}$ be a Hilbert space and let $T \in (\mathcal{H}^*)$. Then there is a unique $h \in \mathcal{H}$ such that $T(f) = <f, h>$ for all $f \in \mathcal{H}$.*

Proof. By Lemma 1, we may select $g \in \mathcal{H}$ such that $\|g\| = 1$ and $T(g) = \|T\|$. By Lemma 2, it suffices to show that

$$\lim_{t \to 0} \frac{\|g + tf\| - \|g\|}{t} = \langle f, g \rangle$$

for all $f \in \mathcal{H}$; then, upon taking $h = g\|T\|$, the result follows.

But the above is an easy computation:

$$\lim_{t \to 0} \frac{\|g + tf\| - \|g\|}{t} = \lim_{t \to 0} \frac{\|g + tf\|^2 - \|g\|^2}{(\|g + tf\| + \|g\|)t}$$

$$= \lim_{t \to 0} \frac{1}{2t}\left(\|g + tf\|^2 - \|g\|^2\right)$$

$$= \lim_{t \to 0} \frac{t}{2}\|f\|^2 + \langle f, g \rangle$$

$$= \langle f, g \rangle.$$

For uniqueness, suppose that $T(f) = \langle f, h \rangle = \langle f, \hat{h} \rangle$ are two different representations of T. Then for all f, $\langle f, h - \hat{h} \rangle = 0$, which implies that $h = \hat{h}$.

Note that this result says that the only linear functionals on $\mathcal{L}_2$ are of the form $T(f) = \int fg \, d\lambda$ for some $g \in \mathcal{L}_2$.

There are many different ways of establishing the foregoing representation theorem, some of them strikingly different from the above (as, for example, Problem 5.23). Some of these alternative proofs are slicker, but we have chosen one which emphasizes geometric considerations. From the very beginning, the study of the $\mathcal{L}_p$ spaces has tended to move in the direction of studying geometric, i.e., global, inequalities based on the norm; one advantage of this approach is that techniques using geometry in $\mathcal{L}_2$ have a good chance of extending to more general spaces. As an example, in the next two sections we prove a representation theorem for $\mathcal{L}_p$ based on the ideas of this section.

PROBLEMS

3.42. Let E be a normed linear space and let $x, y \in E$.

(a) Show that the function f defined on $\Re \setminus \{0\}$ given by

$$f(t) = \frac{\|x + ty\| - \|x\|}{t}$$

is a nondecreasing function. [*Hint:* Show, for $t > 0$, that $f(-t) \leq f(t)$, then treat the cases $t_1 < t_2 < 0$ and $t_2 > t_1 > 0$ separately.]

(b) Show that both of the following limits exist:

$$D_x^+(y) = \lim_{t \to 0^+} \frac{\|x + ty\| - \|x\|}{t} \quad \text{and}$$

$$D_x^-(y) = \lim_{t \to 0^-} \frac{\|x + ty\| - \|x\|}{t}$$

and that $D_x^+(y) \geq D_x^-(y)$.

(c) Show that D_x^+ and D_x^- are bounded linear functionals defined on E.

(d) If E is a Hilbert space, show that $D_x^+ = D_x^- = x/(2\|x\|)$ for all $x \neq 0$ in E.

3.43. Let $\mathcal{H}$ be a Hilbert space; we denote by $B(\mathcal{H})$ the class of all continuous linear mappings from $\mathcal{H}$ into itself.

(a) Show that for each $T \in B(\mathcal{H})$ there is a unique $T^* \in B(\mathcal{H})$ satisfying

$$\langle T(x), y \rangle = \langle x, T^*(y) \rangle$$

for all $x, y \in \mathcal{H}$. (T^* is called the *adjoint* of T.)

(b) Show that $(T_1 + T_2)^* = T_1^* + T_2^*$ for all $T_1, T_2 \in B(\mathcal{H})$ and that $T_1^{**} = T_1$.

(c) Show that $\|TT^*\| = \|T\|^2$ for all $T \in B(\mathcal{H})$ where $\|T\| = \sup\{\|T(x)\| : x \in \mathcal{H}, \|x\| \leq 1\}$ (see 13.I).

(d) Let $T \in B(\mathcal{H})$. Show there is an element $x \in \mathcal{H}$ such that $T(x) = \|T\|$.

(e) Show that, if $T \in B(\mathcal{H})$, then

$$\|T\|^2 = \max\{\lambda : \exists x \in \mathcal{H}, \ x \neq 0 \ \text{for which} \ T^*Tx = \lambda x\}.$$

(If there is an x for which $T^*Tx = \lambda x$, then λ is an *eigenvalue* of T^*T and x is an *eigenvector*.) [Operators and their adjoints in $B(\mathcal{H})$ play an important role in differential equations.]

3.7. CLARKSON'S INEQUALITIES. If $1 < p < \infty$, then the spaces $\mathcal{L}_p$ share many structural properties with general Hilbert spaces. While these properties are always more difficult to establish in $\mathcal{L}_p$, and sometimes more difficult to state, they can generally be deduced using techniques similar to those employed in Hilbert space. This section is devoted to an important set of inequalities, discovered by J. A. Clarkson in 1936,[*] which constitute a generalized "parallelogram law" for these

[*] *Uniformly convex spaces*, Transactions of the American Mathematical Society, **40**, 396-414 (1936).

spaces. In Section 3.8 we will use the inequalities to characterize the dual space of $\mathcal{L}_p$ if $1 < p < \infty$.

The inequalties in question will sharpen the triangle inequality in much the same way the parallelogram law does for Hilbert space. Of course, in general we cannot expand $|f + g|^p$ in an elementary closed form, and so we instead deduce inequality estimates on this quantity. Clarkson's original proof achieved these inequalities by essentially using Taylor's series; luckily, cleaner arguments are now known.

Throughout, we assume that p and q are conjugate pairs.

3.7.1. Lemma. *If* $-1 \le x \le 1$ *and* $1 < p \le 2$, *then*

$$\left(\frac{1+x}{2}\right)^q + \left(\frac{1-x}{2}\right)^q \le \left(\frac{1+|x|^p}{2}\right)^{q-1}. \tag{1}$$

If $2 \le p < \infty$, *then the inequality (1) is reversed.*

Proof. Since (1) is symmetric in x and $-x$, it suffices to prove the lemma for $x \ge 0$. For $0 \le \alpha \le 1$ and $0 \le x \le 1$, set

$$f(\alpha, x) = \left(\frac{1 + \alpha^{1-q}x}{2}\right)\left(\frac{1 + \alpha x}{2}\right)^{q-1} + \left(\frac{1 - \alpha^{1-q}x}{2}\right)\left(\frac{1 - \alpha x}{2}\right)^{q-1}.$$

Notice that $f(1, x)$ is the left-hand side of (1), while, since $(p-1)(q-1) = 1$, $f(x^{p-1}, x)$ is the right-hand side of (1). Thus, to establish the lemma for nonnegative x, it suffices to show that $\frac{\partial f}{\partial \alpha} \le 0$ if $1 < p \le 2$, and $\frac{\partial f}{\partial \alpha} \ge 0$ if $2 \le p < \infty$.

An elementary computation shows that

$$\frac{\partial f}{\partial \alpha} = \frac{(q-1)}{2} x(1 - \alpha^{-q})\left(\left(\frac{1 + \alpha x}{2}\right)^{q-2} - \left(\frac{1 - \alpha x}{2}\right)^{q-2}\right)$$

(you should verify this!). Since $0 \le \alpha \le 1$, $1 - \alpha^{-q} \le 0$. Thus $\frac{\partial f}{\partial \alpha}$ is less than or equal to zero if and only if

$$\left(\frac{1 + \alpha x}{2}\right)^{q-2} - \left(\frac{1 - \alpha x}{2}\right)^{q-2} \ge 0$$

while $\frac{\partial f}{\partial \alpha} \ge 0$ if and only if this inequality is reversed.

But now if $1 < p \le 2$, then $q - 2 \ge 0$, while if $2 \le p < \infty$, then $q - 2 \le 0$, and thus the desired inequalities are immediate, proving the lemma.

3.7.2. Corollary. *If* u *and* v *are any real numbers and* $1 < p \le 2$, *then*

$$\left|\frac{u+v}{2}\right|^q + \left|\frac{u-v}{2}\right|^q \le \left(\frac{|u|^p + |v|^p}{2}\right)^{q-1}.$$

If $2 \le p < \infty$, *the inequality is reversed.*

Proof. We may assume without loss of generality that $0 < |u| \leq |v|$; dividing through by $|u|^q$ reduces to Lemma 3.7.1.

3.7.3. Clarkson's Inequalities. *Let $1 < p \leq 2$ and let f and g be functions in $\mathcal{L}_p$. Then*

$$\left\| \frac{f+g}{2} \right\|_p^q + \left\| \frac{f-g}{2} \right\|_p^q \leq \left(\frac{\|f\|_p^p + \|g\|_p^p}{2} \right)^{q-1}.$$

If $2 \leq p < \infty$, the inequality is reversed.

Proof. We will prove this in the case $1 < p \leq 2$; for the other case all of the inequalities will be reversed.

Note that, if $h \in \mathcal{L}_p$, then $|h|^q \in \mathcal{L}_{p-1}$, since $(p-1)q = p$; moreover,

$$\| |h|^q \|_{p-1} = \|h\|_p^q.$$

We exploit this observation and Minkowski's inequality for $0 < p-1 \leq 1$ as follows:

$$\left\| \frac{f+g}{2} \right\|_p^q + \left\| \frac{f-g}{2} \right\|_p^q = \left\| \left| \frac{f+g}{2} \right|^q \right\|_{p-1} + \left\| \left| \frac{f-g}{2} \right|^q \right\|_{p-1}$$

$$\leq \left\| \left| \frac{f+g}{2} \right|^q + \left| \frac{f-g}{2} \right|^q \right\|_{p-1}$$

$$= \left(\int \left(\left| \frac{f+g}{2} \right|^q + \left| \frac{f-g}{2} \right|^q \right)^{p-1} d\lambda \right)^{\frac{1}{p-1}}.$$

Now apply Corollary 3.7.2 and the fact that $(p-1)^{-1} = q - 1$:

$$\left\| \frac{f+g}{2} \right\|_p^q + \left\| \frac{f-g}{2} \right\|_p^q \leq \left(\int \left(\frac{|f|^p + |g|^p}{2} \right)^{(q-1)(p-1)} d\lambda \right)^{q-1}$$

$$= \left(\frac{\|f\|_p^p + \|g\|_p^p}{2} \right)^{q-1},$$

completing the proof.

Notice that if $p = q = 2$, the inequalities above reduce to the parallelogram law. The following easy reformulation will be of use in the next section.

3.7.4. Corollary. *Let $1 < p \leq 2$ and let φ and ψ be functions in $\mathcal{L}_p$. Then*

$$2^{q-1}\left(\|\varphi\|_p^q + \|\psi\|_p^q \right) \leq \left(\|\varphi + \psi\|_p^p + \|\varphi - \psi\|_p^p \right)^{q-1}.$$

If $2 \leq p < \infty$ the inequality is reversed.

Proof. In 3.7.3, take $f = \varphi + \psi$ and $g = \varphi - \psi$; the result is then immediate.

The ingenious proof above is due to K. O. Friedrichs [*] and is far simpler than Clarkson's original proof; to better appreciate the elegance of Friedrich's approach the student should look up Clarkson's original 1936 paper.

PROBLEMS

3.44.

(a) Suppose $1 < p \leq 2$ and $0 \leq x \leq 1$. Show that

$$H(x) \equiv 2^{(p-2)/p} \frac{(x^p + 1)^{\frac{1}{p}}}{(x^q + 1)^{\frac{1}{q}}} \leq 1.$$

Show the inequality is reversed if $2 \leq p < \infty$. [*Hint:* Compute $H'(x)$.]

(b) Let $x, y \geq 0$ and let $1 < p \leq 2$. Show that

$$2(x^q + y^q)^{p-1} \geq 2^{p-1}(x^p + y^p).$$

Show that the inequality is reversed if $2 \leq p < \infty$.

(c) Deduce the third of Clarkson's Inequalities: If $1 < p \leq 2$ and $f, g \in \ell_p$ then

$$2(\|f\|_p^p + \|g\|_p^p) \geq \|f + g\|_p^p + \|f - g\|_p^p \geq 2^{p-1}(\|f\|_p^p + \|g\|_p^p).$$

If $2 \leq p < \infty$ the inequality is reversed.

3.45. Let C be a closed convex set in $\mathcal{L}_p$, and let $f \in \mathcal{L}_p$. Show there is a unique vector $g \in C$ such that $\|f - g\|_p = \min\{\|f - h\|_p : h \in C\}$. [See part (a) of problem 3.23; 3.23(b) fails unless $p = 2$—indeed, part (b) characterizes Hilbert spaces of dimension larger than 2.]

3.46. Suppose that $2 \leq p < \infty$ and $t = 2^{-n}$ for some integer n. Then for all $f, g \in \mathcal{L}_p$

$$\|(1 - t)f + tg\|_p^p + (2^{p-1} - 1)^{-1} t(1 - t^{p-1})\|f - g\|_p^p$$
$$\leq (1 - t)\|f\|_p^p + t\|g\|_p^p.$$

(Compare with Problem 3.22.)

3.47. Suppose that $1 < p \leq 2$ and $t = 2^{-n}$ for some integer n. Then for all $f, g \in \mathcal{L}_p$

$$\|(1 - t)g + tg\|_p^q + (2^{q-1} - 1)^{-1} t((1 - t^{q-1})\|f - g\|_p^q$$
$$\leq (1 - t)\|f\|_p^q + t\|g\|_p^q.$$

3.48. Let K be a bounded, convex subset of $\mathcal{L}_p$ ($1 < p < \infty$). Define the *diameter* of K to be

$$d \equiv d(K) = \sup\{\|f - g\| : f, g \in K\}$$

[*] *On Clarkson's Inequalities*, Communications in Pure and Applied Mathematics, **23**, 603-607 (1970).

and the *Cebyšev radius* of K to be

$$r \equiv r(K) = \inf\{\varrho > 0 : \cap_{x \in K} B(x; \varrho) \neq \emptyset\}.$$

Show that

(a) If $2 \leq p < \infty$, then $(1 - 2^{1-p})^{-1}r^p \leq d^p$; and

(b) if $1 < p \leq 2$, then $(1 - 2^{1-q})^{-1}r^q \leq d^q$.

(Apply Problems 3.46 and 3.47. If $p = 2$ the estimates above are sharp.)

3.8. RIESZ REPRESENTATION THEOREM FOR $\mathcal{L}_p$.

We now have the machinery in hand to verify the Riesz Representation Theorem for the $\mathcal{L}_p$ spaces. Our approach will imitate Section 3.6, using Clarkson's inequalities in place of the parallelogram law.

3.8.1. Lemma. *Let T be a nonzero bounded linear functional defined on $\mathcal{L}_p$ for some $1 < p < \infty$. Then there is a $g \in \mathcal{L}_p$ such that $\|g\| = 1$ and $T(g) = \|T\|$.*

Proof. We will prove this lemma in the case $1 < p < 2$; the similar proof in the case $2 < p < \infty$ is left as an exercise. Without loss of generality we may assume that $\|T\| \neq 0$. As in Lemma 3.6.1, we select a sequence $\{f_n\} \subset \mathcal{L}_p$ such that $\|f_n\| = 1$ and $\lim |T(f_n)| = \|T\|$. Setting $g_n = \text{sgn}(T(f_n))f_n$, we see that $\|g_n\| = 1$ and $\lim T(g_n) = \|T\|$.

Now fix $\epsilon > 0$ and select $\xi > 0$ so small that $\xi < 2\|T\|$ and

$$\left(2^q - \|T\|^{-q}(2\|T\| - \xi)^q\right)^{\frac{1}{q}} \leq \epsilon$$

where $1/q + 1/p = 1$. Next select N so large that $n \geq N$ implies that

$$T(g_n) \geq \|T\| - \xi/2 > 0.$$

Now, in view of Clarkson's inequalities, if $n, m \geq N$, then

$$\|g_n + g_m\|^q \leq 2^q - \|g_n - g_m\|^q$$

since $\|g_n\| = \|g_m\| = 1$. Consequently,

$$\|T\| \geq \left|\frac{T(g_n + g_m)}{\|g_n + g_m\|}\right| \geq \frac{T(g_n) + T(g_m)}{\left(2^q - \|g_n - g_m\|^q\right)^{\frac{1}{q}}}$$

$$\geq \frac{2\|T\| - \xi}{\left(2^q - \|g_n - g_m\|^q\right)^{\frac{1}{q}}}.$$

Rearranging terms gives, if $n, m \geq N$,

$$\|g_n - g_m\| \leq \left(2^q - \|T\|^{-q}(2\|T\| - \xi)^q \right)^{\frac{1}{q}} \leq \epsilon$$

and hence $\{g_n\}$ is a Cauchy sequence. By the Riesz-Fischer Theorem there is a $g_\infty \in \mathcal{L}_p$ such that $\lim g_n = g_\infty$. It then follows readily that $\|g_\infty\| = 1$ and $T(g_\infty) = \|T\|$.

One of the additional complications in the $\mathcal{L}_p$ spaces is computing the directional derivative of the $\mathcal{L}_p$ norm— something made trivial by the inner product in Hilbert space. Consequently, the argument in 3.6.3 needs to be expanded upon; the Monotone Convergence Theorem and the following simple lemma about the convex function $|ta + b|^p$ combine to give the derivative.

3.8.2. Lemma. *For any real numbers a and b and any $p > 1$, set $\varphi(t) = |ta + b|^p$. Then if $0 \leq t_1 \leq t_2$*

(i)

$$\frac{\varphi(t_1) - \varphi(0)}{t_1} \leq \frac{\varphi(t_2) - \varphi(0)}{t_2},$$

while if $t_2 \leq t_1 < 0$,

(ii)

$$\frac{\varphi(t_1) - \varphi(0)}{t_1} \leq \frac{\varphi(t_2) - \varphi(0)}{t_2}.$$

Proof. Observe that $\varphi'(t) = pa|ta + b|^{p-1}\text{sgn}(ta + b)$ and $\varphi'' = p(p - 1)a^2|ta + b|^{p-2} > 0$. Thus φ is convex; i.e., for any t_1, t_2 and $\lambda \in [0, 1]$,

$$\varphi(\lambda t_1 + (1 - \lambda)t_2) \leq \lambda\varphi(t) + (1 - \lambda)\varphi(t_2).$$

In particular if $0 \leq t_1 \leq t_2$ then

$$\varphi(t_1) \leq \frac{t_2 - t_1}{t_2}\varphi(0) + \frac{t_1}{t_2}\varphi(t_2).$$

Upon rearranging, this yields (i). Inequality (ii) is derived in exactly the same manner.

3.8.3. Lemma. *If $f, g \in \mathcal{L}_p$ and $1 < p < \infty$, then*

$$\lim_{t \to 0} \frac{\|g + tf\|^p - \|g\|^p}{pt} = \int f|g|^{p-1}\text{sgn}(g) \, d\lambda.$$

Proof. First suppose that $t > 0$ and set

$$h_t(x) = |f(x) + g(x)|^p - |g(x)|^p - \frac{|g(x) + tf(x)|^p - |g(x)|^p}{t}.$$

Taking $t_2 = 1$ in (i) of Lemma 3.8.2 implies that $h_t(x) \geq 0$; moreover, (i) implies that $\{h_t(x)\}$ increases as t decreases. Thus the Monotone Convergence Theorem implies that

$$\|f + g\|^p - \|g\|^p - \lim_{t \to 0+} \frac{\|g + tf\|^p - \|g\|^p}{t}$$

$$= \lim_{t \to 0+} \int |f + g|^p - |g|^p - \frac{|g + tf|^p - |g|^p}{t} \, d\lambda$$

$$= \int |f + g|^p - |g|^p - \lim_{t \to 0+} \frac{|g + tf|^p - |g|^p}{t} \, d\lambda$$

$$= \|f + g\|^p - \|g\|^p - \int pf|g|^{p-1}\mathrm{sgn}(g) \, d\lambda,$$

applying L'Hospital's Rule in the last step. Hence

$$\lim_{t \to 0+} \frac{\|g + tf\|^p - \|g\|^p}{pt} = \int f|g|^{p-1}\mathrm{sgn}(g) \, d\lambda.$$

The same procedure, applying (ii) of Lemma 3.8.2 to $-h_t$ gives

$$\lim_{t \to 0-} \frac{\|g + tf\|^p - \|g\|^p}{pt} = \int f|g|^{p-1}\mathrm{sgn}(g) \, d\lambda,$$

completing the proof.

A final lemma is the following analogue of 3.6.2. Since we have actually found the derivative of the pth power of the $\mathcal{L}_p$ norm, the lemma must be modified to take this into account.

3.8.4. Lemma. *Let E be a normed linear space and let $T \in E^*$. Suppose that f and g in E satisfy for some $p \geq 1$*
  (i) $\lim\limits_{t \to 0} \dfrac{\|g + tf\|^p - \|g\|^p}{pt}$ *exists; and*
  (ii) $T(g) = \|T\|$ *and* $\|g\| = 1$.
Then $T(f) = \|T\| \lim\limits_{t \to 0} \dfrac{\|g + tf\|^p - \|g\|^p}{pt}.$

Proof. Observe first by L'Hôpital's Rule that

$$\lim_{t \to 0} \frac{(T(g) + tT(f))^p - (T(g))^p}{pt} = T(f)\|T\|^{p-1}$$

since $T(g) = \|T\|$. Thus we may proceed as in Lemma 3.6.2.

$$\lim_{t \to 0^-} \|T\|^p \left(\frac{\|g + tf\|^p - \|g\|^p}{pt} \right) \leq \lim_{t \to 0^-} \frac{(T(g + tf))^p - (T(g))^p}{pt}$$

$$= T(f)\|T\|^{p-1}$$

$$\leq \lim_{t \to 0^+} \|T\|^p \left(\frac{\|g + tf\|^p - \|g\|^p}{pt} \right)$$

and thus, by (i), the conclusion follows.

3.8.5. Theorem (Riesz Representation Theorem for $\mathcal{L}_p$). *Let $1 < p < \infty$ and let T be a bounded linear functional defined on $\mathcal{L}_p$. Then there is a unique $h \in \mathcal{L}_q$ such that $T(f) = \int fh \, d\lambda$ for all $f \in \mathcal{L}_p$.*

Proof. By Lemma 3.8.1, there is a $g \in \mathcal{L}_p$ such that $\|g\| = 1$ and $T(g) = \|T\|$. By Lemma 3.8.3, 3.8.4(i) holds for all $f \in \mathcal{L}_p$, and thus

$$T(f) = \|T\| \lim_{t \to 0} \frac{\|g + tf\|^p - \|g\|^p}{pt}$$

$$= \|T\| \int f \operatorname{sgn}(g)|g|^{p-1} \, d\lambda.$$

Now since $g \in \mathcal{L}_p$, $|g|^{p-1} \in \mathcal{L}_q$ and thus $h = \|T\|\operatorname{sgn}(g)|g|^{p-1}$ is the desired function in $\mathcal{L}_q$.

For uniqueness, suppose that h and $\hat{h}$ provide representations for T, so that $\int f(h - \hat{h}) \, d\lambda = 0$ for all $f \in \mathcal{L}_p$. Since $(q - 1)p = q$, the function $\operatorname{sgn}(h - \hat{h})|h - \hat{h}|^{q-1} \in \mathcal{L}_p$, and thus

$$\int |h - \hat{h}|^p \, d\lambda = 0.$$

This implies that $h = \hat{h}$, as desired.

Chapter 4

Metric Spaces

4.1. INTRODUCTION. While the Banach spaces, especially those of Section 3.1, are the main spaces of interest in this book, there is a collection of results which are both more natural and more useful when phrased in slightly greater generality. As a rule, these results do not rely on any algebraic structure, but use only a notion of "distance" (which is implicit in the norm for Banach spaces). In this section we introduce the notion of a "metric space" and list some important examples.

4.1.1. Definition. Let M be an abstract set and let d be a mapping from $M \times M$ to $[0, \infty)$. Then d is said to be a *metric* if for all $x, y, z \in M$,

    (i) $d(x, y) = 0$ if and only if $x = y$ and $d(x, y) = d(y, x)$ for all $x, y \in M$.

    (ii) $d(x, y) \leq d(x, z) + d(z, y)$.

The pair (M, d) is called a *metric space*; inequality (ii) is called the *triangle inequality.*

4.1.2. Examples. Of course the most obvious examples of metric spaces are the real numbers and $\Re^n$, n-dimensional Euclidean space. Other examples are:

    (a) If X is a Banach space, and $d(x, y) = \|x - y\|$, then $(X; d)$ is a metric space.

    (b) If X is a Banach space and M is any subset of X, then (M, d) is a metric space with d defined as above.

    (c) If (M, d) is a metric space and $M_1 \subset M$, then (M_1, d) is a metric space.

    (d) It is sometimes possible to define a metric on a Banach space X which is not equivalent to the norm on X. For example, let $\mathcal{H}$ be a separable Hilbert space with complete orthonormal set $\{e_n\}$. For $x, y \in \mathcal{H}$, define

$$d(x, y) = \sum_{1}^{\infty} 2^{-n} |\langle x - y, e_n \rangle|.$$

Note that $d(x, y) \leq \|x - y\|$ by the Cauchy-Schwartz inequality. Also, $d(x, y) = 0$ if and only if $\langle x - y, e_n \rangle = 0$ for all n, which since $\{e_n\}$ is complete implies that

$x - y = 0$. Also if $x, y, z \in \mathcal{X}$, then

$$d(x, y) = \sum_{1}^{\infty} 2^{-n} |\langle x - y, e_n \rangle|$$

$$\leq \sum_{1}^{\infty} (|\langle x - z, e_n \rangle| + |\langle z - y, e_n \rangle|) 2^{-n}$$

$$= d(x, z) + d(z, y),$$

and thus $(\mathcal{X}, d)$ is a metric space. We will return to this example momentarily and show that d and $\| \cdot \|$ induce different convergence properties on $\mathcal{X}$.

(e) We can define other metrics on the real numbers as well. For example, if x and y are real numbers, define $d(x, y)$ to be

$$d(x, y) = |x - y|(1 + |x - y|)^{-1}.$$

This metric has the property of always being bounded above by one.

Since most of our notions of continuity and convergence rely only on the properties (i) and (ii) of 4.1.1, it is possible to extend virtually all of our definitions, with little change, to metric spaces. The next omnibus definition translates many familiar notions to the setting of metric spaces.

4.1.3. Definition. Let (M, d) be a metric space and let $\{x_n\}$ be a sequence in M; then $\{x_n\}$ is a *Cauchy sequence* if for every $\epsilon > 0$ there is an N such that if $m, n \geq N$, then

$$d(x_n, x_m) \leq \epsilon.$$

The sequence $\{x_n\}$ is said to *converge* to x if $\lim d(x_n, x) = 0$; in this case we write $\lim x_n = x$.

It is evident that every convergent sequence is Cauchy; if every Cauchy sequence in a metric space (M, d) is also convergent, then we say that (M, d) is *complete*.

Given $x \in M$ and $\epsilon > 0$, the *ball centered at x* with radius ϵ is the set

$$B(x; \epsilon) = \{y \in M : d(x, y) < \epsilon\}.$$

A set $U \subset M$ is *open* if corresponding to each $x \in U$ there is an $\epsilon > 0$ such that the ball $B(x; \epsilon)$ satisfies

$$B(x; \epsilon) \subset U.$$

A set $F \subset M$ is *closed* if $M \backslash F$ is open.

The following properties of metric spaces are more or less immediate from these definitions; the proofs are identical with those for the real numbers.

4.1.4. Proposition. *Let (M, d) be a metric space. Then*

(a) *if $\{U_\alpha\}$ is any family of open subsets of M, then $\cup U_\alpha$ is an open subset of M;*

(b) *if $\{U_1, \ldots, U_n\}$ is a finite family of open subsets of M, then $\cap_1^n U_j$ is an open subset of M;*

(c) *the sequence $\{x_n\}$ converges to x if and only if for each open set U containing x there is an N such that $n \geq N$ implies that $x_n \in U$;*

(d) *if (M, d) is complete and F is a closed subset of M, then (F, d) is also complete.*

Proof. Problem 4.1.

[Using the metric on $\mathcal{H}$ given in Example 4.1.2(d), $d(e_n, 0) = 2^{-n}$, and so $\lim e_n = 0$ in $(\mathcal{H}, d)$ while $\{e_n\}$ is discrete in the Hilbert space norm; thus $(\mathcal{H}, d)$ has markedly different convergence properties than does $(\mathcal{H}, \|\cdot\|)$—although convergence in the metric induced by the norm implies convergence under the metric d.]

4.1.5. Definition. Let (M, d) be a metric space and let $A \subset M$. The *closure of* A, $Cl(A)$, is the set

$$Cl(A) = \bigcap \{F \subset M : A \subset F \text{ and } F \text{ is closed}\}.$$

4.1.6. Lemma. *Let (M, d) be a metric space and let $A \subset M$. Then $x \in Cl(A)$ if and only if there is a sequence $\{x_n\} \subset A$ such that $\lim x_n = x$.*

Proof. Sufficiency is immediate from 4.1.4(c). For necessity, fix $x \in Cl(A)$, fix $\epsilon > 0$, and suppose that

$$B(x; \epsilon) \cap A = \emptyset.$$

Then if $F = Cl(A) \backslash B(x; \epsilon)$, F is a closed set and $A \subset F$, from which $F \supset Cl(A)$, contradicting $x \in Cl(A)$. Thus, for each n,

$$B(x; 1/n) \cap A \neq \emptyset$$

and thus we may select $x_n \in B(x; 1/n) \cap A$. Clearly, $\{x_n\}$ is the desired sequence.

Our conventional notions of continuity also carry over for mappings between metric spaces.

4.1.7. Definition. Let (M_1, d_1) and (M_2, d_2) be metric spaces and let f be a mapping from M_1 to M_2; f is continuous if $f^{-1}(U)$ is an open subset of M_1 whenever U is an open subset of M_2.

As usual, f is continuous if and only if f and "lim" commute.

4.1.8. Proposition. *Let (M_1, d_1) and (M_2, d_2) be metric spaces and let f be a mapping from M_1 to M_2. The function f is continuous if and only if whenever $\{x_n\} \subset M_1$ converges to $x \in M_1$, then $\lim f(x_n) = f(x)$.*

Proof. We first show necessity. Thus suppose that f is continuous and let $\{x_n\}$ be a convergent sequence in M_1 with limit x. Let U_1 be any open set in M_2 containing $f(x)$ and set $U_2 = f^{-1}(U_1)$. Then U_2 is an open subset of M_1 and $x \in U_2$. This implies that there is an N such that if $n \geq N$, then $x_n \in U_2$, from which $\lim f(x_n) = f(x)$.

For the converse, we suppose that f fails to be continuous; in this case there is an open set $U \subset M_2$ for which $f^{-1}(U)$ fails to be open. This that means there is an $x_0 \in f^{-1}(U)$ for which

$$B(x_0; \epsilon) \not\subset f^{-1}(U)$$

for all $\epsilon > 0$. In particular, for each n there is an $x_n \in M_1$ such that

$$x_n \in B(x_0; 1/n) \text{ but } f(x_n) \notin U.$$

Clearly, $\lim x_n = x_0$ but since $\{f(x_n)\} \cap U = \emptyset$, $\lim f(x_n) \neq f(x_0)$, contradicting our hypothesis.

While most of the definitions carry over to metric spaces with little change, some fundamental properties—especially compactness—require some extra attention. In particular, the equivalence "compact $\leftrightarrow$ closed and bounded" does *not* hold in general metric spaces. Characterizing the compact metric spaces is the topic of the next section.

PROBLEMS

4.1. Prove Proposition 4.1.4.

4.2. Definition. Two metric spaces (M, d) and (M_2, d_2) are *homeomorphic* if there is a continuous one-to-one mapping f of M_1 onto M_2 such that f^{-1} is also continuous.

If (M, d) is a metric space, show that there is a metric ϱ defined on M satisfying
(a) $\varrho(x, y) \leq 1$ for all $x, y \in M$; and
(b) (M, d) and (M, ϱ) are homeomorphic.

4.3. Let $\mathcal{H}$ be a Hilbert space and let d be a metric given by 4.1.2(d). Show that $(\mathcal{H}, \| \cdot \|)$ and $(\mathcal{H}; d)$ are not homeomorphic.

4.4. Let (M, d) be a complete metric space and let $f: M \to M$ satisfy

$$d(f(x), f(y)) \leq k d(x, y).$$

If $0 < k < 1$, show that there is a unique $x_0 \in M$ for which $f(x_0) = x_0$. (This is the *Banach Contraction Mapping Theorem*.)

4.5. Let (M, d) be a complete metric space, let $f : M \to M$ be an arbitrary function, and let $\varphi : M \to [0, \infty]$ be a continuous function. Suppose for each $x \in M$ that

$$d\big(x, f(x)\big) \leq \varphi(x) - \varphi\big(f(x)\big).$$

Show there is an $x_0 \in M$ for which $f(x_0) = x_0$. [Hint: Define a partial order "$\succeq$" on M by $x \succeq y \leftrightarrow d(x, y) \leq \varphi(x) - \varphi(y)$. Apply Zorn's Lemma to find a maximal element. This is a variation of Ekeland's Theorem.[*]]

4.6. Let (M, d) be a complete metric space and denote by $\mathcal{F}(M)$ the collection of all nonempty closed subsets of M. For $E, F \in \mathcal{F}(M)$ and $x \in M$ defined

$$d(x, A) = \inf\{d(x, y) : y \in A\}$$
$$d_A(B) = \sup\{d(x, A) : x \in B\}$$
$$h(A, B) = \max\{d_A(B), d_B(A)\}.$$

Show that $\big(\mathcal{F}(M), h\big)$ is a metric space (h is called the *Hausdorff metric*).

4.7. Let A and B be disjoint closed subsets of the metric space (M, d). Show there is a continuous function $f : M \to [0, 1]$ such that $f(A) = 0$ and $f(B) = 1$.

4.8. Let (M_1, d_1) and (M_2, d_2) be metric spaces and define the functions ϱ_1 and ϱ_2 on $(M_1 \times M_2) \times (M_1 \times M_2)$ by

$$\varrho_1\big((x_1, y_1), (x_2, y_2)\big) = d_1(x_1, x_2) + d_2(y_1, y_2)$$
$$\varrho_2\big((x_1, y_1), (x_2, y_2)\big) = \max\{d_1(x_1, x_2), d_2(y_1, y_2)\}.$$

(a) Show that $\big((M_1 \times M_2), \varrho_1\big)$ and $\big((M_1 \times M_2), \varrho_2\big)$ are metric spaces, and are complete metric spaces if (M_1, d_1) and (M_2, d_2) are complete.

(b) Show that $\big((M_1 \times M_2), \varrho_1\big)$ and $\big((M_1 \times M_2), \varrho_2\big)$ are homeomorphic.

4.9. Let (M, d) be a metric space, let A be a closed subset of M, and let f be a continuous mapping from A to the reals. Show that there is a continuous function $\tilde{f} : M \to \Re$ such that $\tilde{f}(x) = f(x)$ for all $x \in A$. (Compare with 2.5.4.).

4.10. Let (M, d) be a metric space.

(a) Show that there is a *complete* metric space (X, ϱ) such that $d = \varrho$ on $M \times M$ and M is dense in X (X is called the *completion* of M). [Hint: Take Y to be the class of all Cauchy sequences in M; call two sequences $\{x_n\}$ and $\{y_n\}$ *equivalent* if $d(x_n, y_m) \to 0$ as $n, m \to \infty$. Let X be the class of all such equivalence classes.]

(b) Show that $\mathcal{L}_1(\Re)$ is the completion of $C_G(\Re)$ under the $\mathcal{L}_1$ norm. This gives a topological construction of the Lebesgue integral.

---

[*] See, for example, *Sur les problemes variationnels*, Compte Rendus Acad. Sci. Paris, **275**, 1057-1059 (1976).

4.2. COMPACT METRIC SPACES.   To the extent possible, we will try to mimic the definitions of compactness given in Appendix A for the real numbers. However, the theory of compact metric spaces is, in general, much more subtle than the corresponding theory for compact subsets of $\Re$, and so our development will be correspondingly more involved.

4.2.1. Definition. Let (M, d) be a metric space and let $A \subset M$; a point $x_0 \in M$ is a *cluster point* of A if for each open set U containing x_0,

$$U \cap A \setminus \{x_0\} \neq \emptyset.$$

As in Section 3.1, a set $D \subset M$ is *dense* if for each $x \in M$ and each $\epsilon > 0$, $B(x; \epsilon) \cap D \neq \emptyset$.

The idea with a cluster point is to force the set A to "bunch up" at x_0; the point x_0 is omitted from the intersection to make sure that cluster points are not just isolated points in A.

If $A = \{x_n\}$ and x_0 is a cluster point of A, then there is a subsequence of $\{x_n\}$ which converges to x_0. Compactness is one of our main ways of deciding which sequences have convergent subsequences, and the notion of a cluster point is a useful extension of the notion of subsequences.

4.2.2. Lemma. *Let (M, d) be a metric space and suppose that every infinite subset of M has a cluster point. Then M has a countable dense subset.*

Proof. Let r be a fixed positive number. We begin by showing that there is a finite collection $\{x_1, \ldots, x_n\}$ such that $\cup_1^n B(x; r) = M$ (where n possibly depends on r). (This just says that any member of M can be approximated, up to order r, by one of the points $x_1, \ldots, x_n$.) If this is not true, then an easy induction gives a sequence $\{x_n\} \subset M$ for which $x_{n+1} \notin \cup_1^n B(x_k; r)$. By hypothesis, $\{x_n\}$ has a cluster point $x_\infty \in M$, and so there is a subsequence $\{x_{n_j}\}$ converging to x_∞. Now $\{x_{n_j}\}$ must be a Cauchy sequence and so there is an N such that, if $j > k \geq N$ then

$$d(x_{n_j}, x_{n_k}) \leq r/3.$$

Since this implies that $x_{n_j} \in B(x_{n_k}; r/2)$, we have the desired contradiction. Thus for each rational number $r > 0$, there is a finite set $S(r) \subset M$ such that

$$\cup_{x \in S(r)} B(x; r) = M.$$

It is now clear that $\cup_{\substack{r > 0 \\ r \in \mathcal{Q}}} S(r)$ is a countable dense subset of M.

As is the case with compact subsets of $\Re$, the notion of an open cover is especially useful is describing compact sets.

4.2.3. Definition. Let (M, d) be a metric space and let $\{U_\alpha\}$ be a family of subsets of M. If $A \subset M$, the family $\{U_\alpha\}$ is said to *cover* A if $\cup_\alpha U_\alpha \supset A$. If each of the sets U_α are in addition open, then the family $\{U_\alpha\}$ is said to be an *open cover* of A. If $\{U_{\alpha'}\}$ is a subfamily of $\{U_\alpha\}$, then $\{U_{\alpha'}\}$ is a *subcover* for A if $\{U_{\alpha'}\}$ is a cover for A.

The "covering properites" of sets turn out to be very useful in describing other properties (such as compactness). As a first approximation to admitting a finite subcover, we might ask for a countable subcover; the next lemma tells us that the "cluster point" property described in 4.2.2 guarantees at least this weaker covering property.

4.2.4. Theorem (Lindelöf). *Let (M, d) be a metric space and suppose that M has a countable dense subset D. Then every open cover of M admits a countable subcover.*

Proof. Let $\{U_\alpha\}$ be an open cover of M. Denote by $\mathcal{B}$ the collection of subsets of M given by

$$\mathcal{B} = \{B(x; r) : x \in D, r \in \mathcal{Q} \text{ and } r > 0\},$$

so that $\mathcal{B}$ is countable. Denote by $\mathcal{B}'$ the subset of $\mathcal{B}$ given by

$$\mathcal{B}' = \{B(x; r) : B(x; r) \in \mathcal{B} \text{ and } B(x; r) \subset U_\alpha \text{ for some } \alpha\},$$

so that $\mathcal{B}'$ is again a countable set.

Now for each $B \in \mathcal{B}'$, select $U_B \in \{U_\alpha\}$. We claim that the countable subfamily $\{U_B\}$ of $\{U_\alpha\}$ covers M. To see this, fix $x \in M$ and select U_α so that $x \in U_\alpha$. Since U_α is open, there is a number $\epsilon > 0$ such that $B(x; \epsilon) \subset U_\alpha$. Now select $y \in B(x; \epsilon/3) \cap D$ and choose a rational number r in the interval $(\epsilon/3, 2\epsilon/3)$, so that $x \in B(y; r)$. Moreover, if $w \in B(y; r)$, then

$$d(w, x) \le d(w, y) + d(y, x) < 2\epsilon/3 + \epsilon/3 = \epsilon$$

and thus $B(y; r) \subset B(x; \epsilon) \subset U_\alpha$. Consequently, $B(y; r) \in \mathcal{B}'$. Since $x \in B(y; r)$ this implies that $x \in \cup U_B$ and, since $x \in M$ was arbitrary, that $\{U_B\}$ is an open cover for M.

With these preliminary approximations in hand, we can now describe, in terms of open covers, precisely which metric spaces are compact.

4.2.5. Definition. Let (M, d) be a metric space; then M is *compact* if every open cover of M admits a finite subcover.

4.2.6. Theorem. *Let (M, d) be a metric space. Then (M, d) is compact if and only if every infinite subset A of M has a cluster point in M.*

Proof. For necessity, we proceed by contraction and suppose that $A \subset M$ is infinite and has no cluster points. Then for each $x \in M$ there is an open set U_x containing x such that $(U_x \cap A) \setminus \{x\} = \emptyset$. By compactness, the family $\{U_x\}$ admits a finite subcover $\{U_{x_1}, \ldots, U_{x_p}\}$. But then $A \subset \{x_1, \ldots, x_p\}$ since $(A \cap U_{x_i}) \setminus \{x_i\} = \emptyset$, and this contradicts A being infinite.

For sufficiency, let $\mathcal{U}$ be any open cover of M, and extract a countable subcover $\{U_n\}$ using 4.2.2 and 4.2.4. Set $V_n = \cup_1^n U_j$; suppose for contradiction that $V_n \neq M$ for all n. Then, by reindexing if required, we may assume that $V_{n+1} \supset_{\neq} V_n$ for each n. Now select a sequence $\{x_n\}$ so that $x_{n+1} \in V_{n+1} \setminus V_n$ for each n. The sequence $\{x_n\}$ is infinite, and thus must cluster at some $x_\infty \in M$. Now $x_\infty \in U_n$ for some n, and thus x_∞ is in the open set V_n. Thus for $j > n$ sufficiently large, $x_j \in V_n$, which contradicts our construction of $\{x_n\}$.

The property of compact sets which we most frequently wish to exploit is the "cluster point" property described in the result above. As a general rule, the equivalent covering property is difficult to verify in particular spaces and so a more tractable characterization is desirable. (In this sense, the covering properties are only "way stations" leading toward Theorem 4.2.10.) In general metric spaces, bounded sets need *not* be compact, necessitating the following definition.

4.2.7. Definition. Let (M, d) be a metric space and let $A \subset M$; then A is *totally bounded* if for each $\epsilon > 0$, there is a finite collection $\{x_1, \ldots, x_p\} \subset A$ such that $A \subset \cup_1^p B(x_j; \epsilon)$. The set A is *bounded* if there is an element $x \in M$ and an $r > 0$ such that $A \subset B(x; r)$.

Every totally bounded subset is bounded; the converse need not hold (see Problem 4.12). Note that if a set A is totally bounded, then every element $x \in A$ can be approximated, up to order r, by one of $x_1, \ldots, x_n$; this is exactly the property exploited in the proof of 4.2.2.

4.2.8. Proposition. *If A is totally bounded, then A is bounded.*

Proof. Select $\{x_1, \ldots, x_p\}$ such that $A \subset \cup_1^p B(x_i; 1)$; set

$$r = \max\{d(x_1, x_j) : j = 2, \ldots, p\}.$$

Now if $y \in A$, then there is a j such that $y \in B(x_j; 1)$. Thus

$$d(y, x_1) \leq d(y, x_j) + d(x_j, x_1) \leq 1 + r,$$

and thus $A \subset B(x_1; r + 1)$, showing that A is bounded.

Definition 4.2.7 requires that the collection $\{x_1, \ldots, x_n\}$ all be elements of the set A (rather than just elements of M). This will make the proof of 4.2.10 a little easier to write, but necessitates the following lemma.

4.2.9. Lemma. *If A is a totally bounded subset of a metric space (M, d) and if $B \subset A$, then B is totally bounded.*

Proof. Fix $\epsilon > 0$. Since A is totally bounded, there is a collection $\{x_1, \ldots x_p\} \subset A$ such that $A \subset \cup_1^p B(x_j; \epsilon/2)$. Since $B \subset A$, $B \subset \cup_1^p B(x_j; \epsilon/2)$. By deleting sets if needed, we may now assume also that $B \cap B(x_j; \epsilon/2) \neq \emptyset$ for each j. Now select $y_j \in B \cap B(x_j; \epsilon/2)$, and note that $B(x_j; \epsilon/2) \subset B(y_j; \epsilon)$ for each j. To see this, let $z \in B(x_j; \epsilon/2)$ and compute:

$$d(z, y_j) \leq d(z, x_j) + d(x_j, y_j) < \epsilon.$$

Now the collection $\{y_1, \ldots, y_p\}$ is a subset of B and $B \subset \cup_1^p B(x_j; \epsilon/2) \subset \cup_1^p B(y; \epsilon)$, showing that B is totally bounded.

4.2.10. Theorem. *Let (M, d) be a metric space; then M is compact if and only if M is complete and totally bounded.*

Proof. Let M be compact and fix $\epsilon > 0$. Then $\{B(x; \epsilon) : x \in M\}$ is an open cover for M which admits a finite subcover, and thus M is totally bounded. If $\{x_n\}$ is a Cauchy sequence in M, then $\{x_n\}$ must have a cluster point $x_\infty \in M$. Fix $\epsilon > 0$ and select N_1 so large that if $j, k \geq N_1$, then $d(x_j, x_k) \leq \epsilon/2$. Next select N_2 so large that $N_2 \geq N_1$ and $x_{N_2} \in B(x_\infty; \epsilon/2) \cap \{x_n\}$. Then for $j \geq N_2$,

$$d(x_\infty, x_j) \leq d(x_\infty, x_{N_2}) + d(x_{N_2}, x_j) \leq \epsilon,$$

showing that $\lim x_n = x_\infty$, and thus (M, d) is complete as well.

For the converse, let $A \subset M$ be an infinite set; it suffices to show that A has a cluster point in M. Since M is totally bounded, there is a finite collection $\{x_1^1, \ldots, x_{p_1}^1\}$ such that $M = \cup_1^{p_1} B(x_j^1; 1)$. Since A is infinite, it must be the case that $A \cap B(x_j^1; 1)$ is infinite for some j; set $y_1 = x_j^1$. Now $B(y_1; 1)$ is totally bounded, and thus there is a collection $\{x_1^2, \ldots, x_{p_2}^2\} \subset B(y_1; 1)$ such that $B(y_1; 1) \subset \cup_1^{p_1} B(x_j^2; 1/2)$. Thus $A \cap B(x_j^2; 1/2)$ is infinite for some j, and so we set $y_2 = x_j^2$. An easy induction thus gives the existence of a sequence $\{y_j\} \subset M$ such that

$$d(y_n, y_{n+1}) \leq 2^{-n+1} \quad \text{and}$$

$$B(y_n; 2^{-n}) \cap A \quad \text{is infinite}.$$

Since $d(y_n, y_m) \leq \sum_n^{m-1} d(y_j; y_{j+1}) \leq \sum_n^{m+1} 2^{-j+1}$, it follows that $\{y_n\}$ is a Cauchy sequence in M and hence, since M is complete, that $\{y_n\}$ converges to a limit y_∞.

Now fix $\epsilon > 0$ and choose N so large that $n \geq N$ implies that $2^{-n} \leq \epsilon/2$ and $d(y_n, y_\infty) \leq \epsilon/2$. If $z \in B(y_n; 2^{-n})$, then

$$d(z, y_\infty) \leq d(z, y_n) + d(y_n, y_\infty) \leq \epsilon$$

and thus $B(y_n; 2^{-n}) \subset B(y_\infty; \epsilon)$. From this, we may conclude that $B(y_\infty; \epsilon) \cap A$ is infinite, and hence that y_∞ is a cluster point of A.

4.2.11. Corollary. *If A is a closed subset of a compact metric space, then A is compact.*

Continuous functions defined on compact metric spaces have especially regular behavior. By way of example, we deduce here three results which will be useful in later sections. We have already seen their analogues for functions of a real variable.

4.2.12. Theorem. *Let (M, d) be a compact space and let f be a continuous function from M to $\Re$. Then there are elements x_1 and x_2 of M such that*

$$f(x_1) = \inf\{f(t) : t \in M\} \quad \text{and}$$
$$f(x_2) = \sup\{f(t) : t \in M\}.$$

Proof. We show only the existence of x_1, the proof for x_2 being similar. Set $m = \inf\{f(t) : t \in M\}$ and select $\{t_n\} \subset M$ such that $\lim f(t_n) = m$. Since M is compact, $\{t_n\}$ has a cluster point t_∞ in M, and there is a subsequence $\{t_{n_j}\}$ of $\{t_n\}$ so that $\lim t_{n_j} = t_\infty$. Applying continuity of f gives

$$f(t_\infty) = \lim f(t_{n_j}) = \lim f(t_n) = m.$$

4.2.13. Definition. Let (M_1, d_1) and (M_2, d_2) be metric spaces and let f be a mapping from M_1 to M_2; then f is *uniformly continuous* if for each $\epsilon > 0$ there is a $\delta > 0$ such that

$$d_1(x, y) < \delta \quad \text{implies that} \quad d_2\big(f(x), f(y)\big) < \epsilon.$$

4.2.14. Theorem. *Let (M_1, d_1) and (M_2, d_2) be metric spaces and let f be a continuous mapping from M_1 to M_2. If M_1 is compact, then f is uniformly continuous.*

Proof. Fix $\epsilon > 0$. For each $x \in M_1$, there is a $\delta(x) > 0$ such that if $d_1(x, y) < \delta(x)$, then $d_2\big(f(x), f(y)\big) < \epsilon/2$. The open cover $\{B(x; \frac{1}{2}\delta(x))\}$ of M_1 admits a finite cover

$$\left\{ B\left(x_1; \frac{1}{2}\delta(x_1)\right), \cdots, B\left(x_p, \frac{1}{2}\delta(x_p)\right) \right\}$$

since M_1 is compact.
    Set

$$\delta = \tfrac{1}{2} \min\{\delta(x_1), \cdots, \delta(x_p)\}.$$

Now fix x and y in M_1 so that $d_1(x,y) < \delta$. Choose k between 1 and p so that $x \in B(x_k, 1/2\delta(x_k))$; then

$$d_1(y, x_k) \leq d_1(y, x) + d(x, x_k)$$
$$< \delta(x_k)$$

and so $y \in B(x_k, \delta(x_k))$. Finally,

$$d_2(f(x), f(y)) \leq d_2(f(x), f(x_k)) + d_2(f(x_k), f(y))$$
$$< \epsilon,$$

completing the proof.

4.2.15. Definition. Let (M, d) be a metric space and let $\{f_n\}$ be a sequence of functions from M to $\Re$, and let f_∞ be a function from M to $\Re$. The sequence $\{f_n\}$ *converges to f_∞ uniformly* if for each $\epsilon > 0$ there is an N such that $m \geq N$ and $t \in M$ imply that

$$|f_n(t) - f_\infty(t)| \leq \epsilon.$$

As might be expected from Section 1.1, uniform convergence on compact metric spaces has important consequences (see, for example, 4.3.2). Of course, uniform convergence can be quite difficult to test for a particular sequence; Dini's Theorem, which we have already seen for functions of a real variable, remains true in the more general setting. This theorem will play the same crucial role in Chapter 7 that the real variable version played in Chapter 2.

4.2.16. Dini's Theorem. *Let (M, d) be a compact metric space, and let $\{f_n\}$ be a nonincreasing sequence of continuous real-valued functions defined on M. If $\lim f_n(x) = 0$ for each x in M, then $\{f_n\}$ converges to zero uniformly.*

Proof. Fix $\epsilon > 0$. For each $x \in M$ there is an index $N \equiv N(x)$ such that

$$0 \leq f_N(x) \leq \epsilon/2.$$

Since f_N is continuous at x, there is a $\delta(x) > 0$ such that if $y \in B(x; \delta(x))$, then

$$|f_N(y) - f_N(x)| < \epsilon/2,$$

i.e., $0 \leq f_N(y) \leq \epsilon$. Since $\{f_n\}$ is nonincreasing, we have thus selected for each $x \in M$ an index $N(x)$ and an open set $B(x; \delta(x))$ such that if $y \in B(x; \delta(x))$ and $j \geq N(x)$, then

$$0 \leq f_j(y) \leq \epsilon.$$

Now $\{B(x; \delta(x))\}$ is an open cover of M, and so admits a finite subcover

$$\{B(x_1, \delta(x_1)), \ldots, B(x_p, \delta(x_p))\}.$$

Take $m = \max\{N(x_1), \ldots, N(x_p)\}$; fix $y \in M$ and $j \geq m$. Since $y \in B\big(x_j; \delta(x_j)\big)$ for some j it follows that

$$0 \leq f_j(y) \leq f_m(y) \leq f_{N(x_j)}(y) < \epsilon$$

and hence $\{f_n\}$ converges to zero uniformly.

PROBLEMS

4.11. Let (M, d) be a metric space and suppose that $x \in M$ is a cluster point of the set $A \subset M$. Show that there is a sequence $\{x_n\} \subset A$ such that $\lim x_n = x$.

4.12. Give an example of a bounded metric space which is not totally bounded. (Consider $M = \{e_n\} \subset \ell_2$.)

4.13. Prove that every compact metric space is a continuous image of the Cantor set.[*]

4.14. Let (M, d) be a compact metric space and let $f : M \to M$. Suppose whenever $x, y \in M$ and $x \neq y$ that

$$d\big(f(x), f(y)\big) < d(x, y).$$

Show that there is a unique $x_0 \in M$ such that $f(x_0) = x_0$. Show that the conclusion fails if "compact" is weakened to "complete and bounded." (Compare with Problem 4.4.)

4.15. Let (M_1, d_1) and (M_2, d_2) be compact metric spaces. Show that $\big((M_1 \times M_2), \varrho_1\big)$ is a compact metric space with ϱ_1 as defined in Problem 4.8.

4.16. Let (M_1, d_1) and (M_2, d_2) be metric spaces and let f be a continuous map from (M_1, d_1) to (M_2, d_2). If M_1 is compact, show that $f(M_1)$ is a compact subset of M_2.

4.17. Let (M_1, d_1) and (M_2, d_2) be metric spaces and let f be a continuous, one-to-one map from (M_1, d_1) onto (M_2, d_2). Show that f^{-1} is continuous if M is compact.

4.18. A *basis* for a metric space is a family of open sets $\mathcal{G}$ with the property that a set U is open if and only if

$$U = \cup\{G \in \mathcal{G} : G \subset U\}.$$

Show that a metric space has a countable dense subset if and only if it has a countable basis.

4.19. A family of open sets S is a *subbasis* for a metric space M if whenever U is an open set and $x \in U$, there is a finite collection $S_1, \ldots, S_n$ of members of S such that $x \in \cap S_i \subset U$. (Note that if S is enlarged to include all possible finite intersections, then S is a basis.) Show that M is compact if and only if every open cover of M by members of S admits a finite subcover. (This is called the *Alexandroff Subbasis Theorem* and affords one of the more elegant proofs of the Tychonoff Theorem.)

4.20. Show that a metric space M has a countable dense subset if and only if every uncountable subset of M has a cluster point.

---

[*] This is a classical theorem of general topology due to Alexandroff and Urysohn in *Mémoire sur les éspaces topologiques compactes*.

4.21. Let (M, d) be a compact metric space and let $\mathcal{U}$ be an open cover of M. Then there is a number δ greater than zero such that if $A \subset M$ and $\sup\{d(x, y) : x, y \in A\} \leq \delta$, then $A \subset U$ for some U in $\mathcal{U}$. (The number δ is called the *Lebesgue number* for the cover $\mathcal{U}$.)

4.3. THE SPACE $\mathcal{C}(M)$. This section is devoted to the study of the space of continuous real-valued functions defined on a compact metric space (M, d). Many of the elementary properties detailed for $\mathcal{C}[a, b]$ in Section 1.1 carry over easily to this generality. We will further develop some other properties of $\mathcal{C}(M)$ at this time [such as characterizing compact sets and dense subsets of $\mathcal{C}(M)$] which will apply to $\mathcal{C}[a, b]$ as a special case. Some of these results will have further consequences for the spaces $\mathcal{L}_p(\Re)$; these consequences are partly developed in this section and also in the problems at the end of the section.

4.3.1. Definition. Let (M, d) be a compact metric space; we denote by $\mathcal{C}(M)$ the class of all continuous functions from M to $\Re$. If $f \in \mathcal{C}(M)$, we define the norm of f to be

$$\|f\| = \max\{|f(t)| : t \in M\}.$$

It is clear that $\mathcal{C}(M)$ is a vector space and not difficult to show that $\|\cdot\|$ is a norm (Problem 4.22); as in Section 3.1, we are able to infer that $\mathcal{C}(M)$ is a Banach space.

4.3.2. Theorem. *If M is a compact metric space, then $\mathcal{C}(M)$ is a Banach space.*

Proof. Let $\{f_n\}$ be a Cauchy sequence in $\mathcal{C}(M)$; if $t \in M$, then

$$|f_n(t) - f_m(t)| \leq \|f_n - f_m\|$$

and thus $\{f_n(t)\}$ is a Cauchy sequence in $\Re$; set $f(t) = \lim f_n(t)$. We must next verify that this pointwise limit is itself a continuous function.

To see that f is continuous, let $\{t_n\}$ be a sequence in M such that $\lim t_n = t$. Fix $\epsilon > 0$ and select N so large that $\|f_N - f\| \leq \epsilon/3$. Now f_N is continuous and so there is a $\delta > 0$ such that, if $d(s, t) \leq \delta$, then $|f_N(s) - f_N(t)| < \epsilon/3$. Next select k so large that $n \geq k$ implies that $d(t_n, t) \leq \delta$. Then if $n \geq k$,

$$|f(t_n) - f(t)| \leq |f(t_n) - f_N(t_n)| + |f_N(t_n) - f_N(t)| + |f_N(t) - f(t)|$$

$$\leq \|f - f_N\| + \epsilon/3 + \|f_N - f\| \leq \epsilon$$

and thus $\lim f(t_n) = f(t)$, showing that $f \in \mathcal{C}(M)$.

To see that $\lim \|f_n - f\| = 0$, fix $\epsilon > 0$ and select N so large that $n, m \geq N$ implies that $\|f_n - f_m\| \leq \epsilon$. Then for each $t \in M$

$$|f_n(t) - f_m(t)| \leq \|f_n - f_m\| \leq \epsilon.$$

Thus, letting n tend to infinity, we see that $|f(t) - f_m(t)| \leq \epsilon$ if $m \geq N$. Taking the supremum over all $t \in M$ gives, for $m \geq N$, $\|f - f_m\| \leq \epsilon$, whence $\lim \|f_m - f\| = 0$.

If M is infinite, then the unit ball $\{f : \|f\| \leq 1\}$ of $C(M)$ is not compact in the norm topology (Problem 4.23); it is possible, however, to characterize the compact subsets of $C(M)$ using the results of the preceding section.

4.3.3. Definition. Let $\mathcal{F} \subset C(M)$; the family $\mathcal{F}$ is *equicontinuous* if for each $\epsilon > 0$ there is a $\delta > 0$ such that if $t, s \in M$ and $d(t, s) \leq \delta$, then $|f(t) - f(s)| < \epsilon$ for all $f \in \mathcal{F}$.

Of course, since M is compact, it is possible to find such a δ for each $f \in M$; equicontinuity adds the requirement that the same δ work uniformly for all $f \in \mathcal{F}$. It is this uniformity that forces bounded equicontinuous families to have compact closure.

4.3.4. Arzela-Ascoli Theorem. *Let (M, d) be a compact metric space and let $\mathcal{F}$ be a closed subset of $C(M)$. Then $\mathcal{F}$ is compact if and only if $\mathcal{F}$ is equicontinuous and bounded.*

Proof. We first show sufficiency; since $\mathcal{F}$ is a closed subset of the compete metric space $C(M)$, $\mathcal{F}$ is complete, and thus it suffices to show that $\mathcal{F}$ is totally bounded.

The idea of the proof is best visualized if M is an interval. In this case the plan is to place a fine "grid" over the cross-product of M and $\mathcal{F}(M)$, the image of M under $\mathcal{F}$, and then form piecewise linear functions by connecting the interstices of this grid. Every function in $C(M)$ must be "close" to one of these piecewise linear functions if the grid is constructed properly.

Fix $\epsilon > 0$; select $\delta > 0$ so that if $d(t, s) \leq \delta$ then $|f(t) - f(s)| < \epsilon/3$ for all $f \in \mathcal{F}$. Since M is compact, there is a finite collection $\{t_1, \ldots, t_p\} \subset M$ such that $M = \cup_1^p B(t_i; \delta)$. (This sets up the "horizontal" grid points.)

Next set

$$E = \{f(t) \in \Re : t \in M \text{ and } f \in \mathcal{F}\}.$$

Since $\mathcal{F}$ is bounded there is a number M for which $\|f\| \leq M$ for all $f \in \mathcal{F}$; this in turn implies that $|x| \leq M$ for all $x \in E$, and thus E is contained in the compact interval $[-M, M]$. Consequently, there is a collection $\{x_1, \ldots, x_q\} \subset \Re$ satisfying

$$E \subset \cup_1^q (x_j - \epsilon/6, x_j + \epsilon/6);$$

set $I_j = (x_j - \epsilon/6, x_j + \epsilon/6)$. (This sets up the "vertical" grid points. Since M is not an interval, we cannot use piecewise linear functions, so we must actually extract members from $\mathcal{F}$ to serve as our "grid" functions. This is the next step.)

Now for $1 \le i \le p$ and $1 \le j \le q$, set

$$\mathcal{F}(i,j) \;=\; \{f \in \mathcal{F} \;:\; f(t_i) \in I_j\}$$

[note that some of the sets $\mathcal{F}(i,j)$ may be empty]. Set $P = \{(i,j) : 1 \le i \le p$ and $1 \le j \le q\}$, and for a subset $\sigma \subset P$, define

$$\mathcal{F}(\sigma) = \cap\{\mathcal{F}(i,j) : (i,j) \in \sigma\};$$

again, some of the sets $\mathcal{F}(\sigma)$ may be empty (we show below that some, at least, are nonempty). If $\mathcal{F}(\sigma) \neq \emptyset$, we may select $f_\sigma \in \mathcal{F}(\sigma)$. Note that there are at most pq such functions f_σ; we will show that $\mathcal{F} \subset \cup_\sigma B(f_\sigma; \epsilon)$ and thus that $\mathcal{F}$ is totally bounded.

Fix $g \in \mathcal{F}$. For each $i = 1, \ldots, p$, $g(t_i)$ must lie in one of the intervals I_j; take $j(i)$ to be an index so that $g(t_i) \in I_{j(i)}$, so $g \in \mathcal{F}(i, j(i))$. If $\sigma = \{(1, j(1)), (2, j(2)), \ldots, (p, j(p))\}$, then $g \in \mathcal{F}(\sigma)$, and so $\mathcal{F}(\sigma) \neq \emptyset$. We conclude the argument by showing that $\|g - f_\sigma\| \le \epsilon$. To see this, fix $s \in M$; select an index i for which $s \in B(t_i; \delta)$. From this, both

$$|g(t_i) - g(s)| < \epsilon/3 \quad \text{and} \quad |f_\sigma(t_i) - f_\sigma(s)| < \epsilon/3.$$

Moreover, by construction, both $g(t_i)$ and $f_\sigma(t_i)$ are in the interval $I_{j(i)}$, which has length $\epsilon/3$; thus

$$|g(t_i) - f_\sigma(t_i)| < \epsilon/3.$$

Combining these estimates gives

$$\begin{aligned}
|g(s) - f_\sigma(s)| &\le |g(s) - g(t_i)| + |g(t_i) - f_\sigma(t_i)| + |f_\sigma(t_i) - f_\sigma(s)| \\
&< \epsilon/3 + \epsilon/3 + \epsilon/3 \\
&= \epsilon
\end{aligned}$$

and thus, since $s \in M$ was arbitrary, $\|g - f_\sigma\| < \epsilon$, as desired.

We now turn to necessity, which is somewhat easier. Suppose that $\mathcal{F}$ is compact and fix $\epsilon > 0$; then $\mathcal{F}$ is totally bounded (hence bounded) and so we may select $\{f_1, \ldots, f_n\} \subset \mathcal{C}(M)$ so that

$$\mathcal{F} \subset \cup_1^n B(f_i; \epsilon/3).$$

Now each f_i is uniformly continuous on M and so, for each $i = 1, \ldots, n$ there is a $\delta_i > 0$ such that if $d(s, t) \le \delta_i$, then $|f_i(s) - f_i(t)| < \epsilon/3$. Set $\delta = \min\{\delta_i\}$ and suppose that $d(t, s) \le \delta$. If $g \in \mathcal{F}$, we may select f_i so that $\|g - f_i\| \le \epsilon/3$; but then

$$\begin{aligned}
|g(t) - g(s)| &\le |g(t) - f_i(t)| + |f_i(t) - f_i(s)| + |f_i(s) - g(s)| \\
\\
&< 2\|f_i - g\| + \epsilon/3 \\
\\
&\le \epsilon
\end{aligned}$$

and so, since g was arbitrary, $\mathcal{F}$ is equicontinuous.

As a consequence of the Theorem 4.3.4, a uniformly bounded and equicontinuous sequence $\{f_n\} \subset C(M)$ must have a uniformly convergent subsequence. It turns out that this sufficient condition is not necessary.

The remainder of this section is devoted to ascertaining when a subset of $C(M)$ is dense in $C(M)$. The main result, the Stone-Weierstrass Theorem, has many consequences; for example, we will be able to conclude from this that the sequence $\{\cos(nt) + \sin(nt)\}_{|n|=0}^{\infty}$ is a complete orthonormal set in $\mathcal{L}_2[0, 2\pi]$. Other consequences are discussed in the problems. We begin with some definitions.

4.3.5. Definition. Let (M, d) be a compact metric space and let $\mathcal{A} \subset C(M)$. Then

(i) $\mathcal{A}$ is a *linear subspace* if for each $f, g \in \mathcal{A}$ and $\alpha \in \mathfrak{R}$, the functions αf and $f + g$ are each in $\mathcal{A}$;

(ii) $\mathcal{A}$ is an *algebra* if $\mathcal{A}$ is a linear subspace and for each $f, g \in \mathcal{A}$, the product fg is in $\mathcal{A}$;

(iii) $\mathcal{A}$ is a *lattice* if for each $f, g \in \mathcal{A}$, the functions $f \bigvee g = \max(f, g)$ and $f \bigwedge g = \min(f, g)$ are in $\mathcal{A}$;

(iv) $\mathcal{A}$ *separates points* in M if for each $x, y \in M$ with $x \neq y$, there is an $f \in \mathcal{A}$ such that $f(x) \neq f(y)$.

We will prove two versions of the Stone-Weierstrass Theorem. The first version is both more general and easier to deduce than the second; however, the hypotheses of the second are frequently easier to verify in applications.

4.3.6. Stone-Weierstrass Theorem I. *Let (M, d) be a compact metric space and let $\mathcal{A}$ be a linear subspace of $C(M)$, which is also a lattice. If $\mathcal{A}$ separates points in M and if $\mathcal{A}$ contains the constant functions, then $\mathcal{A}$ is dense in $C(M)$.*

Proof. We first observe that if $x, y \in M$ and a and b are any real numbers, then there is a function $g \in \mathcal{A}$ such that $g(x) = a$ and $g(y) = b$ (provided, of course, that $x \neq y$).

To see this, select $f \in \mathcal{A}$ such that $f(x) \neq f(y)$ and define the function $g \in \mathcal{A}$ by

$$g(t) \; = \; a + (b - a)\big(f(t) - f(x)\big)\big(f(y) - f(x)\big)^{-1}.$$

Since $\mathcal{A}$ is a subspace containing the constant functions, $g \in \mathcal{A}$; an easy computation shows that g is the required function.

Next let $x \in M$, $f \in C(M)$, and $\epsilon > 0$ be arbitrary. We will show that there is a function $\varphi_x \in \mathcal{A}$ satisfying:

(1) $\varphi_x(t) \leq f(t) + \epsilon$ for all $t \in M$; and

(2) $\varphi_x(x) = f(x)$.

To see this, select for each $t \in M$ a function $g_t \in A$ so that $g_t(x) = f(x)$ and $g_t(t) = f(t)$. Since both g_t and f are continuous, there is an open set $U_t \subset M$ so that

$$t \in U_t \quad \text{and if} \quad s \in U_t, \quad \text{then}$$
$$g_t(s) \le f(s) + \epsilon.$$

Now the sets $\{U_t\}_{t \in M}$ are an open cover for M and hence admit a finite subcover, $\{U_{t_1}, \ldots, U_{t_n}\}$. Set

$$\varphi_x = \bigwedge_1^n g_{t_i}.$$

Since A is a lattice $\varphi_x \in A$. Since $g_{t_i}(x) = f(x)$ for each i, $\varphi_x(x) = f(x)$. Moreover, if $s \in M$, then $s \in U_{t_i}$ for some i, and $\varphi_x(s) \le g_{t_i}(s) \le f(s) + \epsilon$, so φ_x is the desired function.

Still keeping f and ϵ fixed, select for each $x \in M$ a function $\varphi_x \in A$ satisfying (1) and (2). Since both φ_x and f are continuous and $\varphi_x(x) = f(x)$, we may select for each x an open set $G_x \subset M$ with $x \in G_x$ and satisfying

$$\varphi_x(s) \ge f(s) - \epsilon \quad \text{if} \quad s \in G_x.$$

Once again, the collection $\{G_x\}_{x \in M}$ is an open cover of M and so admits a finite subcover $\{G_{x_1}, \ldots, G_{x_m}\}$. Set $\psi = \bigvee_1^m \varphi_{x_i}$, so $\psi \in A$. Moreover, if $s \in M$, then $s \in G_{x_i}$ for some i and so $\psi(s) \ge \varphi_{x_i}(s) \ge f(s) - \epsilon$. Since $f(s) + \epsilon \ge \bigvee_1^m \varphi_{x_i}(s) = \psi(s)$, it follows that $|f(s) - \psi(s)| \le \epsilon$; since $s \in M$ was arbitrary, this implies that $\|f - \psi\| \le \epsilon$. Since f and ϵ were arbitrary, this shows A is dense in $C(M)$.

The lattice property is often difficult to verify. Consequently, a slightly different statement of this theorem, regarding algebras, is sometimes more useful in applications. We will deduce this second version from 4.3.6; however, we will require some preliminary results and definitions. We will ultimately show that an algebra which separates points and contains the constant functions is in fact a lattice.

4.3.7. Lemma. *Let (M, d) be a compact metric space and let A be an algebra in $C(M)$; then the closure of A is also an algebra in $C(M)$.*

Proof. Fix f, g in the closure of A and select $\{f_n\}$ and $\{g_n\}$ in A so that $\lim f_n = f$ and $\lim g_n = g$, the limits being uniform on M. Then if $\alpha \in \Re$,

$$\lim \alpha f_n = \alpha f$$
$$\lim f_n + g_n = f + g$$
$$\lim f_n g_n = fg;$$

since each of the functions on the left are in A by assumption, 4.1.6 implies that the functions on the right are in the closure of A, and hence A is an algebra.

The following lemma gives an easy test to determine when a closed algebra is a lattice. The simple identities are the same as those used in the proof of 2.1.15.

4.3.8. Lemma. *Let (M, d) be a compact metric space and let A be a closed algebra in $C(M)$. Suppose that whenever $f \in A$ that $|f| \in A$; then A is a lattice.*

Proof. This is immediate from the identities

$$f \vee g = 1/2 (f + g + |f - g|) \quad \text{and} \quad f \wedge g = 1/2 (f + g - |f - g|).$$

We will require one final technical lemma prior to stating our second version of the Stone-Weierstrass Theorem; the lemma will give us a "Taylor's series" approximation for $|x| = (x^2)^{\frac{1}{2}}$ for $x \in \Re$.

4.3.9. Lemma. *Let $\alpha \in \Re$ and define, for $n \geq 0$,*

$$\binom{\alpha}{0} = 1 \quad \text{and} \quad \binom{\alpha}{n} = \frac{\alpha(\alpha - 1) \cdots (\alpha - n + 1)}{n!}.$$

If $-1 < x < 1$, then

$$f_\alpha(x) \equiv \sum_0^\infty \binom{\alpha}{n} x^n \ = \ (1 + x)^\alpha.$$

Proof. First observe that

$$\left| \frac{\binom{\alpha}{n+1} x^{n+1}}{\binom{\alpha}{n} x^n} \right| \ = \ \frac{|\alpha - n|}{n + 1} |x|$$

and so, since $|x| < 1$, the ratio test implies that the series converges on $(-1, 1)$ and converges uniformly on each interval of the form $(-a, a)$, where $0 < a < 1$.
Also observe that

$$\alpha f_{\alpha - 1}(x) \ = \ \alpha \sum_{n=0}^\infty \frac{(\alpha - 1)(\alpha - 2) \cdots (\alpha - 1 - n + 1)}{n!} x^n$$

$$= \ \sum_{n=0}^\infty (n + 1) \binom{\alpha}{n + 1} x^n$$

$$= \ \sum_{m=1}^\infty m \binom{\alpha}{m} x^{m-1}.$$

From this, the series $\sum_{m=0}^{\infty} \left(\binom{\alpha}{m} x^m \right)'$ also converges uniformly on each subinterval of the form $(-a, a)$, where $0 < a < 1$ and thus f_α is differentiable and $f_\alpha'(x) = \alpha f_{\alpha-1}(x)$.

Next observe that

$$(1+x)f_{\alpha-1}(x) = (1+x)\sum_{n=0}^{\infty} \binom{\alpha-1}{n} x^n$$

$$= 1 + \sum_{n=1}^{\infty} \binom{\alpha-1}{n} x^n + \sum_{n=0}^{\infty} \binom{\alpha-1}{n} x^{n+1}$$

$$= 1 + \sum_{n=1}^{\infty} \binom{\alpha-1}{n} x^n + \sum_{n=1}^{\infty} \binom{\alpha-1}{n-1} x^n$$

$$= 1 + \sum_{n=1}^{\infty} \left[\binom{\alpha-1}{n} + \binom{\alpha-1}{n-1} \right] x^n$$

$$= 1 + \cdots$$

$$\cdots \sum_{n=1}^{\infty} \frac{(\alpha-1)(\alpha-2)\cdots(\alpha-1-n+2)\left[\alpha-1-n+1+n\right]}{n!} x^n$$

$$= \sum_{0}^{\infty} \binom{\alpha}{n} x^n = f_\alpha(x).$$

Combining these observations gives

$$(1+x)f_\alpha'(x) = \alpha(1+x)f_{\alpha-1}(x) = \alpha f_\alpha(x).$$

Since $f_\alpha(0) = 1$, this implies that $f_\alpha(x) = (1+x)^\alpha$ upon solving the initial value problem $(1+x)y' = \alpha y$, $y(0) = 1$.

The series above actually converges on $[-1, 1]$, a somewhat more difficult fact to prove; as this is not needed for our application, we omit the proof of this fact. The proof also makes implicit use of the fact that the initial value problem has a *unique* solution—see Problem 4.29.

4.3.10. Stone-Weierstrass Theorem II. *Let (M, d) be a compact metric space and let A be an algebra in $C(M)$ which separates points and contains the constant functions. Then A is dense in $C(M)$.*

Proof. Set $\overline{A} = Cl(A)$; it suffices, in view of 4.3.6 and 4.3.8, to show that if $f \in \overline{A}$, then $|f| \in \overline{A}$.

Let $f \in \overline{A}$; if $\|f\| = 0$, the result is obvious and so we suppose that $\|f\| > 0$. Set $g = \|f\|^{-1}f$. Since $\overline{A}$ is an algebra and $|f| = \|f\|\,|g|$, it suffices to show that $|g| \in \overline{A}$. Now for $\delta \in (0, 1)$ and $t \in M$ define

$$h_\delta(t) = \delta + (1 - \delta)g^2(t).$$

Since $\overline{A}$ contains the constant functions, $h_\delta \in \overline{A}$. Next set

$$\varphi_n(t) = \sum_{k=0}^{n} \binom{\frac{1}{2}}{k}\left(h_\delta(t) - 1\right)^k;$$

note that each $\varphi_n \in \overline{A}$. Also, since

$$0 \leq 1 - h_\delta(t) = (1 - \delta)(1 - g^2(t)) \leq (1 - \delta),$$

$$\left|(h_\delta(t))^{\frac{1}{2}} - \varphi_n(t)\right| = \left|\sum_{k=n+1}^{\infty} \binom{\frac{1}{2}}{k}(1 - h_\delta(t))^k\right|$$

$$\leq \sum_{k=n+1}^{\infty} \binom{\frac{1}{2}}{k}(1 - \delta)^n$$

and so $\{\varphi_n\}$ converges to $(h_\delta)^{\frac{1}{2}}$ uniformly on M. This in turn implies that $(h_\delta)^{\frac{1}{2}} \in \overline{A}$.

Now since the square root function is uniformly continuous on $[0, 1]$, the sequence $\{h_\delta\}$ converges uniformly on M to $(g^2)^{\frac{1}{2}} = |f|$ as δ tends to zero, from which $|g| \in \overline{A}$, as desired.

The student may have been familiar with the following corollary to the Stone-Weierstrass Theorem; 4.3.9 can be thought of as a very special case of 4.3.11, that the absolute value function is approximable by polynomials. In general some weakened form of 4.3.11 appears to be required to deduce 4.3.10.

4.3.11. Corollary. *The polynomial functions are dense in $C[a, b]$.*

4.3.12. Corollary. *The polynomial functions with rational coefficients are dense in $C[a, b]$. In particular, $C[a, b]$ is separable.*

We have already argued that the continuous functions with compact support are dense in $\mathcal{L}_p(\Re)$. Since $C_C(\Re) = \cup C[-n, n]$, it follows from the above that C_C is separable in the uniform norm; it is not hard to extend this conclusion to $\mathcal{L}_p$.

4.3.13. Corollary. *If $1 \le p < \infty$, then the spaces $\mathcal{L}_p(\Re)$ are all separable.*

One of the main consequences of the Stone-Weierstrass Theorem is that the trigonometric polynomials are a dense subset of $\mathcal{L}_2$. As a preliminary result, we first argue that they are dense in the 2π-periodic functions under the uniform norm. Note that we apply the theorem to $C(M)$, where M is not an interval of real numbers.

4.3.14. Corollary. *Denote by $P[0, 2\pi]$ the collection of all functions $f \in C[0, 2\pi]$ for which $f(0) = f(2\pi)$. The span of $\{\cos(nx) + \sin(nx)\}_{|n|=0}^{\infty}$ is dense in $P[0, 2\pi]$.*

Proof. Using polar coordinates on the plane $\Re \times \Re$, we denote by $M = \{(r, \theta) : r = 1, 0 \le \theta \le 2\pi\}$. For $f \in P[0, 2\pi]$ define $\tilde{f}(r, \theta) = f(\theta)$ so $\tilde{f} \in C(M)$ and $\|f\| = \|\tilde{f}\|$. Thus it suffices to argue that the functions $\{\cos(n\theta) + \sin(n\theta)\}_{|n|=0}^{\infty}$ span a dense subset of $C(M)$. If

$$\mathcal{A} = \text{sp}\{\cos(n\theta) + \sin(n\theta)\}_{|n|=0}^{\infty},$$

it suffices to show that $\mathcal{A}$ is an algebra which separates points in M (clearly, $\mathcal{A}$ contains the constant functions).

Since $\cos(-n\theta) + \sin(-n\theta) = \cos(n\theta) - \sin(n\theta)$, both $\cos(n\theta)$ and $\sin(n\theta)$ are in $\mathcal{A}$; from this it is routine to argue that $\mathcal{A}$ separates points in M. Since $\mathcal{A}$ is, by definition, a subspace of $C(M)$, it remains only to show that the product of two elements of $\mathcal{A}$ are again in $\mathcal{A}$.

Setting $f_j = \cos(j\theta) + \sin(j\theta)$, we see that

$$\begin{aligned}
f_n(\theta) f_m(\theta) &= \big(\cos(n\theta) + \sin(n\theta)\big)\big(\cos(m\theta) + \sin(m\theta)\big) \\
&= \cos(-n\theta)\cos(m\theta) - \sin(-n\theta)\sin(m\theta) + \cdots \\
&\quad \cdots + \cos(n\theta)\sin(m\theta) + \sin(n\theta)\cos(m\theta) \\
&= \cos\big((m-n)\theta\big) + \sin\big((n+m)\theta\big) \in \mathcal{A}
\end{aligned}$$

since both $\cos\big((m-n)\theta\big)$ and $\sin\big((n+m)\theta\big)$ are in $\mathcal{A}$. Thus if $\{f_{n_j}\}_{j=1}^{N}$ and $\{f_{m_k}\}_{k=1}^{M}$ and $\{\alpha_1, \ldots \alpha_N\} \subset \Re$, $\{\beta_1, \ldots \beta_M\} \subset \Re$ are arbitrary, then

$$\left(\sum_{j=1}^{N} \alpha_j f_{n_j}\right)\left(\sum_{k=1}^{M} \beta_k f_{n_k}\right) = \sum_{j=1}^{N}\sum_{k=1}^{M} \alpha_j \beta_k f_{n_j} f_{n_k}$$

and hence the product of two elements in $\mathcal{A}$ is again in $\mathcal{A}$, showing that $\mathcal{A}$ is an algebra.

4.3.15. Theorem. *The set* $\{\cos(n\theta) + \sin(n\theta)\}_{|n|=0}^{\infty}$ *is a complete orthonormal set in* $\mathcal{L}_2[0, 2\pi]$.

Proof. In light of 3.4.10(iv) it suffices to show that $\mathcal{A} = \text{sp}\{\cos(n\theta) + \sin(n\theta)\}_{|n|=0}^{\infty}$ is dense in $\mathcal{L}_2[0, 2\pi]$.

To accomplish this, let $f \in \mathcal{L}_2[0, 2\pi]$ and fix $\epsilon > 0$. By 3.2.8 we may select $g \in C[0, 2\pi]$ for which $\|g - f\|_2 \le \epsilon/3$. Let $M = \max\{|g(t)| : 0 \le t \le 2\pi\}$. Next choose $\delta > 0$ so that

$$\delta \le \left(\frac{\epsilon}{12M}\right)^2$$

and define $h \in P[0, 2\pi]$ by

$$h(t) = \begin{cases} \frac{g(\delta) - g(2\pi)}{\delta}t + g(2\pi) & 0 \le t < \delta \\ g(t) & \delta \le t < 2\pi. \end{cases}$$

Then

$$\|h - g\|_2 = \left(\int_0^{\delta} \left(\frac{g(\delta) - g(2\pi)}{\delta}t + g(2\pi) - g(t)\right)^2 dt\right)^{\frac{1}{2}}$$

$$\le \left(\int_0^{\delta} \left(\frac{2M}{\delta}\right)^2 t^2\, dt\right)^{\frac{1}{2}} + \left(\int_0^{\delta} g^2(2\pi)\right)^{\frac{1}{2}} + \left(\int_0^{\delta} g^2(t)\, dt\right)^{\frac{1}{2}}$$

$$\le 2M\delta^{\frac{1}{2}} + M\delta^{\frac{1}{2}} + M\delta^{\frac{1}{2}}$$

$$\le \epsilon/3.$$

Finally, select $\varphi \in \mathcal{A}$ so that the uniform norm $\|\varphi - h\|$ is smaller than $\epsilon/3\sqrt{2\pi}$. Then

$$\|\varphi - f\|_2 \le \|\varphi - h\|_2 + \|h - g\|_2 + \|g - f\|_2$$

$$\le \left(\int_0^{2\pi} |\varphi(t) - h(t)|^2\, dt\right)^{\frac{1}{2}} + \epsilon/3 + \epsilon/3$$

$$\le \epsilon.$$

showing that $\mathcal{A}$ is dense in $\mathcal{L}_2[0, 2\pi]$.

PROBLEMS

4.22. Let (M, d) be a compact metric space. Show that 4.3.1 defines a norm on $C(M)$.

4.23.

(a) Let (M, d) be a metric space and suppose that s, t are in M with $s \neq t$. Show that there is an $f \in C(M)$ such that $f(s) = 0$ and $f(t) = 1$. [Consider $f(u) = \frac{d(u,s)}{d(s,t)}$.]

(b) Let (M, d) be a compact metric space and suppose that M is infinite. Show that the set $\{f \in C(M) : \|f\| \leq 1\}$ is not compact.

4.24. Let g be a continuous function defined on $\Re$ and define a map from $C[0, 1]$ to $C[0, 1]$ by

$$T(f)(x) = \int_0^x g\big(f(t)\big)\, dt.$$

If $B = \{f \in C[0, 1] : \|f\| \leq 1\}$, show that $T(B)$ has compact closure in $C[0, 1]$. (This is a preliminary step in the Peano Existence Theorem for Differential Equations.)

4.25.

(a) Show that the polynomials with rational coefficients are dense in $C[a, b]$.

(b) Prove Corollary 4.3.12.

4.26 Let $f \in C[0, 1]$ and suppose that $\int_0^1 t^n f(t)\, dt = 0$ for $n = 0, 1, \ldots$. Show that $f \equiv 0$.

4.27. Show that if $\lambda(E) > 0$, then $\mathcal{L}_\infty(E)$ is not separable.

4.28. Find an equicontinuous family $\mathcal{F} \subset C[0, 1]$ which does not have compact closure.

4.29. Let g be a differentiable function defined on $\Re$ and suppose, in addition, that g' is continuous. Define a map from $C[0, A]$ to $C[0, A]$ by

$$T(f)(x) = y_0 + \int_0^x g\big(f(t)\big)\, dt,$$

where y_0 is a fixed number. Show that if A is sufficiently small, there is a unique $f_0 \in C[0, A]$ such that $T f_0 = f_0$. [Note that f_0 is differentiable and solves $f' = g(f)$, $f(0) = y_0$. This is the Picard Existence Theorem for Differential Equations.] (*Hint:* apply Problem 4.4.)

4.4. The Baire Category Theorem. This seemingly innocuous theorem has proven to be a very powerful tool in a wide variety of settings in analysis. The theorem asserts that certain kinds of sets are "sparse" in metric spaces, and frequently plays an important role in existence and continuity arguments. We will see an application of the latter in this section, and of the former in Section 4.6.

4.4.1. Definition. Let (M, d) be a metric space; a subset A of M is said to be *nowhere dense* if the closure of A, $Cl(A)$, contains no open sets.

4.4.2. Proposition. *Let (M, d) be a metric space and let A be a subset of M. Then A is nowhere dense if and only if $M \backslash Cl(A)$ is dense (and open).*

Proof. First suppose that E is nowhere dense; fix $x \in M$ and $\epsilon > 0$. Since $B(x; \epsilon) \not\subset Cl(E)$, $B(x; \epsilon) \cap (M \backslash Cl(E))$ is nonempty, showing $M \backslash Cl(E)$ is dense; as $M \backslash Cl(E)$ is clearly open, this completes the proof of necessity.

For sufficiency, suppose that $(M \backslash Cl(E))$ is dense. Then given any $x \in Cl(E)$ and any $\epsilon > 0$, $B(x; \epsilon) \cap ((M \backslash Cl(E)) \neq \emptyset$ and so $B(x; \epsilon) \not\subset Cl(E)$, showing that E is nowhere dense.

4.4.3. Definition. Let (M, d) be a metric space and let A be a subset of M; A is said to be *first category* if there is a collection $\{E_n\}$ of nowhere dense sets such that $A \subset \cup_1^\infty E_n$. If A is not first category, then A is *second category*.

Thus, for example, the rationals are a set of first category in the reals and any line is first category in the plane. The first category sets should be not too large in some sense, and this is the content of the Baire Category Theorem. Before stating the theorem we prove some preliminary lemmas. These lemmas give some intuitive conclusions about intersections of families of dense, open subsets of complete metric spaces; in light of 4.4.2, they also give information about first category sets.

4.4.4. Lemma. *Let (M, d) be a metric space and let $\{U_n\}$ be a family of dense, open subsets of M, and define, for each n, $V_n = \cap_1^n U_j$. Then $\{V_n\}$ is a family of dense, open subsets of M.*

Proof. We proceed by induction, the case $n = 1$ being obvious. Suppose now that $V_n = \cap_1^n U_j$ is dense and open and consider $V_{n+1} = V_n \cap U_{n+1}$; clearly, V_{n+1} is open. To see that V_{n+1} is dense as well, fix $x \in M$ and $\epsilon > 0$. Since V_n is dense, we may select $y \in B(x; \epsilon/2) \cap V_n$; since V_n is open, we may choose $\delta > 0$ so small that $B(y; \delta) \subset V_n$ and $\delta < \epsilon - d(x, y)$. Since U_{n+1} is dense, we may select $w \in B(y; \delta) \cap U_{n+1} \subset V_n \cap U_{n+1}$, so $w \in V_{n+1}$. But $w \in B(x; \epsilon)$, for

$$d(x, w) \leq d(x, y) + d(y, w) \leq d(x, y) + \delta < \epsilon$$

and hence $w \in B(x; \epsilon) \cap V_{n+1}$. Since both x and ϵ were arbitrary, this shows that V_{n+1} is dense.

4.4.5. Lemma. *Let (M, d) be a complete metric space and let $\{U_n\}$ be a family of dense, open subsets of M. Then $\cap_n U_n \neq \emptyset$.*

Proof. For each n, set $V_n = \cap_1^n U_j$, so each V_n is dense and open by the Lemma 4.4.4; it suffices to show that $\cap V_n \neq \emptyset$. To do this we will construct a Cauchy sequence $\{x_n\}$ whose limit x_∞ must be in the intersection.

Select $x_1 \in V_1$ and $r_1 > 0$ so that $B(x_1; 3r_1) \subset V_1$. Now select $x_2 \in B(x_1; r_1) \cap V_2$ and choose $r_2 > 0$ so that $r_2 \leq \frac{1}{2} r_1$ and $B(x_2, 3r_2) \subset V_2$. Continuing in this fashion, an easy induction gives the existence of a sequence $\{x_n\} \subset M$

and a sequence $\{r_n\} \subset (0_1, r_1)$ satisfying, for $n \geq 1$,

(1) $x_{n+1} \in B(x_n, r_n) \cap V_{n+1}$
(2) $B(x_n, 3r_n) \subset V_n$
(3) $r_{n+1} \leq \frac{1}{2} r_n$.

Note that (3) implies, if $j \leq k$, that $r_k \leq 2^{j-k} r_j$. Thus if $n < m$, applying (1) gives

$$d(x_n, x_m) \leq \sum_{j=n}^{m-1} d(x_j, x_{j+1})$$

$$\leq \sum_{j=n}^{m-1} r_j$$

$$\leq \sum_{j=n}^{m-1} 2^{n-j} r_n$$

$$\leq 2r_n$$

$$\leq 2(2^{1-n}) r_1.$$

Thus $\{x_n\}$ is a Cauchy sequence in M and since M is complete, $\{x_n\}$ must converge to a limit $x_\infty \in M$. From the inequalities above we see that if $n < m$, then $d(x_n, x_m) \leq 2r_n$; letting m tend to infinity then yields $d(x_n, x_\infty) \leq 2r_n$, and so, by (2), $x_\infty \in B(x_n; 3r_n) \subset V_n$ for each n, whence $x_\infty \in \cap_n V_n$.

The Baire Category Theorem is now easy to state and prove.

4.4.6. Baire Category Theorem. *Let* (M, d) *be a complete metric space; then* M *is second category.*

Proof. Suppose for contradiction that $M = \cup_1^\infty E_n$ where each E_n is nowhere dense. If $U_n = M \backslash Cl(E_n)$, then U_n is dense and open and moreover,

$$\cap_n U_n = \cap_n (M \backslash Cl(E_n)) = M \backslash \cup_n Cl(E_n) = \emptyset,$$

in contradiction to 4.4.5.

We next give a very important application to Banach spaces.

4.4.7. Principle of Uniform Boundedness. *Let* X *be a Banach space and* let Y be a normed linear space. Suppose that $\mathcal{F}$ is a family of continuous linear mappings from X to Y and that for each $x \in X$,

$$\sup\{\|F(x)\| : F \in \mathcal{F}\} < \infty. \qquad (*)$$

Then there is a constant M such that for all $x \in X$ with $\|x\| \leq 1$

$$\sup\{\|F(x)\| : F \in \mathcal{F}\} \leq M.$$

Proof. For each $F \in \mathcal{F}$ and each natural number m define

$$E(m, F) = \{x \in X : \|F(x)\| \leq m\}$$

and set $E_m = \cap_{F \in \mathcal{F}} E(m, F)$. Since each F is continuous, $E(m, F)$ is closed for each (m, F) and hence E_m is closed. We next claim that $X = \cup E_m$; to see this, let $x \in X$. In view of $(*)$, there is a natural number M_x such that

$$\sup\{\|F(x)\| : F \in \mathcal{F}\} \leq M_x,$$

i.e., $x \in E(M_x, F)$ for all $F \in \mathcal{F}$. This implies that $x \in E_{M_x}$ and hence that $X = \cup E_m$ as desired. Since X is complete, the Baire Category Theorem implies that at least one of the sets E_m must contain an open set U.

Fix $x \in U$ and select $\epsilon > 0$ so that $B(x; 2\epsilon) \subset U \subset E_m$. Note that if $y \in B(x; 2\epsilon)$, then $\|F(y)\| \leq m$ for all F.

Now let $F \in \mathcal{F}$ and $h \in X$ be arbitrary with $\|h\| \leq 1$ [so that $x + \epsilon h \in B(x; 2\epsilon)$]. Then

$$\|F(h)\| \leq \tfrac{1}{\epsilon}\big(\|\epsilon F(h) + F(x)\| + \|F(x)\|\big)$$

$$\leq \tfrac{1}{\epsilon}\big(\|F(x + \epsilon h)\| + m\big)$$

$$\leq 2m/\epsilon.$$

Thus

$$\sup\{\|F(h)\| : \|h\| \leq 1\} \leq 2m/\epsilon,$$

Since F was arbitrary, this establishes the result.

PROBLEMS

4.30. Show that the irrationals are second category in $\Re$.

4.31. Show that $\Re^2$ cannot be written as the countable union of straight lines.

4.32. *Banach-Steinhaus Theorem.* Let E be a Banach space and let $\{f_n\} \subset E^*$. Suppose for each $x \in E$ that

$$f(x) \equiv \lim_{n \to \infty} f_n(x)$$

exists and is finite. Show that $f \in E^*$.

4.33. Let X and Y be Banach spaces and let T be a continuous linear mapping from X *onto* Y. Show that T maps open subsets of X onto open subsets of Y. (This result, known as the *Open Mapping Theorem*, is one of the three basic pillars upon which linear functional analysis rests. The Principle of Uniform Boundedness is another; the third, the Hahn-Banach Theorem, appears in Chapter 8.)

4.34. Let $\{x_n\}$ be a fixed sequence and let p be a fixed number satisfying $1 \leq p < \infty$. Suppose for all sequences $\{y_n\} \in \ell_p$ that

$$\sum_1^\infty |x_n y_n| < \infty;$$

show that $\{x_n\} \in \ell_q$ where p and q are conjugate pairs.

4.35. *Discontinuous functions on $\Re$. Prove or find a counterexample.* There is a function $f : \Re \to \Re$ which is continuous at each rational and discontinuous at each irrational.

Chapter 5

Differentiation

5.1. INTRODUCTION. In 1806, A. M. Ampère published a paper purporting to show that every continuous function is everywhere differentiable. By all accounts this paper was met with calm acceptance by the other leading mathematicians of the day, even though Ampère's claim turns out to be one of the more spectacular "errors" of nineteenth-century mathematics. The point of this anecdote is not that Ampère was stupid (on the contrary), but rather to indicate something of the state of the mathematics of the era. At this time there was not a clear understanding or even formulation of such notions as "continuity," "differentiability," and "function." Indeed, there was resistance to overly formal arguments which devoted attention to rigor at the expense of intuition. Eventually, however, errors and inconsistencies began to accumulate and the need for more structured arguments became evident. Ampère's (wrong) claim and its eventual refutation were a principal impetus for the rigorous investigations which helped to coalesce the basic notions of Newton's calculus into a unified structure and ultimately gave birth to modern analysis.

This section is devoted primarily to the construction of a continuous, nowhere differentiable function. Before turning to this example, we introduce some of the basic definitions which will be used throughout this chapter.

5.1.1. Definitions. Let I be an open interval and let f be a real-valued function defined on I. The *Dini-derivates* of f at $x \in I$ are the numbers given by:

$$D^+ f(x) = \overline{\lim}_{h \to 0+} \frac{f(x+h) - f(x)}{h}$$

$$D_+ f(x) = \underline{\lim}_{h \to 0+} \frac{f(x+h) - f(x)}{h}$$

$$D^- f(x) = \overline{\lim}_{h \to 0+} \frac{f(x-h) - f(x)}{-h}$$

$$D_- f(x) = \underline{\lim}_{h \to 0+} \frac{f(x-h) - f(x)}{-h}.$$

If $D^+ f(x) = D_+ f(x)$, then we say that f has a *right derivative* at x and write

$$f'_+(x) \equiv D^+ f(x) = D_+ f(x).$$

Similarly, if $D^- f(x) = D_- f(x)$, then we say that f has a *left derivative* at x and write

$$f'_-(x) \equiv D^- f(x) = D_- f(x).$$

If $f'_+(x)$ and $f'_-(x)$ both exist and are equal, then we say that f has a *derivative* $f'(x)$ at x. Note that if $f'(x)$ exists, then as usual

$$f'(x) = \lim_{h \to 0} \frac{f(x+h) - f(x)}{h},$$

but $f'(x)$ may be infinite. If $f'(x)$ exists and is finite, then we will say that f is *differentiable* at x.

We distinguish between "functions having a derivative" and "differentiable functions"; this technicality is partly a notational convenience to ease our discussion in Section 5.3. Our notion of a "differentiable function" agrees with the classical notion.

Our study of differentiability will, initially, concentrate on the nondecreasing functions. Some of the elementary notions from measure theory will play a role—we will show that such functions are differentiable a.e.—and so it will be important to know when the Dini derivates are themselves measurable functions. The following proposition provides the necessary information.

5.1.2. Proposition. *If f is a nondecreasing function, then all four of the Dini derivates are measurable functions.*

Proof. For any number r the set $\{x : f(x) \leq r\}$ is an interval since f is nondecreasing, and so f is measurable. This implies that the difference quotients involved in the Dini derivates are measurable functions. Finally, since the derivates are now seen to be the limits of measurable functions, they themselves must be measurable.

We now turn to the main topic of this section—the construction of a continuous, nowhere differentiable function. The first such example was apparently discovered by Bolzano around 1825, although he was denied the right to publish his work by the Austrian government. The first published example, due to Weierstrass, appeared in 1875. The simpler example we present below is due to van der Waerden.

5.1.3. Example. A function which is continuous everywhere and differentiable nowhere.

We denote by $g_0(x)$ the distance from the real number x to the integer nearest x. Clearly, g_0 is continuous— indeed the graph of g_0 is a "sawtooth" consisting of line segments having slope either $+1$ or -1. Next, for $k \geq 1$, set

$$g_k(x) = 2^{-k} g_0(2^k x) \text{ and}$$

$$f(x) = \sum_{k=0}^{\infty} g_k(x). \tag{1}$$

Since $|g_k(x)| \leq 2^{-k}$ for all x, the series $\sum_0^{\infty} g_k$ converges uniformly on each compact interval $[-N, N]$ and thus f is everywhere continuous.

The proof that f is nowhere differentiable rests on two elementary observations.

5.1.4. Lemma. *Let j and N be integers with $N \geq 1$ and let I denote the interval $I = \left[\frac{j}{2N}, \frac{j+1}{2N}\right]$. Then*

(a) Each of the functions $\{g_0, \ldots, g_{N-1}\}$ is linear on I; and

(b) If t is in the interior of I and $0 \leq k \leq N - 1$, then $|g_k'(t)| = +1$.

Proof. Since g_0 is linear on intervals of the form $[j/2, (j+1)/2]$, the conclusions of the lemma are certainly valid for g_0. To obtain g_k we increase the period of g_0 by 2^k and decrease the amplitude of g_0 by the same factor. The former guarantees that g_k is linear on intervals of the form $\left[\frac{j}{2^{k+1}}, \frac{j+1}{2^{k+1}}\right]$ while the latter guarantees that the slope of g_k is either $+1$ or -1. A more formal argument, using induction and distinguishing between the cases when j is even and j is odd, can be given, but a picture is probably most convincing (see Problem 5.1).

When we think of the derivative of a function at a number x, we generally think in terms of the difference quotient $h^{-1}[f(x + h) - f(x)]$; however, if f is differentiable, there is no need to fix x in the difference quotient, as the next lemma shows.

5.1.5. Lemma. *Let φ be a function which is differentiable at the real number x and let $\{u_n\}$ and $\{v_n\}$ be sequences of real numbers satisfying*

(a) $u_n \leq x < v_n$; and

(b) $\lim u_n = \lim v_n = x$.

Then

(c) $\lim \dfrac{\varphi(u_n) - \varphi(v_n)}{u_n - v_n} = \varphi'(x)$.

Proof. Set

$$t_n = \frac{u_n - x}{u_n - v_n}.$$

Then

$$0 \leq t_n \leq 1, \quad (1 - t_n) = \frac{v_n - x}{v_n - u_n}$$

and

$$\frac{\varphi(u_n) - \varphi(v_n)}{u_n - v_n} = t_n \frac{\varphi(u_n) - \varphi(x)}{u_n - x} + (1 - t_n)\frac{\varphi(v_n) - \varphi(x)}{v_n - x}. \tag{2}$$

Now if (c) fails, we may select a subsequence $\left\{\frac{\varphi(u_{n'}) - \varphi(v_{n'})}{u_{n'} - v_{n'}}\right\}$ which does converge and is bounded away from $\varphi'(x)$.

Since $0 \leq t_{n'} \leq 1$ we may assume, by passing to subsequences again if needed, that $\{t_{n'}\}$ converges to some number t_∞. But then (2) implies that

$$\lim \frac{\varphi(u_{n'}) - \varphi(v_{n'})}{u_{n'} - v_{n'}} = t_\infty \varphi'(x) + (1 - t_\infty)\varphi'(x),$$

contradicting our original selection of the subsequence.

Completion of Example 5.1.3. Fix $x \in \Re$. For each N select an integer j so that

$$\frac{j}{2^N} \leq x < \frac{j+1}{2^N}.$$

Set $u_N = j2^{-N}$ and $v_N = (j+1)2^{-N}$, so that the sequences $\{u_N\}$ and $\{v_N\}$ satisfy 5.1.5 (a) and (b). Also, if $k \geq N$, then both $2^k u_N$ and $2^k v_N$ are integers and hence $g_k(u_N) = 0 = g_k(v_N)$. Now if f were differentiable at x, then 5.1.5 (c) would imply that the sequence

$$\left\{ \frac{f(u_N) - f(v_N)}{u_N - v_N} \right\}$$

is Cauchy and, in particular, if N is sufficiently large, that

$$\left(\frac{f(u_N) - f(v_N)}{u_N - v_N} - \frac{f(u_{N+1}) - f(v_{N+1})}{u_{N+1} - v_{N+1}} \right) < 1/2.$$

We will show that the above difference must always be either $+1$ or -1, obtaining the desired contradiction. First observe by 5.1.4 that, for $k = 0, \ldots, N-1$,

$$\frac{g_k(u_N) - g_k(v_N)}{u_N - v_N} = \text{slope of } g_k \text{ on } [u_N, v_N]$$

$$= \frac{g_k(u_{N+1}) - g_k(v_{N+1})}{u_{N+1} - v_{N+1}}$$

since $[u_{N+1}, v_{N+1}] \subset [u_N, v_N]$. Consequently,

$$\frac{f(u_N) - f(v_N)}{u_N - v_N} - \frac{f(u_{N+1}) - f(v_{N+1})}{u_{N+1} - v_{N+1}}$$

$$= \sum_{k=0}^{N-1} \frac{g_k(u_N) - g_k(v_N)}{u_N - v_N} - \sum_{k=0}^{N} \frac{g_k(u_{N+1}) - g_k(v_{N+1})}{u_{N+1} - v_{N+1}}$$

$$= \frac{g_N(u_{N+1}) - g_N(v_{N+1})}{u_{N+1} - v_{N+1}}$$

$$= \text{slope of } g_N \text{ on } [u_{N+1}, v_{N+1}]$$

$$= \pm 1,$$

completing the proof.

For a while in the nineteenth century, finding continuous, nondifferentiable functions became something of a fad, causing at least one mathematician to lament at the profligacy of such examples. However, these examples continued to fascinate many mathematicians well into the current century. In particular, the examples of Weierstrass

$$w(x) \; = \; \sum_{n=0}^{\infty} \frac{\cos(3^n x)}{2^n}$$

and Riemann

$$r(x) \; = \; \sum_{n=1}^{\infty} \frac{\sin(n^2 \pi x)}{n^2}$$

have been of historical importance. The former of these has the interesting property that $w'_+(x) \; = \; -\infty$ and $w'_-(x) \; = \; +\infty$ for all x. Of course, all of these functions have the property that they "wiggle" too fast to be differentiable.

PROBLEMS

5.1.   Give a formal proof of Lemma 5.1.4.

5.2.   Let f be the function defined in 5.1.3. Show that $f'_+(1/2) = -\infty$ and $f'_-(1/2) = +\infty$.

5.3.   Show that 5.1.5 may fail if φ is assumed only to have a derivative rather than to be differentiable.

5.4.   Let

$$f(x) \; = \; \sum_{k=0}^{\infty} b^k \cos(a^k \pi x),$$

where a is an odd, positive integer, b is a real number in the interval $(0,1)$ and $ab > 1 + \frac{3\pi}{2}$. Show that $f'_+(x) = -\infty$ and $f'_-(x) = +\infty$ for all x. [*]

5.5.   Suppose that f is continuous on $[a,b]$ and $D^+ f(x) \geq 0$ for all x in $[a,b)$. Show that $f(a) \leq f(b)$.

5.2. CATEGORY AND DIFFERENTIATION.
Since continuous, nowhere differentiable functions are rather difficult to construct explicitly, many nineteenth century mathematicians suspected that they were peculiarities, and that "most" continuous functions were differentiable at "most" points. Thus the discovery that

---

[*] This is an example of a function which is continuous everywhere but has a derivative nowhere. In 1916, G. H. Hardy (*Trans. Amer. Math. Soc.* **17**, pp. 301-325) gave a detailed analysis of both this function and Riemann's example, $\sum_1^{\infty} n^{-2} \sin(n^2 \pi x)$.

the differentiable functions are first category in $C[a, b]$ came as another shock to the mathematical community.

In this section, which is something of a digression from our main development, we shall prove the following:

5.2.1. Theorem. *In the Banach space $C[a, b]$, take $\mathcal{D}$ to be the collection of all functions $f \in C[a, b]$ corresponding to which there is an $x \in [a, b)$ for which both $D^+ f(x)$ and $D_+ f(x)$ are finite (x may depend on f). Then $\mathcal{D}$ is first category in $C[a, b]$.*

5.2.2. Corollary. *In the Banach space $C[a, b]$, take*

$$\mathcal{E} = \{f \in C[a, b] : |D^+ f(x)| + |D_+ f(x)| = \infty \text{ for all } x \in [a, b)\}.$$

Then $\mathcal{E}$ is second category in $C[a, b]$.

Notice that $\mathcal{E}$ is the class of all functions with the property that at least one of the four Dini derivates is infinite at *every* point; despite the fact that such functions are difficult to construct explicitly, "most" functions fall into this class. This also gives an indirect proof that the class of nowhere differentiable continuous functions is nonempty by showing that its complement is first category. "Generic" results of this type have evolved into an important tool with wide applications in both analysis and topology.

Proof of 5.2.1. For each integer $n > 1$, we set

$$\mathcal{D}_n = \{f \in C[a, b] : \exists\, x \in [a, b - \frac{1}{n}] \text{ for which}$$

$$\sup_{0 < h \le \frac{1}{n}} \left(\frac{f(x + h) - f(x)}{h} \right) \le n\}.$$

Our proof has three major components: First, we show $\mathcal{D} = \cup \mathcal{D}_n$; second, we show that each $\mathcal{D}_n$ is a closed subset of $C[a, b]$; finally, we show that no $\mathcal{D}_n$ contains an open subset of $C[a, b]$. Thus $\mathcal{D}$, being a countable union of nowhere dense sets, must be first category.

I. $\mathcal{D} = \cup \mathcal{D}_n.$

Clearly, $\cup \mathcal{D}_n \subset \mathcal{D}$; for the reverse containment, fix $f \in \mathcal{D}$. Then there is an $x \in [a, b)$ and a constant $M > 0$ such that both

$$|D^+ f(x)| \le M \quad \text{and} \quad |D_+ f(x)| \le M.$$

Thus if $\delta > 0$ is sufficiently small,

$$\sup_{0 < h \le \delta} \left(\frac{f(x + h) - f(x)}{h} \right) \le 2M.$$

Now if n is an integer selected so that

$$n \geq \max\{2M, \; 1/\delta, \; (b-x)^{-1}\},$$

then $x \in [a, b - 1/n]$ and

$$\sup_{0 < h \leq \frac{1}{n}} \left(\frac{f(x+h) - f(x)}{h} \right) \leq n,$$

so $f \in \mathcal{D}_n$, showing that $\mathcal{D} \subset \cup \mathcal{D}_n$, as desired.

II. $\mathcal{D}_n$ is closed.

Fix n and let $\{f_k\}$ be a Cauchy sequence in $\mathcal{D}_n$ with uniform limit f in $\mathcal{C}[a, b]$. For each k we next select $x_k \in [a, b - 1/n]$ so that

$$\sup_{0 < h \leq \frac{1}{n}} \left(\frac{f_k(x_k + h) - f_k(x_k)}{h} \right) \leq n.$$

Since $[a, b - 1/n]$ is compact, by passing to subsequences if needed we may assume that $\{x_k\}$ converges to a number x in $[a, b - 1/n]$. Now fix $h \in (0, 1/n]$ and $\epsilon > 0$, and choose k so large that

$$\begin{aligned}
\|f - f_k\| &\leq h\epsilon/4, \\
|f(x_k) - f(x)| &\leq h\epsilon/4, \quad \text{and} \\
|f(x_k + h) - f(x + h)| &\leq h\epsilon/4.
\end{aligned}$$

Then

$$\left(\frac{f(x+h) - f(x)}{h} \right) \leq \frac{1}{h} \big(|f(x+h) - f(x_k + h)| + \cdots$$
$$\cdots + |f(x_k + h) - f_k(x_k + h)| + \cdots$$
$$\cdots + |f_k(x_k + h) - f_k(x_k)| + \cdots$$
$$\cdots + |f_k(x_k) - f(x_k)| + |f(x_k) - f(x)| \big)$$
$$\leq n + \epsilon.$$

Since ϵ was arbitrary, this shows that $f \in \mathcal{D}_n$ and hence that $\mathcal{D}$ is closed.

III. $\mathcal{D}_n$ contains no open sets.

Suppose for some n that $\mathcal{D}_n$ contains a nonempty open set. Then there is an $f \in \mathcal{D}_n$ and an $\epsilon > 0$ so that the ball

$$B(f; \epsilon) \equiv \{g \in \mathcal{C}[a, b] : \|f - g\| \leq \epsilon\} \subset \mathcal{D}_n.$$

Now the polynomial functions are dense in $\mathcal{C}[a, b]$ and so we may select a polynomial function p so that $\|p - f\| \leq \epsilon/3$, and hence

$$B(p; \epsilon/3) \subset B(f; \epsilon) \subset \mathcal{D}_n.$$

Being a polynomial, p is differentiable everywhere and $p' \in C[a, b]$. Next select $g \in C[a, b]$ so that

$$g'_+(x) \text{ exists for all } x \in [a, b)$$
$$|g'_+(x)| > n + \|p'\| \text{ and}$$
$$\|g\| \le \epsilon/3$$

(see Problem 5.8 for the existence of such a function g).

Then, on the one hand,

$$p + g \in B(p; \epsilon/3) \subset \mathcal{D}_n$$

while, on the other hand,

$$|(p + g)'_+(x)| = |p'(x) + g'_+(x)|$$
$$\ge |g'_+(x)| - \|p'\|$$
$$> n$$

for all $x \in [a, b)$, a contradiction. Thus $\mathcal{D}_n$ contains no nonempty open sets and the proof is complete.

PROBLEMS

5.7. Prove Corollary 5.2.2.

5.8. Let α and β be arbitrary numbers. Construct a piecewise linear function $g \in C[a, b]$ so that $g'_+(x)$ exists for all $x \in [a, b)$, $|g'_+(x)| > \beta$ for all $x \in [a, b)$ and $\|g\| \le \alpha$. (Make g a "sawtooth" with "high frequency" and 'low amplitude').

5.9. Prove that there is no real-valued function f defined on $\Re$ which is continuous at each rational but discontinuous at each irrational.

5.3. LEBESGUE'S DIFFERENTIATION THEOREM. In this section we shall prove that a nondecreasing function has a finite derivative almost everywhere. This theorem was proved by Lebesgue in 1904 as a final consequence of his new integration theory, and the proof we present is fairly close to his original. While other authors have since found proofs which are somewhat shorter, they are either far more technical or far less intuitive (or both).

We begin our discussion with a lemma which shows that if f' exists, then it must be finite almost everywhere. As the proof of the basic theorem is modeled after the proof of this lemma, the student should take special care in understanding the techniques employed.

5.3.1. Lemma. *Let f be a nondecreasing function defined on an interval $[a, b]$ and let*

$$E = \{x \in (a, b) : f'(x) \text{ exists and } f'(x) = +\infty\}.$$

Then $\lambda(E) = 0$.

Proof. For each n, set $E_n = \{x \in (a, b) : f'(x) \text{ exists and } f'(x) \geq n\}$, so $E = \cap E_n$. Since $E_{n+1} \subset E_n$ and $\lambda(E_1) < \infty$, it suffices to show that $\lim \lambda(E_n) = 0$.

Select a closed set $F_n \subset E_n$ such that $\lambda(E_n \backslash F_n) \leq 1/n$. For each $x \in F_n$, select an open interval $I_x \equiv (a_x, b_x)$ about x so that

(1) $I_x \subset (a, b)$ and

(2) $\inf \left\{ \frac{f(y) - f(x)}{y - x} : a_x \leq y \leq b_x \right\} \geq \frac{n}{2}$.

Since F_n is compact, there is a finite collection $I_{x_1}, \ldots, I_{x_p}$ which covers F_n. Reindexing, if needed, we may assume that $x_j \leq x_{j+1}$ for $j = 1, \ldots, p-1$. Further, by shrinking some of the intervals if needed, we may assume that $I_{x_j} \cap I_{x_k} = \emptyset$ if $j \neq k$.

Observe that (2) implies that

$$f(b_{x_j}) - f(a_{x_j}) = f(b_{x_j}) - f(x_j) - (f(a_{x_j}) - f(x_j))$$

$$\geq \frac{n}{2}(b_{x_j} - x_j) + \frac{n}{2}(x_j - a_{x_j})$$

$$= \frac{n}{2}\ell(I_{x_j}).$$

Consequently,

$$f(b) - f(a) \geq f(b_{x_p}) - f(a_{x_1})$$

$$\geq \sum_{j=1}^{p} f(b_{x_j}) - f(a_{x_j})$$

$$\geq \sum_{j=1}^{p} \frac{n}{2}\ell(I_{x_j}).$$

This implies that

$$\sum_{j=1}^{p} \ell(I_{x_j}) \leq \frac{2(f(b) - f(a))}{n}.$$

Thus

$$\lambda(E_n) \le \lambda(F_n) + \lambda(E_n \setminus F_n)$$

$$\le \lambda\left(\cup_1^p I_{x_j}\right) + \frac{1}{n}$$

$$\le \sum_1^p \ell(I_{x_j}) + \frac{1}{n}$$

$$\le \frac{2(f(b) - f(a))}{n} + \frac{1}{n}$$

and hence $\lim \lambda(E_n) = 0$.

The simple technique employed in the proof above, subject to some technical modifications, will also yield the differentiation theorem. We will use essentially the same arguments to show that $D^+(f)(x) = D_+(f)(x)$ a.e.; however, since these derivates involve only one-sided limits, the intervals we can construct will be of the form $(x, x + \delta(x))$, and thus will no longer be an open cover of the closed set F_n [since $x \notin (x, x + \delta(x))$]. However, they are "almost" an open cover, and "almost" admit a finite subcover, as the next lemma shows. This lemma and its corollary can be thought of as applications of Littlewood's three principles. Note that the lemma is proved by applying exactly the same technique as that used to show that closed bounded intervals are compact.

5.3.2. Lemma. *Let F be a compact set and let δ be an arbitrary mapping from F to $(0, \infty)$. Then for each $\epsilon > 0$ there is a finite collection $\{x_1, \ldots, x_p\} \subset F$ satisfying:*

(a) $\lambda\left(F \setminus \cup_1^p (x_j, x_j + \delta(x_j))\right) < \epsilon$; and

(b) $(x_j, x_j + \delta(x_j)) \cap (x_k, x_k + \delta(x_k)) = \emptyset$ if $j \ne k$.

Proof. Set $\alpha = \inf(F)$ and $\beta = \sup(F)$; if $\alpha = \beta$ the result is trivial, and so we assume that $\alpha < \beta$.

For $t \in [\alpha, \beta]$, set $F_t = F \cap [\alpha, t]$. Let S denote the class of all numbers $t \in [\alpha, \beta]$ for which the conclusion of the lemma holds with $F = F_t$, i.e., $t \in S$ if for each $\epsilon > 0$, there is a finite collection $\{x_1, \ldots, x_p\} \subset F_t$ such that (b) holds and

$$\lambda\left(F_t \setminus \cup_1^p (x_j, x_j + \delta(x_j))\right) < \epsilon.$$

Since $\alpha \in S$, $S \ne \emptyset$; also, if $t \in S$ and $\alpha \le s < t$, then $s \in S$.

Now set $\sigma = \sup(S)$; since $\min(\alpha + \delta(\alpha), \beta) \in S$, it follows that $\sigma > \alpha$. We will show that $\sigma \in S$. To see this, fix $\epsilon > 0$ and set $\tau = \max\{\alpha, \sigma - \epsilon/2\}$. Since $\tau < \sigma$ it follows that $\tau \in S$ and so there is a collection $\{x_1, \ldots, x_p\} \subset F_\tau$ such that

(b) holds and

$$\lambda\left(F_\tau\backslash\cup_1^p\big(x_j,x_j+\delta(x_j)\big)\right) < \epsilon/2.$$

Since $\sigma-\tau<\epsilon/2$, it follows that

$$\lambda\left(F_\sigma\backslash\cup_1^p\big(x_j,x_j+\delta(x_j)\big)\right)$$

$$\leq \lambda\left(F_\tau\backslash\cup_1^p\big(x_j,x_j+\delta(x_j)\big)\right) + \lambda(F\cap[\tau,\sigma])$$

$$\leq \epsilon/2+\epsilon/2 = \epsilon,$$

proving that $\sigma\in S$.

The proof will be complete if we can show that $\sigma=\beta$; thus, suppose for contradiction that $\sigma<\beta$, set $\overline{x}=\inf(F\cap[\sigma,\beta])$, and set $\gamma=\min\{\beta,\overline{x}+\delta(\overline{x})\}$, so $\gamma>\sigma$. We will show that $\gamma\in S$, obtaining the desired contradiction.

Fix $\epsilon>0$ and select $\{x_1,\dots,x_p\}\subset F_\sigma$ such that (b) holds and

$$\lambda\left(F_\sigma\backslash\cup_1^p\big(x_j,x_j+\delta(x_j)\big)\right) < \epsilon.$$

Note that if $\overline{x}\in\cup_1^p\big(x_j,x_j+\delta(x_j)\big)$, then $\overline{x}+\xi\in S$ for some $\xi>0$ sufficiently small, contradicting our choice of σ. Thus $\big(\overline{x},\overline{x}+\delta(\overline{x})\big)\cap\big(x_j,x_j+\delta(x_j)\big)=\emptyset$ if $j=1,\dots,p$. Thus if $x_{p+1}=\overline{x}$, (b) holds for the collection $\{x_1,\dots,x_{p+1}\}$ and moreover,

$$\lambda\left(F_\gamma\backslash\cup_1^{p+1}\big(x_j,x_j+\delta(x_j)\big)\right)$$

$$\leq \lambda\left(F_\sigma\backslash\cup_1^{p+1}\big(x_j,x_j+\delta(x_j)\big)\right) + \lambda\left(F\cap[\sigma,\gamma]\backslash\big(\overline{x},\overline{x}+\delta(\overline{x})\big)\right)$$

$$\leq \epsilon+\lambda\left([\overline{x},\gamma]\backslash\big(\overline{x},\overline{x}+\delta(\overline{x})\big)\right) = \epsilon,$$

showing that $\gamma\in S$, as desired.

Since bounded measurable sets are "almost" compact, we can replace "compact" with "bounded and measurable" in 5.3.2.

5.3.3.   Corollary. *Let E be a bounded, measurable set and let δ be an arbitrary function from E to $(0,\infty)$. Then for each $\epsilon>0$ there is a finite collection $\{x_1,\dots,x_p\}\subset E$ satisfying*

(a) $\lambda\left(E\backslash\cup_1^p\left(x_j, x_j + \delta(x_j)\right)\right) < \epsilon$; and

(b) $\left(x_j, x_j + \delta(x_j)\right) \cap \left(x_k, x_k + \delta(x_k)\right) = \emptyset$ if $j \neq k$.

Proof. Fix $\epsilon > 0$ and select a closed set $F \subset E$ such that $\lambda(E\backslash F) \le \epsilon/2$. Select $\{x_1, \ldots, x_p\} \subset F$ so that (b) holds and

$$\lambda\left(F\backslash\cup_1^p(x_j, x_j + \delta(x_j))\right) < \epsilon/2.$$

Then

$$\lambda\left(E\backslash\cup_1^p(x_j, x_j + \delta(x_j))\right)$$

$$\le \lambda(E\backslash F) + \lambda\left(F\backslash\cup_1^p(x_j, x_j + \delta(x_j))\right) \le \epsilon.$$

The next result is a fundamental step in the proof; Corollary 5.3.3 is employed in place of the compactness arguments of 5.3.1. The other estimates (while more complicated) are motivated by those given for $\lambda(F_n)$ in 5.3.1.

5.3.4. Theorem. *Let f be a nondecreasing function defined on an interval $[a, b]$. Then $D_+(f)(x) = D^+(f)(x)$ a.e.*

Proof. Let $E = \{x \in [a, b] : D_+(f)(x) < D^+(f)(x)\}$ and for $p, q \in \mathcal{Q}$ with $p < q$, set

$$E(p, q) = \{x \in [a, b] : D_+(f)(x) < p < q < D^+(f)(x)\}.$$

Since $E = \cup\{E(p, q) : p, q \in \mathcal{Q}, p < q\}$, and since this is a countable union, it suffices to show that $\lambda(E(p, q)) = 0$ for each p, q.

Thus we fix a pair $p < q$ and suppose that $u \equiv \lambda(E(p, q)) > 0$; select $\epsilon > 0$ so that

$$\epsilon < \frac{u(q-p)}{p + 2q}$$

and select an open set $U \supset E(p, q)$ so that $\lambda(U\backslash E(p, q)) < \epsilon$. Now for each $x \in E(p, q)$ there is a number $\delta(x) > 0$ such that $x + \delta(x) \in U$ and

$$f(x + \delta(x)) - f(x) < p\delta(x). \tag{1}$$

By Corollary 5.3.3 there is a finite, disjoint collection of intervals $\{(x_j, x_j + \delta(x_j))\}_{j=1}^n$ for which if $V = \cup_1^n(x_j, x_j + \delta(x_j))$,

$$\lambda(E(p, q)\backslash V) < \epsilon.$$

Since $V \subset U$,

$$\sum_1^n \delta(x_j) = \lambda(V) \leq \lambda(U) \leq \lambda(E(p,q)) + \epsilon = u + \epsilon,$$

and so, by (1),

$$\sum_{j=1}^n f(x_j + \delta(x_j)) - f(x_j) < p(u + \epsilon). \tag{2}$$

Now for each $y \in E(p,q) \cap V$ there is a number $\gamma(y) > 0$ such that $y + \gamma(y) \in V$ and

$$f(y + \gamma(y)) - f(y) > q\gamma(y). \tag{3}$$

Again applying Corollary 5.3.3, we obtain a disjoint family of intervals $\{(y_j, y_j + \gamma(y_j))\}_{j=1}^m$ for which

$$\lambda\left(E(p,q) \cap V \setminus \cup_1^m (y_j, y_j + \gamma(y_j))\right) < \epsilon.$$

Thus

$$u = \lambda(E(p,q)) \leq \lambda(E(p,q) \setminus V) + \lambda(E(p,q) \cap V)$$

$$\leq \epsilon + \lambda\left(E(p,q) \cap V \setminus \cup_1^m (y_j, y_j + \gamma(y_j))\right) + \sum_1^m \gamma(y_j)$$

$$\leq 2\epsilon + \sum_1^m \gamma(y_j).$$

From this and (3),

$$q(u - 2\epsilon) \leq q\sum_1^m \gamma(y_j) \leq \sum_1^m f(y_j + \gamma(y_j)) - f(y_j). \tag{4}$$

Now $\cup_1^m (y_j, y_j + \gamma(y_j)) \subset \cup_1^n (x_j, x_j + \delta(x_j))$ and so, since f is nondecreasing,

$$\sum_1^m f(y_j + \gamma(y_j)) - f(y_j) \leq \sum_1^n f(x_j + \delta(x_j)) - f(x_j).$$

This inequality, together with (2) and (4), implies that $q(u - 2\epsilon) \leq p(u + \epsilon)$, which contradicts our choice of ϵ.

The corresponding statement for $D_-(f)$ and $D^-(f)$ is also true. We need to modify 5.3.2 and 5.3.3 slightly for this case.

5.3.5.   Lemma. *Let E be a bounded, measurable set and let δ be an arbitrary function from E to $(0, \infty)$. Then for each $\epsilon > 0$ there is a finite collection $\{x_1, \ldots, x_p\} \subset E$ satisfying* *(a)* $\lambda\left(E\backslash \cup_1^p\left(x_j-\delta(x_j),x_j\right)\right) < \epsilon;$ *and*

(b) $\left(x_j-\delta(x_j),x_j\right)\cap\left(x_k-\delta(x_k),x_k\right)=\emptyset$ *if* $j\neq k$.

Proof. As with 5.3.3, it suffices to prove the result in the case E is compact. The argument is very similar to 5.3.2, and so we omit many of the details.

Set $\alpha=\inf(E)$ and $\beta=\sup(E)$, and, without loss, suppose that $\alpha<\beta$. For $t\in[\alpha,\beta]$, set $E_t=E\cap[t,\beta]$, and denote by S the collection of all $t\in[\alpha,\beta]$ for which the conclusion of the lemma holds for $E=E_t$.

If $\sigma=\inf(S)$, essentially the same arguments as in 5.3.2 show that $\sigma\in S$. Thus we need only show that $\sigma=\alpha$ to complete the proof. However, if $\sigma>\alpha$, then we may take $\bar{x}=\sup(F\cap[\alpha,\sigma])$ and show, as in 5.3.2, that there is a $\gamma>0$ such that $\bar{x}-\gamma\in S$, contradicting the definition of σ.

Note that the only difference between 5.3.5 and 5.3.2 is that 5.3.5 argues from "right to left." Making similar modifications to the proof of 5.3.4 yields the following version for the left-hand derivates.

5.3.6. Theorem. *Let f be a nondecreasing function defined on $[a,b]$. Then $D_-(f)(x)=D^-(f)(x)$ a.e.*

Proof. Problem 5.10.

The basic differentiation theorem is now an easy consequence.

5.3.7. Lebesgue's Differentiation Theorem. *Let f be a nondecreasing function defined on an interval $[a,b]$. Then f has a finite derivative almost everywhere. Moreover, $f'\in\mathcal{L}_1[a,b]$ and $\int_a^b f'(t)\,dt\leq f(b)-f(a)$.*

Proof. Combining 5.3.4 and 5.3.6, f'_- and f'_+ exist almost everywhere; let $E=\{x\in(a,b):$ both $f'_-(x)$ and $f'_+(x)$ exist$\}$. [Note that if $x\in E$, both $f'_-(x)$ and $f'_+(x)$ may be infinite]. Let $A=\{x\in E:f'_+<f'_-(x)\}$; we begin by showing that A is at most countable (and so has measure zero) by defining an injection from A to $Q\times Q\times Q$.

For each $x\in A$, select $r(x)\in Q$ such that $f'_+(x)<r(x)<f'_-(x)$. In addition we may select rational numbers $s(x)$ and $t(x)$ such that $a<s(x)<x<t(x)<b$ and

$$\frac{f(y)-f(x)}{y-x}>r(x)\ \ \text{if}\ \ s(x)<y<x;$$

$$\frac{f(y)-f(x)}{y-x}<r(x)\ \ \text{if}\ \ x<y<t(x).$$

Combining these inequalities gives, if $s(x)<y<t(x)$ and $y\neq x$, that

$$f(y)-f(x)<r(x)(y-x). \tag{1}$$

Now define $\varphi(x) = \big(r(x), s(x), t(x)\big)$, so $\varphi : A \to Q \times Q \times Q$. To see that φ is an injection, suppose that there are numbers $u, v, \in A$ for which $\varphi(u) = \varphi(v)$. Then both u and v are in the interval $\big(s(u), t(u)\big) = \big(s(v), t(v)\big)$ and so (1) implies that

$$f(u) - f(v) < r(v)(u - v) \quad \text{and}$$
$$f(v) - f(u) < r(u)(v - u).$$

Since $r(u) = r(v)$, combining these inequalities gives $0 < 0$, a contradiction. Thus we have shown that A is countable.

Essentially the same argument shows that if $B = \{x \in E : f'_+(x) > f'_-(x)\}$, then B is countable. Thus $\{x \in E : f'_+(x) \neq f'_-(x)\}$ is countable, and so the set $F = \{x \in [a, b] : f'(x) \text{ exists}\}$ satisfies $\lambda(E \backslash F) = 0$. Since $\lambda([a, b] \backslash E) = 0$, this implies that $\lambda([a, b] \backslash F) = 0$. By 5.3.1 the set

$$D = \{x \in [a, b] : f'(x) \text{ exists and is finite}\}$$

satisfies $\lambda(F \backslash D) = 0$, and so f is differentiable almost everywhere.

All that remains is to verify the estimate $\int_a^b f'(t) \, dt \leq f(b) - f(a)$. Extend f to the right of b by saying that $f(x) = f(b)$ if $x \geq b$ and set

$$g_n(t) = \frac{f(t + \frac{1}{n}) - f(t)}{(1/n)}.$$

Then, since f is nondecreasing, each $g_n \geq 0$ and $\lim g_n = f'$ a.e. Thus, g is measurable and, applying Fatou's Lemma, we obtain

$$\int_a^b f'(t) \, dt \leq \underline{\lim} \int_a^b g_n(t) \, dt$$

$$= \underline{\lim} \, n \left(\int_a^b f\left(t + \frac{1}{n}\right) \, dt - \int_a^b f(t) \, dt \right)$$

$$= \underline{\lim} \, n \left(\left(- \int_a^{a + \frac{1}{n}} f(t) \, dt + \int_b^{b + \frac{1}{n}} f(t) \, dt \right) \right)$$

$$\leq f(b) - f(a)$$

again using the fact that $f(a) \leq f(t)$ if $t \geq a$. Consequently, $\int_a^b f'(t) \, dt < \infty$, so $f' \in \mathcal{L}_1[a, b]$.

In general, it need not be the case that $\int_a^b f' \, d\lambda = f(b) - f(a)$ (see Problem 5.11; this happens for only a special class of continuous functions which we will study in Section 5.5.)

PROBLEMS

5.10. Prove Lemma 5.3.6.

5.11. Give an example of a nondecreasing function f defined on $[0, 1]$ with the property that $f' = 0$ a.e. but $f(x) \neq \int_0^x f'(t) \, dt$ for any x in $(0, 1]$.

5.12. Definition. Let $E \subset \Re$; a family $\mathcal{V}$ of closed intervals of positive length is a *Vitali Cover* of E if for each $x \in E$ and each $\epsilon > 0$ there is an interval $I \in \mathcal{V}$ such that $x \in I$ and $\lambda(I) < \epsilon$.

Prove *Vitali's Theorem*: Let E be a bounded measurable set and let $\mathcal{V}$ be a Vitali Cover of E. Then for each $\epsilon > 0$ there is a finite family $\{I_1, \ldots, I_p\} \subset \mathcal{V}$ such that

$$\lambda\big(E \backslash (\cup_1^p I_k)\big) \leq \epsilon.$$

5.13. Use Vitali's Theorem to prove 5.3.4 and 5.3.6.

5.14. Let f be a nondecreasing function defined on the interval $[a, b]$. Show that f has at most countably many discontinuities. (We will need this observation in Chapter 6.)

5.4. THE SPACE $BV[a, b]$. The results of the preceding section apply to a fairly broad class: the functions of bounded variation. This initial discussion of these functions will be greatly amplified upon in the next chapter, especially in Section 6.2.

5.4.1. Definition. Let f be a real-valued function defined on the interval $[a, b] \subset \Re$; the (*total*) *variation* of f over $[a, b]$ is the number

$$V_f(a, b) = \sup\{ \sum_{n=1}^{m} |f(x_k) - f(x_{k-1})| : m \in \mathcal{N} \text{ and}$$

$$a = x_0 \leq x_1 \leq \cdots \leq x_m = b\}.$$

If $V_f(a, b) < \infty$, then f is said to be of *bounded variation* over $[a, b]$. The collection of all functions f which are of bounded variation over $[a, b]$ is denoted by $BV[a, b]$.

Intuitively, the variation measures measures how much a function "wiggles"; if the variation is infinite, then the function "wiggles" too much to be differentiable [consider $f(x) = x \sin(\frac{1}{x})$]. Observe that if f is nondecreasing, then $V_f(a, b) = f(b) - f(a)$. We will show shortly every function of bounded variation decomposes into the difference of nondecreasing functions, and hence must be differentiable a.e.

Some more elementary properties are evident:

5.4.2. Proposition. *Let $f \in BV[a, b]$ and suppose that $a \leq c \leq b$. Then*
(a) $V_f(a, c) + V_f(c, b) = V_f(a, b)$.
(b) Each of the following functions are nonnegative and nondecreasing:

$$V_f(a, x); \quad V_f(a, x) + f(x); \quad V_f(a, x) - f(x);$$

$$V_f(a, x) - f(x) + f(a); \quad and \quad V_f(a, x) + f(x) - f(a).$$

Proof. We prove only the assertions regarding $V_f(a, x) - f(x) + f(a)$, and leave the other similar proofs to the exercises. First observe that

$$V_f(a, x) \geq |f(x) - f(a)| \geq f(x) - f(a)$$

and hence $V_f(a, x) - f(x) + f(a) \geq 0$. Also, if $a \leq x \leq y \leq b$, then, by (a),

$$\begin{aligned}
V_f(a, x) - f(x) &+ f(a) - \left(V_f(a, y) - f(y) + f(a)\right) \\
&= V_f(x, y) + \left(f(y) - f(x)\right) \\
&\geq |f(y) - f(x)| + f(y) - f(x) \\
&\geq 0
\end{aligned}$$

and thus $V_f(a, x) - f(x) + f(a)$ is nondecreasing.

The above leads at once to the following important decomposition theorem and its corollary.

5.4.3. Jordan Decomposition Theorem. *If $f \in BV[a, b]$, then there are nondecreasing functions g and h defined on $[a, b]$ so that*

$$f = g - h \quad \text{on} \quad [a, b].$$

Proof. Set

$$g(x) = \tfrac{1}{2}\left(V_f(a, x) + f(x)\right) \quad \text{and}$$
$$h(x) = \tfrac{1}{2}\left(V_f(a, x) - f(x)\right).$$

By 5.4.2 each of g and h are nondecreasing functions and thus $f = g - h$ is the desired decomposition.

5.4.4. Lebesgue Differentiation Theorem. *Let $f \in BV[a, b]$; then f is differentiable almost everywhere and $f' \in \mathcal{L}_1[a, b]$.*

Proof. This is immediate from 5.4.3 and 5.3.7.

While the functions of bounded variation are of interest for reasons independent of 5.4.4 (see, for example, Chapter 6), this result alone justifies their further study. We begin by defining a norm on $BV[a, b]$ under which the collection becomes a Banach space; a preliminary and transparent lemma is first needed:

5.4.5. Lemma. *Let $f, g \in BV[a, b]$ and let $\alpha \in \Re$. Then the functions αf and $f + g$ are of bounded variation and, moreover,*

$$V_{\alpha f}(a, b) = |\alpha| V_f(a, b)$$
$$V_{(f+g)}(a, b) \leq V_f(a, b) + V_g(a, b).$$

Proof. Problem 5.16.

5.4.6. Theorem. *The space BV[a, b] is a Banach space with norm*

$$\|f\| = V_f(a, b) + |f(a)|.$$

Proof. By 5.4.5., $BV[a, b]$ is a linear space; it is routine to verify that $\| \cdot \|$ is a norm (Problem 5.17) and so we prove only that $BV[a, b]$ is complete under the given norm.

Thus, let $\{f_n\}$ be a Cauchy sequence in $BV[a, b]$ under the norm above and let $x \in [a, b]$. Then

$$|f_n(x) - f_m(x)| \leq V_{(f_n - f_m)}(a, b) + |f_n(a) - f_m(a)| = \|f_n - f_m\|$$

and consequently $\{f_n(x)\}$ is a Cauchy sequence for each x. Let $f(x) = \lim f_n(x)$.

Now fix $\epsilon > 0$ and select N so large that $n, m \geq N$ implies that $\|f_n - f_m\| \leq \frac{\epsilon}{2}$. Let $a = x_0 \leq x_1 \leq \cdots \leq x_p = b$ be any partition of $[a, b]$ and observe that for n, $m \geq N$,

$$\sum_{j=1}^{p} |f_n(x_j) - f_m(x_j) - (f_n(x_{j-1}) - f_m(x_{j-1}))| \leq V_{f_n - f_m}(a, b) \leq \frac{\epsilon}{2}.$$

Letting m tend to infinity in the above gives

$$\sum_{j=1}^{p} |f_n(x_j) - f(x_j) - (f_n(x_{j-1}) - f(x_{j-1}))| \leq \frac{\epsilon}{2}.$$

Since the partition was arbitrary, this shows that $V_{f_n - f}(a, b) \leq \frac{\epsilon}{2}$, and thus $(f_n - f) \in BV[a, b]$. Since $f = f_n - (f_n - f)$, it follows that $f \in BV[a, b]$.

Also, since $|f_n(a) - f_m(a)| \leq \|f_n - f_m\| \leq \frac{\epsilon}{2}$, it follows that

$$|f_n(a) - f(a)| \leq \frac{\epsilon}{2}$$

if $n \geq N$, and hence

$$\|f_n - f\| \leq V_{f_n - f}(a, b) + |f_n(a) - f(a)| \leq \epsilon,$$

so $\lim \|f_n - f\| = 0$ and $BV[a, b]$ is a Banach space.

One of the most important questions addressed in Chapter 2 was the interchange of integration and pointwise limits. In general the limit of differentiable functions need not be differentiable (as the example in Section 4.5 shows); however, there are conditions when "the derivative of the limit is the limit of the derivatives." We close this section with a partial description of these conditions for nondecreasing functions; a more complete description for functions of bounded variation is given in the problems. As was the case with convergence theorems for integrals, it is more convenient to work with series than with sequences.

5.4.7. Fubini's Differentiation Theorem. *Let $\{f_n\}$ be a sequence of nondecreasing functions defined on the interval $[a,b]$ and define $f(x) \equiv \sum_1^\infty f_n(x)$. If $|f(x)| < \infty$ for $a \le x \le b$, then*

$$f'(x) \;=\; \sum_1^\infty f_n'(x) \quad \text{a.e.}$$

Proof. First observe that f is a nondecreasing function, and hence is differentiable a.e. Moreover, by considering the series $\sum_1^\infty f_n(x) - f_n(a)$, we may without loss of generality assume that each f_n (and hence f) is nonnegative (the constants will drop out when we differentiate).

Now set

$$s_n(x) \;=\; \sum_{k=1}^n f_k(x)$$

so that each s_n is nondecreasing and, by the previous assumption, nonnegative. Moreover, if $h > 0$, then

$$\frac{s_n(x+h) - s_n(x)}{h} - \frac{f(x+h) - f(x)}{h}$$

$$= \frac{\sum_{k=n+1}^\infty f_k(x) - \sum_{k=n+1}^\infty f_k(x+h)}{h} \;\le\; 0;$$

letting h tend to zero then implies that

$$s_n'(x) \;\le\; f'(x)$$

wherever both functions are defined. Clearly, $s_n'(x) \le s_{n+1}'(x)$ [since $f_{n+1}'(x) \ge 0$], and thus the nondecreasing sequence $\{s_n'(x)\}$ is almost everywhere bounded above by $f'(x)$, showing that $\sum_1^\infty f_n'(x)$ converges a.e. and

$$\sum_1^\infty f_n'(x) \;\le\; \left(\sum_1^\infty f_n(x) \right)' \quad \text{a.e.}$$

To see that we actually have equality a.e., it suffices to find a subsequence $\{s_{n_j}\}$ of $\{s_n\}$ so that

$$\lim s_{n_j}'(x) \;=\; f'(x) \quad \text{a.e.}$$

since $\{s_n'(x)\}$ converges almost everywhere.

Since $\lim s_n(b) = f(b)$, we may select n_j so that $|f(b) - s_{n_j}(b)| \le 2^{-j}$, i.e., so that

$$\sum_{j=1}^\infty f(b) - s_{n_j}(b) \;<\; \infty.$$

But now for $a \leq x \leq y \leq b$ and for each n_j

$$0 \leq f(x) - s_{n_j}(x) = \sum_{k=n_j+1}^{\infty} f_k(x)$$

$$\leq \sum_{k=n_j+1}^{\infty} f_k(y) = f(y) - s_{n_j}(y).$$

Thus (taking $y = b$), $0 \leq \sum_{j=1}^{\infty} f(x) - s_{n_j}(x) < \infty$ for all $x \in [a, b]$ and each of the functions $(f - s_{n_j})$ are nondecreasing. By the argument of the preceding paragraph,

$$\sum_{j=1}^{\infty} \left(f'(x) - s'_{n_j}(x) \right) < \infty \text{ a.e.,}$$

from which

$$\lim s'_{n_j}(x) = f'(x) \text{ a.e.}$$

PROBLEMS

5.15. Let $f \in BV[a, b]$ and suppose $a \leq c \leq b$. Show that
  (a) $V_f(a, c) + V_f(c, b) = V_f(a, b)$.
  (b) $V_f(a, x)$ is nondecreasing in x.
  (c) $V_f(a, x) + f(x) - f(a)$ is nondecreasing in x.

5.16. Prove Lemma 5.4.5.

5.17. Verify that the formula in 5.4.6 gives a norm on $BV[a, b]$.

5.18. Prove or disprove: Let $\{f_n\} \subset BV[0, 1]$ and suppose that $f_n \to f_\infty$ uniformly on $[0, 1]$. Then $f_\infty \in BV[0, 1]$. [*Hint:* consider

$$f(x) = \begin{cases} x \sin(\frac{\pi}{2x}) & x > 0 \\ 0 & x = 0. \end{cases}]$$

5.19. Let $\{f_n\} \subset BV[a, b]$ and suppose that both

$$\sum_1^{\infty} V_{f_n}(a, b) < \infty$$

and

$$\sum_1^{\infty} |f_n(a)| < \infty.$$

Then $f(x) \equiv \sum_1^{\infty} f_n(x)$ is an element of $BV[a, b]$ and

$$f'(x) = \sum_1^{\infty} f'_n(x) \text{ a.e.}$$

[*Hint:* Consider $g_n = \frac{1}{2}(V_{f_n} + f_n)$ and $h_n = \frac{1}{2}(V_{f_n} - f_n)$].

5.20. Use 5.19 to give another proof of 5.4.6, using the Riesz-Fisher Theorem as a model.

5.21. Let $f \in BV[a,b]$. Show that $V_f' = |f'|$ a.e.

5.22.

(a) If $f \in BV[a,b]$, show that
$$\lim_{x \to a^+} f(x)$$
exists.

(b) Show that
$$\|f\| \equiv \sup |f(x)| + \lim_{x \to a^+} f(x)$$
defines a norm on $BV[a,b]$.

5.23. Let E be an arbitrary set of real numbers and let λ^* denote outer measure (see Problem 2.21). A real number x is a *density point* for E if
$$\lim_{h,t \to 0^+} \frac{\lambda^* (E \cap (x - h, x + t))}{h + t} = 1.$$

Show that almost all points of E are density points of E. (*Hint:* First consider the case that E is open and bounded.)

5.5. ABSOLUTELY CONTINUOUS FUNCTIONS. In this section we classify those functions F for which

$$F(x) = F(a) + \int_a^x F' \, d\lambda$$

for all x, i.e., those functions which are indefinite integrals. We begin by showing that if $f \in \mathcal{L}_1[a,b]$, then

$$\left(\int_a^x f \, d\lambda \right)' = f(x) \quad \text{a.e. in } [a,b].$$

5.5.1. Theorem. *Let $f \in \mathcal{L}_1[a,b]$ and set*
$$F(x) = \int_a^x f \, d\lambda.$$
Then $F \in C[a,b] \cap BV[a,b]$ and
$$F' = f \quad \text{a.e.}$$

Proof. Proposition 2.6.5 shows that $F \in C[a,b]$. If $a = x_0 \le x_1 \le \cdots \le x_n = b$ is any partition of $[a,b]$, then

$$\sum_1^n |F(x_k) - F(x_{k-1})| \le \sum_1^n \int_{x_{k-1}}^{x_k} |f| \, d\lambda = \|f\|_1,$$

and thus $V_F(a,b) \leq \|f\|_1$, showing that $F \in BV[a,b]$.

For the final conclusion, select functions $g, h \in \mathcal{L}^+$ so that $f = g - h$ a.e. and both $\int_a^b g \, d\lambda < \infty$ and $\int_a^b h \, d\lambda < \infty$. Setting $G(x) = \int_a^x g \, d\lambda$ and $H(x) = \int_a^x h \, d\lambda$, it follows that $F = G - H$, and thus it suffices to show that

$$G' = g \quad \text{and} \quad H' = h \text{ a.e.}$$

We will verify the foregoing assertion for G, the proof for H being identical.

Since $g \in \mathcal{L}^+$, there is a nondecreasing sequence $\{g_n\}$ of continuous nonnegative functions defined on $[a,b]$ such that

$$\lim g_n = g \text{ a.e.}$$

Now set $\varphi_0 = g_0$, and for $n \geq 1$, set $\varphi_n = g_n - g_{n-1}$. Next define

$$\Phi_n(x) = \int_a^x \varphi_n(t) \, dt;$$

each φ_n is continuous, so the fundamental theorem of calculus as deduced in undergraduate analysis implies that Φ_n is differentiable and $\Phi_n'(x) = \varphi_n(x)$ for all x. Since $\varphi_n \geq 0$, it follows that Φ_n is nondecreasing, and so, by Fubini's Differentiation Theorem, $\sum_0^\infty \Phi_n$ is differentiable and

$$\left(\sum_0^\infty \Phi_n\right)'(x) = \sum_0^\infty \Phi_n'(x) \text{ a.e.} \tag{1}$$

But $\sum_0^N \Phi_n(x) = \int_a^x \sum_0^n \varphi_n \, d\lambda = \int_a^x g_n \, d\lambda$, and thus by definition

$$G(x) = \sum_0^\infty \Phi_n(x);$$

similarly, $\sum_0^\infty \Phi_n'(x) = \lim g_n(x) = g(x)$ a.e., and thus (1) implies that

$$G'(x) = g(x) \text{ a.e.,}$$

completing the proof.

Lebesgue's singular function (Problem 1.17) provides an example of a function which is continuous but not absolutely continuous. The functions F to which 5.5.1 applies do have one critical property which is not shared by Lebesgue's singular function, however, and we now concentrate on this property.

5.5.2. Definition. A real-valued function f defined on the interval $[a,b]$ is *absolutely continuous* if, corresponding to each $\epsilon > 0$, there is a $\delta > 0$ such that whenever $\{(x_j, y_j)\}_{j=1}^p$ is a finite family of disjoint subintervals of $[a,b]$ satisfying

$$\sum_1^p (y_j - x_j) < \delta,$$

then

$$\sum_{1}^{p} |f(y_j) - f(x_j)| \leq \epsilon.$$

The class of all absolutely continuous functions defined on $[a, b]$ is denoted by $AC[a, b]$.

5.5.3. Proposition. *If $f \in \mathcal{L}_1[a, b]$ and $F(x) = \int_a^x f \, d\lambda$, then F is absolutely continuous.*

Proof. This is, once again, a special case of 2.6.5.

Clearly, an absolutely continuous function is continuous; Lebesgue's singular function fails to be absolutely continuous. The following proposition gives a slightly more intuitive view of absolute continuity.

5.5.4. Proposition. *Let f be a continuous, real-valued function of bounded variation defined on the interval $[a, b]$. Then $f \in AC[a, b]$ if and only if*

$$\left(E \subset [a, b] \text{ and } \lambda(E) = 0\right) \implies \lambda\left(f(E)\right) = 0.$$

Proof. Problem 5.25.

5.5.5. Proposition. *If $f \in AC[a, b]$, then $f \in BV[a, b]$; in particular, if $f \in AC[a, b]$, then f is differentiable a.e., and $f' \in \mathcal{L}_1[a, b]$.*

Proof. Fix $f \in AC[a, b]$ and choose δ corresponding to $\epsilon = 1$ in Definition 5.5.2. Select the least integer N so that $N \geq (b - a)/\delta$, and let $a = x_0 \leq x_1 \leq \cdots \leq x_N = b$ be a subdivision of $[a, b]$ satisfying $x_k - x_{k-1} \leq \frac{b-a}{N}$ for all k [for example, $x_k = a + k(b-a)/\delta$]. Then since $x_k - x_{k-1} \leq \delta$, it follows from our choice of δ that

$$V_f(x_{k-1}, x_k) \leq 1$$

and hence

$$V_f(a, b) = \sum_{1}^{N} V_f(x_{k-1}, x_k) \leq N$$

showing that $f \in BV[a, b]$. The remaining conclusions follow from 5.4.4.

We next verify that for $f \in AC[a, b]$, $f(x) = f(a) + \int_a^x f' \, d\lambda$; a critical preliminary step in the proof of this fact is the following.

5.5.6. Lemma. *Let $f \in AC[a, b]$ and suppose that $f' = 0$ a.e.; then f is a constant function.*

Proof. Fix $\epsilon > 0$ and $t \in [a, b]$; we will show that $|f(t) - f(a)| \leq 2\epsilon$, from which, since ϵ is arbitrary, $f(t) = f(a)$.

Corresponding to our fixed ϵ, select δ as in Definition 5.5.2. Set

$$E = \{x \in (a, t) : f'(x) = 0\},$$

so that $\lambda(E) = t - a$. Corresponding to each $x \in E$ there is a number $\delta(x) > 0$ such that $x + \delta(x) < t$ and

$$|f(x + \delta(x)) - f(x)| \leq \epsilon\delta(x)(t - a)^{-1} \tag{2}$$

Applying 5.3.3, there is a finite collection $\{x_1, \cdots, x_p\} \subset E$ satisfying

$$\lambda\left(E \setminus \cup_1^p (x_j, x_j + \delta(x_j))\right) < \delta \quad \text{and}$$

$$(x_j, x_j + \delta(x_j)) \cap (x_k, x_k + \delta(x_k)) = \emptyset \quad \text{if } j \neq k;$$

without loss of generality we assume that

$$a < x_1 < x_2 < \cdots < x_p < t.$$

Since

$$t - a = \lambda(E) \leq \delta + \sum_{j=1}^{p} \delta(x_j)$$

it follows that the sum of the lengths of the intervals

$$(a, x_1), \ (x_1 + \delta(x_1), x_2), \ \ldots, \ (x_p + \delta(x_p), t),$$

which are complementary to $\cup_1^p (x_j, x_j + \delta x_j)$ must be less than δ. Consequently, our choice of δ implies that

$$|f(x_1) - f(a)| + \sum_{j=1}^{p-1} |f(x_j + \delta(x_j)) - f(x_{j+1})| + \cdots \tag{3}$$

$$\cdots + |f(x_p + \delta(x_p)) - f(t)| < \epsilon$$

Combining the estimates (2) and (3) now gives

$$|f(t) - f(a)| \leq |f(x_1) - f(a)| + \cdots$$

$$\cdots + \sum_{j=1}^{p-1} |f(x_j + \delta(x_j)) - f(x_{j+1})| + \cdots$$

$$\cdots + |f(x_p + \delta(x_p)) - f(t)| + \sum_{j=1}^{p} |f(x_j + \delta(x_j)) - f(x_j)|$$

$$< \epsilon + \sum_{j=1}^{p} \frac{\epsilon\delta(x_j)}{t - a}$$

$$\leq 2\epsilon$$

since $\sum_1^p \delta(x_j) \le (t - a)$. Thus $f(t) = f(a)$, and f is constant on $[a, b]$.

5.5.7. Theorem. *Let $f \in AC[a, b]$; then $f' \in \mathcal{L}_1[a, b]$ and*

$$f(x) = f(a) + \int_a^x f' \, d\lambda$$

for all x in $[a, b]$.

Proof. We showed in 5.5.5 that $f' \in \mathcal{L}_1[a, b]$. Set $g(x) = \int_a^x f' \, d\lambda$, so $g \in AC[a, b]$ and $g' = f'$ a.e. by 5.5.1. Thus if $h = f - g$, then $h \in AC[a, b]$ and $h' = 0$ a.e. It then follows from 5.5.6. that

$$h(t) \ = \ h(a) \ = \ f(a)$$

for all $t \in [a, b]$. Rearranging gives

$$f(t) \ = \ f(a) + g(t) \ = \ f(a) + \int_a^t f' \, d\lambda,$$

as desired.

The following decomposition theorem for functions of bounded variation, due to Lebesgue, is a special case of a more general and powerful result that we will deduce in Chapter 7.

5.5.8. Theorem. *Let $f \in BV[a, b]$. Then there are functions g and h in $BV[a, b]$ satisfying*

$$g \in AC[a, b]; \quad h' = 0 \ \text{a.e.}; \quad f = g - h.$$

Proof. Set $g(x) = \int_a^x f' \, d\lambda$ and $h = g - f$.

We conclude this chapter by using the foregoing to show that $(\mathcal{L}_1[0, 1])^* = \mathcal{L}_\infty[0, 1]$.

5.5.9. Riesz Representation Theorem for $\mathcal{L}_1[0, 1]$. *Let $T \in (\mathcal{L}_1[0, 1])^*$; then there is a function $g \in \mathcal{L}_\infty[0, 1]$ such that*
   (4) $T(f) = \int_0^1 fg \, d\lambda$ *for all $f \in \mathcal{L}_1[0, 1]$.*
Moreover, $\|T\| = \|g\|_\infty$.

Proof. For $0 \le s \le 1$, let e_s be the characteristic function of the interval $[0, s]$, and define the real-valued function φ on $[0, 1]$ by

$$\varphi(s) \ = \ T(e_s).$$

We begin by showing that $\varphi \in AC[0, 1]$.

Fix $\epsilon > 0$ and set $\delta = \epsilon/\|T\|$. Let $\{(x_j, y_j)\}_{j=1}^p$ be any finite family of disjoint subintervals of $[0, 1]$ satisfying

$$\sum_{j=1}^p (y_j - x_j) \leq \delta.$$

Then

$$\sum_{j=1}^p |\varphi(y_j) - \varphi(x_j)| = \sum_{j=1}^p |T(e_{y_j} - e_{x_j})|$$

$$= T\left(\sum_{j=1}^p (e_{y_j} - e_{x_j}) \mathrm{sgn}(\varphi(y_j) - \varphi(x_j)) \right).$$

Now if $f = \sum_{j=1}^p (e_{y_j} - e_{x_j}) \mathrm{sgn}(\varphi(y_j) - \varphi(x_j))$, then

$$\|f\|_1 = \int_0^1 |f| \, d\lambda = \int_0^1 \sum_{j=1}^p (e_{y_j} - e_{x_j}) \, d\lambda$$

$$= \sum_{j=1}^p (y_j - x_j) \leq \delta$$

and thus

$$\sum_{j=1}^p |\varphi(y_j) - \varphi(x_j)| = T(f) \leq \|T\| \|f\|_1 = \epsilon,$$

showing that $\varphi \in AC[0, 1]$, as desired.

Thus $\varphi(s) = \int_0^s \varphi' \, d\lambda$ for all s in $[0, 1]$. Defining $g \equiv \varphi'$, this implies that

$$T(\chi_{[0,s]}) = \int_0^1 g\chi_{[0,s]} \, d\lambda$$

for all s in $[0, 1]$. Now every step function σ can be written (except possibly for a finite number of points) as a linear combination

$$\sigma = \sum_1^n \mu_j \chi_{[0,s_j]};$$

thus (4) holds whenever f is a step function.

Next let $\psi \in C[0, 1]$ and select a sequence $\{\sigma_n\}$ of step functions so that $|\sigma_n| \leq |\psi|$ and $\lim \|\sigma_n - \psi\|_\infty = 0$. Since this implies that $\lim \|\sigma_n - \psi\|_1 = 0$, it follows that

$$T(\psi) = \lim T(\sigma_n) = \lim \int_0^1 g\sigma_n \, d\lambda = \int_0^1 g\psi \, d\lambda,$$

applying Dominated Convergence in the last step. This shows that (4) holds whenever f is continuous.

Now let f be a bounded measurable function and use Lusin's Theorem (actually, 2.5.5) to select $\{\psi_n\} \subset C[0,1]$ so that $\|\psi_n\|_\infty \leq \|f\|_\infty$ and $\lim \|\psi_n - f\|_1 = 0$. Exactly as before, this implies that (4) holds whenever f is a bounded measurable function.

Before finally verifying (4) for all $f \in \mathcal{L}_1[0,1]$, we at this point show that $g \in \mathcal{L}_\infty[0,1]$ and $\|g\|_\infty \leq \|T\|$. Fix $\epsilon > 0$ and set $E = \{t \in [0,1] : |g(t)| \geq \|T\| + \epsilon\}$. Then $f = \operatorname{sgn}(g)\chi_E$ is a bounded measurable function and $\|f\|_1 = \lambda(E)$. Thus

$$\lambda(E)\|T\| \geq T(f) = \int_0^1 fg\,d\lambda$$
$$= \int_E |g|\,d\lambda$$
$$\geq \lambda(E)(\|T\| + \epsilon),$$

which implies that $\lambda(E) = 0$ and hence that $g \in \mathcal{L}_\infty[0,1]$ (and that $\|g\|_\infty \leq \|T\|$).

Now for arbitrary $f \in \mathcal{L}_1[0,1]$, select a sequence $\{\xi_n\}$ of simple functions so that $|\xi_n| \leq |f|$ for all n and $\lim \|\xi_n - f\|_1 = 0$ (using 2.4.12 and Dominated Convergence). Then since $|\xi_n g| \leq |f|\,\|g\|_\infty$, Dominated Convergence implies that

$$T(f) = \lim T(\xi_n) = \lim \int_0^1 \xi_n g\,d\lambda = \int_0^1 fg\,d\lambda.$$

For the final conclusion, notice that

$$|T(f)| \leq \|f\|_1 \|g\|_\infty$$

and so $\|T\| \leq \|g\|_\infty$. Since we already argued that $\|g\|_\infty \leq \|T\|$, this completes the proof.

PROBLEMS

5.24.

(a) Suppose that $f \in C[0,1] \cap BV[0,1]$ and that for every $\epsilon > 0$, that $f \in AC[\epsilon, 1]$. Prove that $f \in AC[0,1]$.

(b) Use (a) to show that if $f(x) = \sqrt{x}$, then $f \in AC[0,1]$.

5.25. Prove 5.5.4. [*]

5.26. Let $f, g \in AC[a,b]$. Verify the integration-by-parts formula

$$\int_a^b gf'\,d\lambda = f(b)g(b) = f(a)g(a) - \int_a^b fg'\,d\lambda.$$

(*Hint:* First show $fg \in AC[a,b]$).

5.27. Definition. Let f be a real-valued function defined on $(-\infty, \infty)$. If

$$\lim_{a \to \infty} V_f(-a, a) < \infty,$$

[*] This relationship was discovered by Banach and published in 1925 in *Fund. Math.*, pp. 225-236.

then f is of globally bounded variation.

Let f be a real-valued function defined on $(-\infty, \infty)$. Show that

$$f(x) = \int_{-\infty}^{x} f' \, d\lambda$$

if and only if each of the following hold:

(a) $f \in AC[-t, t]$ for all $t > 0$;
(b) f is of globally bounded variation; and
(c) $\lim\limits_{x \to -\infty} f(x) = 0$.

5.28. Let p be fixed with $1 < p < \infty$. Using the technique of 5.5.9, show that $(\mathcal{L}_p[0,1])^* = \mathcal{L}_q[0,1]$. (This gives a proof of the Riesz Representation Theorem for $\mathcal{L}_p[0,1]$ which is independent of Clarkson's Inequalities.)

Chapter 6

Stieltjes Integrals

6.1. THE RIEMANN-STIELTJES INTEGRAL. This integral is of great practical importance in connection with many problems arising in mechanics, physics, and probability theory. As examples, line and surface integrals and various moments associated with nonhomogeneous mass distributions are related to Stieltjes integrals (and their generaliztions to higher-dimensional spaces). Stieltjes introduced them in 1894 in an important paper dealing with continued fractions and the problem of moments (which is discussed in the exercises); some of Stieltjes' work was anticipated by Julius König.

The Stieltjes integral shares many of the basic properties—and flaws—of the Riemann Integral. This integral can be extended to an integral of Lebesgue type, and this extension will be one of the principal topics of the next chapter. The main purpose of the current chapter is to use this integral to describe the dual space of $C[a, b]$. Along the way, we will deduce some of the convergence properties of the integral. We begin by defining the Stieltjes integral for absolutely continuous functions; eventually we will extend this definition to include all bounded Borel functions relative to an arbitrary function of bounded variation.

6.1.1. Definition. If α is an absolutely continuous function and if f is a bounded measurable function, then we define the *Riemann-Stieltjes integral of f relative to α* to be

$$\int_a^b f(t)\, d\alpha(t) = \int_a^b f\, d\alpha = \int_a^b f\alpha'\, d\lambda.$$

Notice that if $\alpha(t) = t$, then 6.1.1 just reduces to the Lebesgue integral. While this definition is underwhelming to say the least, it is strong enough to yield an important integration-by-parts formula:

6.1.2. Proposition. *If α and f are both absolutely continuous functions, then*

$$\int_a^b f\, d\alpha = f(b)\alpha(b) - f(a)\alpha(a) - \int_a^b \alpha\, df.$$

Proof. This is immediate from the identity

$$f(b)\alpha(b) - f(a)\alpha(a) = \int_a^b (f\alpha)' \, d\lambda$$

$$= \int_a^b f\alpha' + f'\alpha \, d\lambda$$

$$= \int_a^b f \, d\alpha + \int_a^b \alpha \, df.$$

Now notice that $\int_a^b \alpha \, df$ makes sense whenever f is absolutely continuous and α is a bounded measurable function; in particular, if $\alpha \in BV[a, b]$, 6.1.2 gives a way to define the Riemann-Stieltjes integral relative to all $f \in AC[a, b]$.

6.1.3. Definition Let $\alpha \in BV[a, b]$ and let $f \in AC[a, b]$; then we define the *Riemann-Stieltjes integral of f relative to α* to be

$$\int_a^b f(t) \, d\alpha(t) \equiv \int_a^b f \, d\alpha = f(b)\alpha(b) - f(a)\alpha(a) - \int_a^b \alpha \, df$$

$$\left(= f(b)\alpha(b) - f(a)\alpha(a) - \int_a^b \alpha f' \, d\lambda \right).$$

Because of 6.1.2, this is consistent with 6.1.1.

Some elementary properties of the Riemann-Stieltjes integral are now more or less evident.

6.1.4. Proposition. *Suppose that $\alpha, \beta \in BV[a, b]$ and $f, g \in AC[a, b]$. Then*

(i) $\int_a^b f + g \, d\alpha = \int_a^b f \, d\alpha + \int_a^b g \, d\alpha$;

(ii) if ξ is any real number, $\int_a^b \xi f \, d\alpha = \xi \int_a^b f \, d\alpha$;

(iii) if ξ is any real number, $\int_a^b f \, d(\xi\alpha) = \xi \int_a^b f \, d\alpha$;

(iv) $\int_a^b f \, d(\alpha + \beta) = \int_a^b f \, d\alpha + \int_a^b f \, d\beta$;

(v) $\alpha \in AC[a, b] \Rightarrow$

$$\int_a^b f \, d\alpha = f(b)\alpha(b) - f(a)\alpha(a) - \int_a^b \alpha \, df;$$

(vi) if α is nondecreasing and if $f \leq g$, then

$$\int_a^b f \, d\alpha \leq \int_a^b g \, d\alpha.$$

Proof. We will prove only (vi), the remaining conclusions being elementary exercises.

Set $h = g - f$; it suffices to show that $\int_a^b h \, d\alpha \geq 0$; i.e., that

$$\int_a^b \alpha h' \, d\lambda \leq h(b)\alpha(b) - h(a)\alpha(a).$$

If $h' \geq 0$, then

$$\int_a^b h'\alpha \, d\lambda \leq \alpha(b)\left(h(b) - h(a)\right)$$

$$\leq \alpha(b)\left(h(b) - h(a)\right) + h(a)\left(\alpha(b) - \alpha(a)\right)$$

$$= \alpha(b)h(b) - \alpha(a)h(a).$$

On the other hand, if $h' \leq 0$, then

$$\int_a^b h'\alpha \, d\lambda \leq \alpha(a)\left(h(b) - h(a)\right)$$

$$\leq \alpha(b)h(b) - \alpha(a)h(a).$$

In particular, if h is monotone on the interval, then (vi) holds. Now if h is a polynomial, then there are numbers $a = x_0 \leq x_1 \leq \cdots \leq x_n = b$ so that h is monotone on each of the intervals $[x_i, x_{i+1}]$. Thus

$$\int_a^b h'\alpha \, d\lambda = \sum_{i=0}^{n-1} \int_{x_i}^{x_{i+1}} h'\alpha \, d\lambda$$

$$\leq \sum_{i=0}^{n-1} \alpha(x_{i+1})h(x_{i+1}) - \alpha(x_i)h(x_i)$$

$$= \alpha(b)h(b) - \alpha(a)h(a),$$

establishing (vi) for polynomial functions.

To establish (vi) for arbitrary $h \in AC[a, b]$, fix $\epsilon > 0$ and suppose for the moment that $\alpha \geq 0$. Select a polynomial p' so that $\|p' - h'\|_1 \leq \epsilon$ and set

$$p(t) = \int_a^t p' \, d\lambda + h(a) + \epsilon.$$

Since

$$\left| \int_a^t p' \, d\lambda + h(a) - h(t) \right| = \left| \int_a^t p' - h' \, d\lambda \right|$$

$$\leq \|p' - h'\|_1$$

$$\leq \epsilon,$$

it follows that $\|p - h\|_\infty \leq 2\epsilon$ and $0 \leq h \leq p$. Thus

$$\int_a^b h'\alpha \, d\lambda = \int_a^b p'\alpha \, d\lambda + \int_a^b (h' - p')\alpha \, d\lambda$$
$$\leq p(b)\alpha(b) - p(a)\alpha(a) + \epsilon\alpha(b)$$
$$\leq (h(b) + 2\epsilon)\alpha(b) - (h(a) - 2\epsilon)\alpha(a) + \epsilon\alpha(b)$$
$$\leq h(b)\alpha(b) - h(a)\alpha(a) + 3\epsilon\alpha(b) + 2\epsilon\alpha(a).$$

Since $\epsilon > 0$ was arbitrary, this establishes (vi) for arbitrary $h \in AC[a, b]$ and $\alpha \geq 0$ and nondecreasing.

To complete the proof, set $\hat{\alpha} = \alpha + \alpha(a)$, so that $\hat{\alpha} \geq 0$. Moreover,

$$\int_a^b \alpha h' \, d\lambda + \alpha(a) \left(h(b) - h(a)\right) =$$
$$= \int_a^b \hat{\alpha} h' \, d\lambda$$
$$\leq \hat{\alpha}(b)h(b) - \hat{\alpha}(a)h(a)$$
$$= \alpha(b)h(b) - \alpha(a)h(a) + \alpha(a)\left(h(b) - h(a)\right),$$

proving the general case.

Of course, it is highly restrictive to require that the integrand f in 6.1.2 be an absolutely continuous function. If we now proceed as with the Riemann integral (see Section 1.2), we can extend the Riemann-Stieltjes integral to include a much broader class of integrands. In order to exploit 6.1.4(vi), we initially will restrict our attention to nondecreasing α.

6.1.5. Definition. Suppose that α is a nondecreasing function defined on $[a, b]$ and that f is a bounded function defined on $[a, b]$. Suppose further that, corresponding to each $\epsilon > 0$, there are functions $f_1, f_2 \in AC[a, b]$ with the following properties:

(i) $f_1 \leq f \leq f_2$; and
(ii) $\int_a^b f_2 - f_1 \, d\alpha \leq \epsilon$.

Then we will say that f *is Riemann-Stieltjes integrable with respect to* α and write

$$\int_a^b f \, d\alpha = \sup \left\{ \int_a^b f_1 \, d\alpha : f_1 \leq f \text{ and } f_1 \in AC[a, b] \right\}$$
$$\left(= \inf \left\{ \int_a^b f_2 \, d\alpha : f_2 \geq f \text{ and } f_2 \in AC[a, b] \right\} \right).$$

Immediate from the definition above is an analogue of 6.1.4.

6.1.6. Proposition. *Suppose that α, β are nondecreasing functions and f and g are each Riemann-Stieltjes integrable with respect to both α and β. Then*
   (i) $f + g$ is Riemann-Stieltjes integrable with respect to α and

$$\int_a^b f + g\,d\alpha = \int_a^b f\,d\alpha + \int_a^b g\,d\alpha;$$

   (ii) if ξ is any real number, $\int_a^b \xi f\,d\alpha = \xi \int_a^b f\,d\alpha;$

   (iii) if ξ is any real number, $\int_a^b f\,d(\xi\alpha) = \xi \int_a^b f\,d\alpha;$
   (iv) f is integrable with respect to $\alpha + \beta$ and

$$\int_a^b f\,d(\alpha + \beta) = \int_a^b f\,d\alpha + \int_a^b f\,d\beta;$$

(v) $\alpha \in AC[a, b] \Rightarrow$

$$\int_a^b f\,d\alpha = f(b)\alpha(b) - f(a)\alpha(a) - \int_a^b \alpha\,df;$$

   (vi) If α is nondecreasing and if $f \leq g$, then

$$\int_a^b f\,d\alpha \leq \int_a^b g\,d\alpha.$$

In view of the above, 6.1.5 now extends to include arbitrary $\alpha \in BV[a, b]$.

6.1.7. Definition Suppose that $\alpha \in BV[a, b]$ and that $\alpha = \alpha_1 - \alpha_2$ where both α_1 and α_2 are nondecreasing functions. If f is Riemann-Stieltjes integrable with respect to both α_1 and α_2, then we define the *Riemann-Stieltjes integral of f relative to α* to be

$$\int_a^b f\,d\alpha = \int_a^b f\,d\alpha_1 - \int_a^b f\,d\alpha_2.$$

Because of 6.1.6, this definition is independent of the decompositon of α into nondecreasing functions.

   Our next goal is to show that if $f \in C[a, b]$ and $\alpha \in BV[a, b]$, then $\int_a^b f\,d\alpha$ exists. Before doing this, we will need to prove some lemmas.

6.1.8. Lemma. *Let $\alpha \in BV[a, b]$. For $x \in [a, b]$ each of the limits*

$$\alpha(x+) \equiv \lim_{h \downarrow 0} \alpha(x + h) \ \text{and}$$

$$\alpha(x-) \equiv \lim_{h \uparrow 0} \alpha(x + h)$$

exist. For $\epsilon > 0$ set $E_\epsilon = \{x : |\alpha(x+) - \alpha(x-)| \geq \epsilon\}$. Then E_ϵ is finite; in particular, α has at most countably many discontinuities.

Proof. It suffices to consider only the case that α is nondecreasing. For nondecreasing functions, notice that $\alpha(x+)$ and $\alpha(x-)$ both exist and are finite. Moreover, $\alpha(x+) - \alpha(x-) \geq 0$ and

$$\alpha(x+) - \alpha(x-) = 0 \leftrightarrow \alpha \text{ is continuous at } x.$$

Next set

$$E_n = \left\{ x : \alpha(x+) - \alpha(x-) \geq \frac{1}{n} \right\}$$

so that

$$E \equiv \{ x : \alpha \text{ is not continuous at } x \} = \cup_n E_n.$$

We will show that each E_n is finite.

Fix E_n; let N be any integer larger than $\frac{\alpha(b) - \alpha(a)}{n}$; we claim that E_n must have fewer that N elements. For contradiction, suppose that we can find $N+1$ elements $\{x_1, \ldots, x_{N+1}\}$ in E_n. Then

$$\alpha(b) - \alpha(a) \geq \sum_{i=1}^{N+1} \alpha(x_i+) - \alpha(x_i-)$$

$$\geq \sum_{i=1}^{N+1} \frac{1}{n}$$

$$\geq (N+1)n$$

$$> \alpha(b) - \alpha(a),$$

a contradiction.

6.1.9. Lemma. *Suppose that $\alpha \in BV[a,b]$, that $a < c < b$, and that α is continuous at c. Then*

(i) *$g = \chi_{[c,b]}$ and $h = \chi_{(c,b]}$ are Riemann-Stieltjes integrable with respect to α; and*

(ii) *$\int_a^b g \, d\alpha = \int_a^b h \, d\alpha = \alpha(b) - \alpha(c)$.*

Proof. As usual, it suffices to consider only the case that α is nondecreasing.

The idea is to approximate g and h with absolutely continuous functions; while there are many possible choices, we elect to use cubic splines: for $\delta > 0$ define

$$g_\delta(t) = \begin{cases} 0 & t < 0 \\ -2\left(\frac{t}{\delta}\right)^3 + 3\left(\frac{t}{\delta}\right)^2 & 0 \leq t \leq \delta \\ 1 & \delta < t. \end{cases}$$

Then, as is readily checked, g_δ is absolutely continuous, is nondecreasing, and $\int_0^\delta g_\delta'(t) \, dt = 1$.

For fixed δ, set

$$f_1(t) = g_\delta(t - c) \quad \text{and} \quad f_2(t) = g_\delta(t - c + \delta),$$

so that $f_1 \leq g \leq h \leq f_2$.

Moveover, if δ is sufficiently small,

$$\int_a^b \alpha(t) f_1'(t) \, dt = \int_{a-c}^{b-c} \alpha(t + c) g_\delta'(t) \, dt$$
$$= \int_0^\delta \alpha(t + c) g_\delta'(t) \, dt.$$

In particular, since $\int_0^\delta g_\delta'(t) \, dt = 1$,

$$\alpha(c) \leq \int_a^b \alpha(t) f_1'(t) \, dt \leq \alpha(c + \delta).$$

Similarly,

$$\alpha(c - \delta) \leq \int_a^b \alpha(t) f_2'(t) \, dt \leq \alpha(c).$$

Thus

$$0 \leq \int_a^b f_2 - f_1 \, d\alpha$$
$$= f_2(b)\alpha(b) - f_1(b)\alpha(b) - f_2(a)\alpha(a) + f_1(a)\alpha(a) - \cdots$$
$$\cdots - \int_a^b \alpha(t) \left(f_2'(t) - f_1'(t) \right) dt$$
$$= 0 - \int_a^b \alpha(t) f_2'(t) \, dt + \int_a^b \alpha(t) f_1'(t) \, dt$$
$$\leq \alpha(c + \delta) - \alpha(c - \delta).$$

Since by assumption α is continuous at c, this proves that both g and h are Riemann-Stieltjes integrable with respect to α. The remaining conclusion is evident from the estimates above.

The following corollaries are immediate from Lemma 6.1.9.

6.1.10. Corollary.   *Suppose that $\alpha \in BV[a, b]$, that $a \leq c \leq d \leq b$, and that α is continuous at both c and d. Then*

(i) All four of the functions

$$f_1 = \chi_{[c,d]}, \quad f_2 = \chi_{(c,d]}, \quad f_3 = \chi_{[c,d)}, \quad f_4 = \chi_{(c,d)}$$

are Riemann-Stieltjes integrable with respect to α; and

(ii) $\int_a^b f_i \, d\alpha = \alpha(d) - \alpha(c) \quad (i = 1, \ldots, 4)$.

6.1.11. Definition. If I is an interval having endpoints $a \le b$, if $\alpha \in BV[a,b]$, and if α is continuous at a and b, then we can define the α-length, $\ell_\alpha(I)$, of I to be $\ell_\alpha(I) = \alpha(b) - \alpha(a)$. This definition of length extends the notion of "length of an interval" introduced in Section 1.2 and enables us to extend the Riemann-Stieltjes integral to include step functions. (Note that the α-length need *not* be positive!)

6.1.12. Corollary *Suppose that $\alpha \in BV[a,b]$ and that $s = \sum_1^n \xi_i \chi_{I_i}$ is a step function defined on $[a,b]$. If α is continuous at the endpoints of the intervals $\{I_i\}$, then s is Riemann-Stieltjes integrable with respect to α and*

$$\int_a^b s \, d\alpha = \sum_1^n \xi_i \ell_\alpha(I_i).$$

Finally, with 6.1.12 in hand, we can show that if $\alpha \in BV[a,b]$ and if $f \in C[a,b]$, then f is Riemann-Stieltjes integrable with respect to α.

6.1.13. Theorem. *If $\alpha \in BV[a,b]$ and if $f \in C[a,b]$, then f is Riemann-Stieltjes integrable with respect to α.*

Proof. Fix $\epsilon > 0$; once again we consider only the case that α is nondecreasing. Since f is uniformly continuous on $[a,b]$, we may select $\delta > 0$ so that if $s,t \in [a,b]$ and $|s - t| \le \delta$, then

$$|f(s) - f(t)| \le \frac{\epsilon}{3(\alpha(b) - \alpha(a))}.$$

Next select numbers

$$a = t_0 < t_1 < \cdots < t_n = b$$

so that α is continuous at $t_1, \ldots, t_{n-1}$, and so that $|t_j - t_{j-1}| \le \delta$, $j = 1, \ldots, n$. For $j = 1, \ldots, n-1$, set $I_j = [t_{j-1}, t_j)$ and set $I_n = [t_{n-1}, b]$. Now define for $j = 1, \ldots, n$

$$m_j = \min\{f(t) : t \in I_j\} \text{ and}$$
$$M_j = \max\{f(t) : t \in I_j\}$$

so that

$$0 \le M_j - m_j \le \frac{\epsilon}{3(\alpha(b) - \alpha(a))}.$$

Finally, set

$$s_1 = \sum_1^n m_j \chi_{I_j} \text{ and}$$

$$s_2 = \sum_1^n M_j \chi_{I_j}.$$

Then $s_1 \leq f \leq s_2$ and

$$\int_a^b s_2 - s_1 \, d\alpha \leq \sum_1^n (M_j - m_j) \, \ell_\alpha(I_j)$$

$$\leq \frac{\epsilon}{3(\alpha(b) - \alpha(a))} \sum_1^n \ell_\alpha(I_j)$$

$$\leq \frac{\epsilon}{3}.$$

Now each of s_1 and s_2 are Riemann-Stieltjes integrable, and so there are absolutely continuous functions $f_1 \leq s_1 \leq f_2$ and $g_1 \leq s_2 \leq g_2$ so that

$$\int_a^b f_2 - f_1 \, d\alpha \leq \frac{\epsilon}{3} \text{ and}$$

$$\int_a^b g_2 - g_1 \, d\alpha \leq \frac{\epsilon}{3}.$$

Note that $f_1 \leq f \leq g_2$ and that since $s_2 - g_1 \geq 0$ and $f_2 - s_1 \geq 0$,

$$\int_a^b g_2 - f_1 \, d\alpha \leq \int_a^b g_2 - g_1 + s_2 + f_2 - s_1 - f_1 \, d\alpha$$

$$= \int_a^b g_2 - g_1 \, d\alpha + \int_a^b s_2 - s_1 \, d\alpha + \int_a^b f_2 - f_1 \, d\alpha$$

$$\leq \epsilon,$$

showing that f_1 and g_2 fulfill the requirements of 6.1.5. Hence f is Riemann-Stieltjes integrable with respect to α.

Another standard property of the Riemann-Stieltjes integral is embodied in the following:

6.1.14. Proposition. *If $\alpha \in BV[a,b]$ and if $f \in C[a,b]$, then f and $|f|$ are Riemann-Stieltjes integrable with respect to both α and V_α, the variation of α. Moreover,*

$$\left| \int_a^b f \, d\alpha \right| \leq \int_a^b |f| \, dV_\alpha.$$

Proof. Recall (see 5.4.2) that if

$$\beta(t) = \frac{1}{2} \left(V_\alpha(a, t) + \alpha(t) - \alpha(a) \right) \text{ and}$$

$$\gamma(t) = \frac{1}{2} \left(V_\alpha(a, t) - \alpha(t) + \alpha(a) \right),$$

then $\alpha = \beta - \gamma - \alpha(a)$, $V_\alpha = \beta + \gamma$ and $\beta \geq 0$, $\gamma \geq 0$. Then [applying 6.1.6(vi)]

$$\left| \int_a^b f \, d\alpha \right| = \left| \int_a^b f \, d(\beta - \gamma) \right| = \left| \int_a^b f \, d\beta - \int_a^b f \, d\gamma \right|$$

$$\leq \left| \int_a^b f \, d\beta \right| + \left| \int_a^b f \, d\gamma \right|$$

$$\leq \int_a^b |f| \, d\beta + \int_a^b |f| \, d\gamma$$

$$= \int_a^b |f| \, d(\beta + \gamma)$$

$$= \int_a^b |f| \, dV_\alpha,$$

completing the proof.

By combining the ideas in 6.1.9 and 6.1.13, it is possible to show the following (see Problem 6.4).

6.1.15. Theorem. *If $\alpha_1, \alpha_2 \in BV[a, b]$, if $f \in C[a, b]$, and if $\alpha_1 - \alpha_2$ is constant on a dense subset of $[a, b]$, which includes the endpoints $\{a, b\}$, then $\int_a^b f \, d\alpha_1 = \int_a^b f \, d\alpha_2$.*

The import of 6.1.15 is that the values $\alpha \in BV[a, b]$ assumes at its discontinuities in (a, b) have no impact on the value of $\int_a^b f \, d\alpha$.

6.1.16. Definition. If $\alpha \in BV[a, b]$, then α is said to be *normalized* if $\alpha(a) = 0$ and

$$\alpha(t) = \frac{1}{2} \left(\alpha(t+) - \alpha(t-) \right)$$

for all $t \in (a, b)$.

Since $\alpha \in BV[a, b]$ can have at most countably many discontinuities, α has a normalization $\hat{\alpha}$ with the property $\alpha = \hat{\alpha}$ on a dense subset of $[a, b]$ (including the endpoints of the interval). Relative to this normalization,

$$\int_a^b f \, d\alpha = \int_a^b f \, d\hat{\alpha}$$

for all $f \in C[a, b]$.

PROBLEMS

6.1. Set $f(t) = t^2$ and

$$\alpha(t) = \begin{cases} t & 0 \le t \le 1 \\ 3 - t & 1 < t \le 2. \end{cases}$$

Find $\int_0^2 f \, d\alpha$.

6.2. Suppose that $\alpha \in BV[a, b]$, $f \in C[a, b]$ and that $a < c < b$. Show that

$$\int_a^b f \, d\alpha = \int_a^c f \, d\alpha + \int_c^b f \, d\alpha.$$

6.3. *True or False:* If $\alpha \in BV[a, b]$, if f is Riemann-Stieltjes integrable relative to α, and if $a < c < b$, then

$$\int_a^b f \, d\alpha = \int_a^c f \, d\alpha + \int_c^b f \, d\alpha.$$

6.4. Prove Theorem 6.1.15.

6.5 Suppose that $\alpha \in BV[a, b] \cap C[a, b]$; if $I_1, \ldots, I_n \subseteq [a, b]$ are disjoint intervals, show that

$$\ell_\alpha \left(\cup_1^n I_j \right) = \sum_1^n \ell_\alpha(I_j).$$

6.6. Suppose that $f \in C[a, b]$ and that $\alpha \in BV[a, b]$; set

$$h(x) = \int_a^x f \, d\alpha.$$

Show that $h \in BV[a, b]$ and that if α is continuous at x, then so is h.

6.7. Suppose that $f \in C[a, b]$ and that α is a nondecreasing function; show that there is a point $\xi \in [a, b]$ with the property that

$$\int_a^b f \, d\alpha = f(\xi) \left(\alpha(b) - \alpha(a) \right).$$

Does the conclusion hold for arbitrary $\alpha \in BV[a, b]$?

6.8. Let α be a nondecreasing function and let $f \in \mathcal{L}_1[a, b]$. Show that there is a number $\xi \in [a, b]$ so that

$$\int_a^b \alpha f \, d\lambda = \alpha(a) \int_a^\xi f \, d\lambda + \alpha(b) \int_\xi^b f \, d\lambda.$$

6.9. Let α be a nondecreasing function. Show that a bounded function f is Riemann-Stieltjes integrable relative to α if and only if each of the following hold:

(a) f is continuous at each of the points of discontinuity of α;

(b) For each $\epsilon > 0$ there are points $a = t_0 < t_1 < \cdots < t_n = b$ so that α is continuous at each t_i, $i = 1, \cdots, n-1$, and

$$\sum_1^n \omega\left(f;(t_{i-1},t_i)\right)\left(\alpha(t_i) - \alpha(t_{i-1})\right) \leq \epsilon.$$

6.10. Use Problem 6.9 to show that f is Riemann integrable if and only if f is continuous almost everywhere.

6.2.   Convergence Theorems.   Because Stieltjes integrals involve two functions there are more possible types of convergence theorems. In particular, we could examine any of the following sequences:

$$\left\{\int f_n \, d\alpha\right\} \quad \left\{\int f \, d\alpha_n\right\} \quad \left\{\int f_n \, d\alpha_m\right\}.$$

The primary goal of the next chapter will be to examine the first of the sequences above; the third is probably better considered in a somewhat more abstract setting and so we will not consider it in this book. The middle sequence is the main topic of the current section.

While examination of the second sequence is our primary goal, we can give a simple preliminary result with respect to the first; together with Dini's Theorem, this will provide the basis for the Lebesgue-type convergence theorems derived in the next chapter for Stieltjes Integrals.

6.2.1.   Theorem. *Let $\alpha \in BV[a,b]$, let $\{f_n\} \subset C[a,b]$, and suppose that $\{f_n\}$ converges to f uniformly on $[a,b]$. Then*

$$\lim \int_a^b f_n \, d\alpha \;=\; \int_a^b f \, d\alpha.$$

Proof. Fix $\epsilon > 0$ and choose N so large that $n \geq N$ implies that

$$\|f_n - f\| \;\leq\; \epsilon\left(V_\alpha(a,b)\right)^{-1}$$

Then, for $n \geq N$,

$$\left|\int_a^b f_n \, d\alpha - \int_a^b f \, d\alpha\right| \leq \int_a^b \|f_n - f\| \, dV_\alpha(a,x)$$

$$\leq \; \epsilon.$$

Before considering the sequences $\{\int f\,d\alpha_n\}$ we need to establish some preliminary lemmas. The first of these is of independent importance; the diagonalization technique it exploits can be of great use in a variety of settings.

6.2.2. Selection Principle. *Let* $\{a(m,n)\}_{m,n\in\mathcal{N}}$ *be a doubly indexed sequence of nonnegative real numbers and let* $A > 0$. *Suppose for all* m *and* n *that*

$$|a(m,n)| \leq A.$$

Then there is a set of positive integers

$$n_0 < n_1 < n_2 < \cdots \tag{1}$$

and a set of numbers $\{a_0, a_1, \ldots\}$ *such that*

$$\lim_{i\to\infty} a(m, n_i) = a_m \quad \text{for each} \quad m = 0, 1, 2, \ldots. \tag{2}$$

Proof. Since $\{a(0,n)\}_{n=0}^\infty$ is a bounded sequence there is at least one cluster point a_0. Hence there are integers

$$n_0^0 < n_1^0 < n_2^0 < \cdots \tag{3}$$

such that

$$\lim_{i\to\infty} a(0, n_i^0) = a_0.$$

Now the sequence $\{a(1, n_i^0)\}$ is bounded and also must have at least one cluster point a_1. Thus we may find a subsequence of (3),

$$n_0^1 < n_1^1 < n_2^1 < \cdots$$

such that

$$\lim_{i\to\infty} a(1, n_i^1) = a_1.$$

Proceeding in this manner, we obtain for each m a number a_m and a sequence

$$n_0^m < n_1^m < n_2^m < \cdots,$$

which is a subsequence of its predecessor

$$n_0^{m-1} < n_1^{m-1} < n_2^{m-1} < \cdots$$

and which satisfies

$$\lim_{i\to\infty} a(m, n_i^m) = a_m.$$

Now set $n_i = n_i^i$. Since

$$n_i = n_i^i < n_{i+1}^i \leq n_{i+1}^{i+1} = n_{i+1}$$

the set $\{n_i\}$ satisfies (1). Moreover, for each fixed m, the sequence $\{n_i\}_{i=m}^{\infty}$ is a subsequence of $\{n_i^m\}_{i=1}^{\infty}$, and hence

$$\lim_{i \to \infty} a(m, n_i) = \lim_{i \to \infty} a(m, n_i^m) = a_m,$$

showing that (2) is fulfilled as well.

In analyzing the behavior of $\{\int f \, d\alpha_n\}$ we will first examine when a sequence $\{\alpha_n\}$ of functions of bounded variation has a subsequence converging pointwise to a function of bounded variation. It turns out that is happens under *very* mild restrictions on the sequence $\{\alpha_n\}$. Recall that the Arzela-Ascoli Theorem gave fairly stringent conditions for a sequence of continuous functions to have a continuous limit—the difference between these conditions is what makes $\{\int f \, d\alpha_n\}$ converge more "easily" than $\{\int f_n \, d\alpha\}$.

As a first approximation, we examine when a sequence of nondecreasing functions has a subsequence which converges pointwise. The following lemma is a "continuous" version of the Selection Principle.

6.2.3. Lemma. *Let $\{\alpha_n\}$ be a sequence of nondecreasing functions defined on the closed bounded interval $[a, b]$. Suppose that for some $A > 0$, for each $x \in [a, b]$, and for each n*

$$|\alpha_n(x)| \leq A$$

Then there is a set of integers

$$n_0 < n_1 < n_2 < \cdots$$

and a nondecreasing function α defined on $[a, b]$ satisfying

$$\lim_{i \to \infty} \alpha_{n_i}(x) = \alpha(x) \qquad\qquad (x \in [a, b]).$$

Proof. Let $\{r_m\}_{m=0}^{\infty}$ be an enumeration of the rational numbers in $[a, b]$. By the Selection Principle, there is a sequence of integers

$$n_0^0 < n_1^0 < n_2^0 < \cdots \qquad\qquad\qquad (4)$$

and a sequence of numbers $\{\alpha(r_m)\}$ such that

$$\lim_{i \to \infty} \alpha_{n_i^0}(r_m) = \alpha(r_m).$$

Next set

$$\overline{\alpha}(x) = \overline{\lim}_{i \to \infty} \alpha_{n_i^0}(x) \quad \text{and} \quad \underline{\alpha}(x) = \underline{\lim}_{i \to \infty} \alpha_{n_i^0}(x).$$

We claim that both $\overline{\alpha}$ and $\underline{\alpha}$ are nondecreasing functions. We will show this for $\overline{\alpha}$, the proof for $\underline{\alpha}$ being similar. Fix $x_1 < x_2$ in $[a, b]$ and fix $\epsilon > 0$; select i so large that

$$\sup_{j \geq i} \alpha_{n_j^0}(x_2) \leq \overline{\alpha}(x_2) + \epsilon/2$$

and select $k \geq i$ so that

$$\sup_{j \geq i} \alpha_{n_j^0}(x_1) \leq \alpha_{n_k^0}(x_1) + \epsilon/2.$$

Then

$$\overline{\alpha}(x_1) \leq \sup_{j \geq i} \alpha_{n_j^0}(x_1)$$

$$\leq \alpha_{n_k^0}(x_1) + \epsilon/2$$

$$\leq \alpha_{n_k^0}(x_2) + \epsilon/2$$

$$\leq \sup_{j \geq i} \alpha_{n_j^0}(x_2) + \epsilon/2$$

$$\leq \overline{\alpha}(x_2) + \epsilon;$$

since ϵ was arbitrary this shows that $\overline{\alpha}(x_1) \leq \overline{\alpha}(x_2)$, establishing our claim.
Set

$$D = \{t \in [a, b] : \text{ both } \overline{\alpha} \text{ and } \underline{\alpha} \text{ are continuous at } t\};$$

since $([a, b] \backslash D)$ is at most countable, D is a dense subset of $[a, b]$. The next step of the proof is to show that $\overline{\alpha} = \underline{\alpha}$ on the set D. To this end, fix $t \in D$ and select $\{r_{m_k}\} \subset Q \cap [a, b]$ so that $\lim r_{m_k} = t$. Then, using the definition of $\alpha(r_{m_k})$,

$$\overline{\alpha}(t) = \lim_{k \to \infty} \overline{\alpha}(r_{m_k})$$

$$= \lim_{k \to \infty} \overline{\lim}_{i \to \infty} \alpha_{n_i^0}(r_{m_k})$$

$$= \lim_{k \to \infty} \alpha(r_{m_k})$$

$$= \lim_{k \to \infty} \underline{\lim} \alpha_{n_i^0}(r_{m_k})$$

$$= \lim_{k \to \infty} \underline{\alpha}(r_{m_k}) = \underline{\alpha}(t)$$

and thus, on D, $\overline{\alpha} = \underline{\alpha}$.

Finally, let $\{s_m\}$ be an enumeration of $D\backslash[a,b]$. By the Selection Principle, applied to $\{\alpha_{n_i^0}(s_m)\}$ there is a subsequence of (4),

$$n_0 < n_1 < n_2 < \cdots,$$

and a sequence of numbers $\{a_m\}$ such that

$$\lim_{i\to\infty} \alpha_{n_i}(s_m) = a_m.$$

Now define

$$\alpha(x) = \begin{cases} a_m & \text{if } x = s_m \\ \bar{\alpha}(x) & \text{if } x \in D. \end{cases}$$

Clearly, $\lim \alpha_{n_i}(x) = \alpha(x)$ for each x. Since the limit of nondecreasing functions is again nondecreasing, this completes the proof.

We are now in a position to prove:

6.2.4. Helly's Theorem. *Let $\{\alpha_n\}$ be a norm-bounded sequence in $BV[a,b]$. Then there is a subsequence $\{\alpha_{n_i}\}$ and an $\alpha \in BV[a,b]$ such that*

$$\lim \alpha_{n_i}(x) = \alpha(x)$$

for each x in $[a,b]$.

The above is a very strong and startling compactness result—certainly much stronger than the corresponding result we deduced for $C[a,b]$. Recall that norm bounded sequences in $BV[a,b]$ are both of uniformly bounded variation and are pointwise bounded.

Proof of 6.2.4. For each $x \in [a,b]$, set

$$g_n(x) = 1/2\big(V_{\alpha_n}(a,x) + \alpha_n(x) - \alpha_n(a)\big)$$
and
$$h_n(x) = 1/2\big(V_{\alpha_n}(a,x) - \alpha_n(x) + \alpha_n(a)\big).$$

We have previously seen that both g_n and h_n are nondecreasing; it is evident that

$$\alpha_n(x) = g_n(x) - h_n(x) + \alpha_n(a)$$

and

$$V_{\alpha_n}(a,x) = g_n(x) + h_n(x).$$

Also,

$$\alpha_n(a) - \alpha_n(x) \le |\alpha_n(a) - \alpha_n(x)| \le V_{\alpha_n}(a,x),$$

and hence $g_n(x) \ge 0$; similarly, $h_n(x) \ge 0$. Thus if $\|\alpha_n\| \le M$, then

$$|g_n(x)| = g_n(x) \le g_n(x) + h_n(x) = V_{\alpha_n}(a,x) \le M$$

and also $|h_n(x)| \leq M$. Hence, by 6.2.3, there is a nondecreasing function g and a subsequence

$$n_0^0 < n_1^0 < n_2^0 < \cdots \tag{5}$$

so that $\lim g_{n_i^0}(x) = g(x)$. A second application of 6.2.3 to $\{h_{n_i^0}\}$ gives a nondecreasing function h and a subsequence of (5)

$$n_0^1 < n_1^1 < n_2^1 < \cdots \tag{6}$$

so that $\lim h_{n_i^1}(x) = h(x)$. Finally, since $\{\alpha_{n_i^1}(a)\}$ is bounded, there is a number A and a subsequence of (6)

$$n_0 < n_1 < n_2 < \cdots$$

so that $\lim \alpha_{n_i}(a) = A$. Consequently,

$$\lim_{i \to \infty} \alpha_{n_i}(x) = \lim_{i \to \infty} g_{n_i}(x) - h_{n_i}(x) + \alpha_{n_i}(a)$$

$$= g(x) - h(x) + A.$$

Setting $\alpha = g - h + A$ gives the result.

If we have a sequence $\{\alpha_n\} \subseteq BV[a, b]$ which converge pointwise to a function $\alpha \in BV[a, b]$ and if $\|\alpha_n\|$ is bounded, then $\int_a^b f \, d\alpha_n \to \int_a^b f \, d\alpha$ as $n \to \infty$.

6.2.5. Helly-Bray Theorem. *Let $\{\alpha_n\}$ be a norm-bounded sequence in $BV[a, b]$ and suppose that there is an $\alpha \in BV[a, b]$ such that $\lim \alpha_n(x) = \alpha(x)$. Then for all $f \in C[a, b]$,*

$$\lim_{n \to \infty} \int_a^b f \, d\alpha_n = \int_a^b f \, d\alpha.$$

Proof. Set

$$M = \sup\{\|\alpha_n\|, V_\alpha(a, b)\}$$

and fix $\epsilon > 0$ and $f \in C[a, b]$. Select $\delta > 0$ so small that if $|x - y| \leq \delta$, then $|f(x) - f(y)| \leq \epsilon/2M$. Select points $a = t_0 < t_1 < \cdots < t_m = b$ so that α and all of the functions $\{\alpha_n\}$ are continuous at $\{t_1, \ldots, t_{m-1}\}$ and so that $t_i - t_{i-1} \leq \delta$.

For $i = 0, \ldots, m-1$ select x_i between t_i and t_{i+1}. Then

$$\left| \int_a^b f \, d\alpha_n - \int_a^b f \, d\alpha \right| =$$

$$= \left| \sum_{k=0}^{m-1} \int_{t_k}^{t_{k+1}} f \, d\alpha_n \mp \sum_{k=0}^{m-1} \int_{t_k}^{t_{k+1}} f(x_k) \, d\alpha_n(t) \pm \cdots \right.$$

$$\cdots \pm \sum_{k=0}^{m-1} \int_{t_k}^{t_{k+1}} f(x_k) \, d\alpha(t) - \cdots$$

$$\left. \cdots - \sum_{k=0}^{m-1} \int_{t_k}^{t_{k+1}} f \, d\alpha \right|$$

$$\leq \sum_{k=0}^{m-1} \int_{t_k}^{t_{k+1}} |f(t) - f(x_k)| \, dV_{\alpha_n}(a, t) + \cdots$$

$$\cdots + \left| \sum_{k=0}^{m-1} \int_{t_k}^{t_{k+1}} f(x_k) \, d\big(\alpha_n(t) - \alpha(t)\big) \right| + \cdots$$

$$\cdots + \sum_{k=0}^{m-1} |f(t) - f(x_k)| \, dV_\alpha(a, t)$$

$$\leq \epsilon/2 + \sum_{k=0}^{m-1} |f(x_k)| \cdots \big(\alpha_n(x_{k+1}) - \alpha(x_{k+1}) - \cdots$$

$$\cdots - \alpha_n(x_k) + \alpha(x_k)\big) + \frac{\epsilon}{2}.$$

Now taking the limit superior as $n \to \infty$ yields

$$\overline{\lim} \left(\int_a^b f \, d\alpha_n - \int_a^b f \, d\alpha \right) \leq \epsilon.$$

Since ϵ was arbitrary, this establishes the result.

Combining the foregoing two results yields the following very important theorem. We will see the same theorem from a different viewpoint in Chapter 8 (the Banach-Alaoglu Theorem).

6.2.6. Corollary. *Let $\{\alpha_n\}$ be a norm-bounded sequence in $BV[a,b]$. Then there is an $\alpha \in BV[a,b]$ and a subsequence $\{\alpha_{n_j}\}$ such that*

$$\lim \alpha_{n_j}(x) = \alpha(x) \qquad (x \in [a, b]).$$

Moreover, for all $f \in C[a,b]$,

$$\lim \int_a^b f \, d\alpha_{n_j} = \int_a^b f \, d\alpha.$$

PROBLEMS

6.11. Let α be a normalized function of bounded variation and suppose that

$$\int_0^1 t^n \, d\alpha(t) = 0$$

for all natural numbers n. Show that $\alpha(t) \equiv 0$.

6.12. Let $g \in C[0, 1]$ and suppose that

$$\int_0^1 g(t) \, dt^n = 0$$

for all natural numbers n. Show that g is a constant function.

6.13. (See Problem 5.26.) Only *one* of Helly's Theorem and the Helly-Bray Theorem remains true for functions of globally bounded variation defined on $(-\infty, \infty)$. Using the $[a, b]$ case, *prove* the one that is true and find a counterexample to the one that fails.

6.14. Let α be a nondecreasing function and suppose for some $s_0 > 0$ that

$$\sup_{u \geq 0} \int_0^u e^{-s_0 t} \, d\alpha(t) < \infty.$$

Then for every $s \geq s_0$

$$\lim_{R \to \infty} \int_0^R e^{-st} \, d\alpha(t)$$

exists. Moreover, if $\beta(t) = \int_0^t e^{-st} \, d\alpha(t)$, then

$$\lim_{R \to \infty} \int_0^R e^{-st} \, d\alpha(t) = \lim_{R \to \infty} \int_0^R e^{-st} \beta(t) \, dt.$$

6.15. Let $\alpha \in BV[a, b]$ and for $s > 0$ define

$$f(s) = \int_a^b e^{-st} \, d\alpha(t).$$

Show that f is differentiable and that

$$f'(s) = -\int_a^b t e^{-st} \, d\alpha(t).$$

6.16. Let $\alpha \in BV[a, b]$ and define a linear functional L on $C[a, b]$ by

$$L(f) = \int_a^b f \, d\alpha.$$

Show that L is continuous.

6.3. THE DUAL SPACE OF $C[a,b]$. Fix $\alpha \in BV[a,b]$; the functional L defined for $f \in C[a,b]$ by

$$L(f) = \int_a^b f \, d\alpha$$

is an element of $(C[a,b])^*$ (see Problem 6.16). The main topic of this section is to prove that the converse to this problem is also true; along the way, we will incidentally extend the Stieltjes integral to include all bounded Borel measurable functions. Our first result shows that every bounded linear functional defined on $C[a,b]$ decomposes into a positive and negative part; because we will need this result in Chapter 7 for $C(M)$ where M is an arbitrary compact metric space, we deduce it here in that generality.

6.3.1. Definition. A function $f \in C(M)$ is *positive* if $f(t) \geq 0$ for all t; a linear functional $L \in (C(M))^*$ is *positive* if $L(f) \geq 0$ whenever f is positive.

6.3.2. Jordan Decomposition Theorem. *For each $L \in (C(M))^*$ there are positive linear functionals P and N in $(C(M))^*$ for which $L = P - N$.*

Proof. Fix $L \in (C(M))^*$; define the *positive cone* of $C[a,b]$ to be the set

$$P^+ = \{f \in C(M) : f(t) \geq 0 \text{ for all } t\}.$$

We begin by defining P and N on P^+; for $f \in P^+$, set

$$P(f) = \sup\{L(\varphi) : \varphi \in C(M), \ 0 \leq \varphi \leq f\}.$$

If $0 \leq \varphi \leq f$, then $|L(\varphi)| \leq \|L\|\,\|\varphi\| \leq \|L\|\,\|f\|$, and so $P(f) < \infty$ for all f. Moreover,

$$P(f) \geq L(0) = 0 \quad \text{and} \quad P(f) \geq L(f).$$

Thus if $N(f) = P(f) - L(f)$, $N(f) \geq 0$ on P^+. It is further evident that for $\lambda \geq 0$, $\lambda P(f) = P(\lambda f)$ and $P(f + g) \geq P(f) + P(g)$ when $f,\, g \in P^+$. The reverse inequality also holds, for suppose that $0 \leq \sigma \leq f + g$ for some $\sigma \in C(M)$. If we set

$$\varphi_1 = f \wedge \sigma \quad \text{and} \quad \varphi_2 = \sigma - \varphi_1,$$

then $0 \leq \varphi_1 \leq f$ and, since $\varphi_1 \leq \sigma$, $\varphi_2 \geq 0$. Also,

$$\varphi_2(x) - g(x) = \sigma(x) - \varphi_1(x) - g(x) = \begin{cases} \sigma(x) - f(x) - g(x) & \text{if } f(x) < \sigma(x) \\ 0 - g(x) & \text{if } f(x) \geq \sigma(x) \end{cases}$$

and thus $\varphi_2 - g \leq 0$, showing that $0 \leq \varphi_2 \leq g$. Now fix $\epsilon > 0$ and choose $\sigma \in C(M)$ so that $0 \leq \sigma \leq f + g$ and

$$P(f + g) \leq L(\sigma) + \epsilon.$$

Then with φ_1 and φ_2 defined as above,

$$
\begin{aligned}
P(f+g) &\le L(\sigma) + \epsilon \\
&= L(\varphi_1) + L(\varphi_2) + \epsilon \\
&\le P(f) + P(g) + \epsilon.
\end{aligned}
$$

Since ϵ was arbitrary, this gives $P(f+g) = P(f) + P(g)$ [and $N(f+g) = N(f) + N(g)$] for all $f, g \in P^+$.

Now any $f \in C(M)$ may be written as the difference of two functions in P^+ (for example, $f = f^+ - f^-$); we would like to extend P (and hence $N \equiv P - L$) to all of $C(M)$ using this fact. As usual, we must guarantee that this extension is independent of the decomposition of f. Thus fix $f \in C(M)$ and suppose that

$$
f = f_1 - f_2 = g_1 - g_2
$$

for some functions f_1, f_2, g_1, and $g_2 \in P^+$. We wish to show that

$$
P(f_1) - P(f_2) = P(g_1) - P(g_2);
$$

by symmetry it suffices to show only "$\le$." Select functions σ_1, σ_2, s_1, and s_2 so that

$$
0 \le \sigma_1 \le f_1; \quad 0 \le \sigma_2 \le f_2; \quad 0 \le s_1 \le g_1; \quad \text{and} \quad 0 \le s_2 \le g_2.
$$

Then

$$
\sigma_1 \le f_1 = f + f_2 = g_1 - g_2 + f_2 \le g_1 - s_2 + f_2,
$$

i.e.,

$$
\sigma_1 + s_2 \le g_1 + f_2.
$$

Thus

$$
L(\sigma_1 + s_2) \le P(g_1 + f_2) = P(g_1) + P(f_2).
$$

Taking the supremum over all such σ_1, s_2 gives

$$
P(f_1) + P(g_2) \le P(g_1) + P(f_2);
$$

rearranging gives the desired inequality.

Thus we may extend P (and N) to all of $C(M)$. We next argue that the extended versions of P and N are linear. If $f, g \in C(M)$, then

$$
P(f+g) = P((f^+ + g^+) - (f^- + g^-)) = P(f) + P(g),
$$

so P is additive. Since $P(0) = 0$,

$$
0 = P(f - f) = P(f) + P(-f)
$$

and so $P(-f) = -P(f)$. If $\lambda \ge 0$,

$$
P(\lambda f) = P(\lambda f^+ - \lambda f^-) = \lambda P(f)
$$

while if $\lambda < 0$,

$$
P(\lambda f) = -P(-\lambda f) = \lambda P(f)
$$

and thus each of P and N is linear. Finally since, on $\mathcal{P}^+$, $|P(f)| \leq \|L\| \|f\|$, for any $\varphi \in C(M)$

$$|P(\varphi)| \leq |P(\varphi^+)| + |P(\varphi^-)| \leq 2\|L\| \|\varphi\|,$$

it follows that P (and hence $N = P - L$) are each bounded linear functionals. As we have already verified that P and N are positive, this completes the proof.

The basic idea of the proof of our representation of $(C[a,b])^*$ is the same as in the proof that $(\mathcal{L}_1)^* = \mathcal{L}_\infty$ given in 5.5.9; to apply this idea, we need to extend a given functional defined on $C[a,b]$ to a slightly larger class.

6.3.3. Definition. We denote by $\mathcal{B}[a,b]$ the collection of all bounded functions f corresponding to which there is a nondecreasing sequence $\{g_n\} \subset C[a,b]$ satisfying

$$\lim g_n(x) = f(x)$$

for all $x \in [a,b]$.

It can be argued (Problem 6.17) that $\mathcal{B}[a,b]$ is precisely the class of bounded, Borel measurable functions defined on $[a,b]$. Note that $\mathcal{B}[a,b]$ is a normed linear space under the supremum norm

$$\|f\| = \sup\{|f(t)| : a \leq t \leq b\}$$

and that for each $x \in [a,b]$ the characteristic function $\chi_{[a,x]}$ is a member of $\mathcal{B}[a,b]$.

6.3.4. Lemma. *Let L be a positive, bounded linear functional defined on $C[a,b]$. Then there is a positive bounded linear functional $\tilde{L}$ defined on $\mathcal{B}[a,b]$ such that*

$$L(f) = \tilde{L}(f) \text{ for all } f \in C[a,b].$$

Moreover, there is a nondecreasing function α defined on $[a,b]$ such that for all $f \in C[a,b]$

$$\int_a^b f \, d\alpha = L(f).$$

Proof. Let $f \in \mathcal{B}[a,b]$ and select a nondecreasing sequence $\{\varphi_n\} \subset C[a,b]$ such that

$$\lim \varphi_n(x) = f(x) \qquad (a \leq x \leq b).$$

Since L is positive, $L(\varphi_n - \varphi_{n+1}) \leq 0$, and hence $\lim L(\varphi_n)$ exists by the linearity of L. Since $\varphi_n(t) \leq \|f\| < \infty$ for all t, $L(\varphi_n) \leq L(\|f\|)$ (L applied to the constant function $\|f\|$), and thus $\lim L(\varphi_n) < \infty$. We would like to define $\tilde{L}(f)$ to be this limit; as usual, we must ensure that the value of the limit is independent of the particular approximating sequence $\{\varphi_n\}$ we happen to select. The proof is basically the same as in 2.1.4.

Let f_1, $f_2 \in \mathcal{B}[a,b]$ and suppose that $f_1(t) \leq f_2(t)$ for all t. Select nondecreasing sequences $\{\varphi_n\}$ and $\{\psi_n\}$ in $C[a,b]$ so that

$$\lim \varphi_n(t) = f_1(t) \quad \text{and} \quad \lim \psi_n(t) = f_2(t)$$

for all t. We will show that

$$\lim L(\varphi_n) \ \leq \ \lim L(\psi_n). \tag{1}$$

Fix n and for each k, set

$$\xi_k(t) \ = \ [\varphi_n(t) - \psi_k(t)]^+.$$

Now $\{\xi_k\}$ is a nonincreasing sequence which converges pointwise to zero. By Dini's Theorem, $\{\xi_k\}$ converges uniformly to zero, and hence $\lim L(\xi_k) = 0$. Since

$$\varphi_n(t) - \psi_k(t) \ \leq \ [\varphi_n(t) - \psi_k(t)]^+$$

it follows from positivity of L that

$$\lim_{k \to \infty} L(\varphi_n(t) - \psi_k(t)) \ \leq \ 0$$

and hence that

$$L(\varphi_n) \leq \lim_{k \to \infty} L(\psi_k).$$

Letting $n \to \infty$ establishes (1).

In view of (1) we may now define for $f \in \mathcal{B}[a,b]$

$$\tilde{L}(f) \ = \ \lim L(\varphi_n),$$

where $\{\varphi_n\}$ is any nondecreasing sequence in $C[a,b]$ satisfying $\lim \varphi_n = f$. Note that (1) also implies that, if $f, g \in \mathcal{B}[a,b]$ and $f \leq g$, then $L(f) \leq L(g)$. Since the limit process is linear

$$\tilde{L}(f+g) = \tilde{L}(f) + \tilde{L}(g) \quad \text{and}$$
$$\tilde{L}(\mu f) = \mu \tilde{L}(f)$$

for all $f, g \in \mathcal{B}[a,b]$ and all $\mu \geq 0$. Moreover,

$$0 = L(0) = \tilde{L}(0) = \tilde{L}(f + (-f)) = \tilde{L}(f) + \tilde{L}(-f),$$

and so if $\mu < 0$ and $f \in \mathcal{B}[a,b]$,

$$\tilde{L}(\mu f) = \tilde{L}((-\mu)(-f)) = (-\mu)(-\tilde{L}(f)),$$

showing that $\tilde{L}$ is linear.

To see that $\tilde{L}$ is bounded, denote by u the characteristic function of $[a,b]$; since, for all $f \in \mathcal{B}[a,b]$,

$$-\|f\|u(t) \leq f(t) \leq \|f\|u(t),$$

it follows from positivity of $\tilde{L}$ that

$$-\|f\|\tilde{L}(u) \leq \tilde{L}(f) \leq \|f\|\tilde{L}(u)$$

and hence $\|\tilde{L}\| \le |L(u)|$.

All that remains is to find a nondecreasing function α solving the integral equation

$$L(f) = \int_a^b f \, d\alpha$$

for all $f \in \mathcal{C}[a, b]$. To find α, set $e_a(t) = 0$ and for $s \in (a, b]$, set

$$e_s(t) = \chi_{[a,s]}(t).$$

Define $\alpha(s) = \tilde{L}(e_s)$, so that α is a nonnegative, nondecreasing function defined on $[a, b]$.

Now suppose without loss that $L \ne 0$ and fix $\epsilon > 0$ and $f \in \mathcal{C}[a, b]$. Select $\delta > 0$ so small that

$$|x - y| \le \delta \text{ implies that } |f(x) - f(y)| < \max\left\{ \frac{\epsilon}{2\|L\|}, \frac{\epsilon}{2V_\alpha(a, b)} \right\}. \tag{2}$$

Next select points $a = x_0 < x_1 < \cdots < x_n = b$ so that α is continuous at each x_i, $i = 1, \ldots, n - 1$, $|x_i - x_{i+1}| \le \delta$ $i = 0, \ldots n - 1$, and select any ξ_i between x_i and x_{i+1}. Then

$$\left| L(f) - \int_a^b f \, d\alpha \right| \le |L(f) - \sum_{k=0}^{n-1} f(\xi_k)\big(\alpha(x_{k+1}) - \alpha(x_k)\big)|$$

$$+ |\sum_{k=0}^{n-1} f(\xi_k)\big(\alpha(x_{k+1}) - \alpha(x_k)\big) - \int_a^b f \, d\alpha|$$

$$\le |\tilde{L}\big(f - \sum_{k=0}^{n-1} f(\xi_k)(e_{x_{k+1}} - e_{x_k})\big)| + \epsilon/2. \tag{3}$$

Now for each $s \in [a, b]$

$$\sum_{k=0}^{n-1} \big(e_{x_{k+1}}(s) - e_{x_k}(s)\big) = e_{x_n}(s) = 1$$

and hence

$$|f(s) - \sum_{k=0}^{n-1} f(\xi_k)\big(e_{x_{k+1}}(s) - e_{x_k}(s)\big)|$$

$$\le \sum_{k=0}^{n-1} |f(s) - f(\xi_k)| \, |e_{x_{k+1}}(s) - e_{x_k}(s)|$$

$$\le \epsilon/(2\|L\|).$$

Substituting this estimate into (3) gives

$$|L(f) - \int_a^b f \, d\alpha| \le \epsilon/2 + \epsilon/2 = \epsilon.$$

Since ϵ was arbitrary, this shows that

$$L(f) = \int_a^b f \, d\alpha.$$

6.3.5. Riesz Representation Theorem for $C[a,b]$. *Let L be a bounded linear functional defined on $C[a,b]$. Then there is an $\alpha \in BV[a,b]$ such that*

$$L(f) = \int_a^b f \, d\alpha \qquad (4)$$

for all $f \in C[a,b]$. If α_1 and α_2 are two normalized functions in $BV[a,b]$ satisfying (4), then $\alpha_1 = \alpha_2$.

Proof. The existence of α is immediate from 6.3.2 and 6.3.5. In general, α need not be unique (α plus any constant works equally well to represent L). We can show, however, that α is unique up to normalization.

Let α_1 and α_2 be two normalized functions satisfying (4) and let s be a point of continuity of α_1 and α_2. Then

$$\alpha_1(s) - \alpha_1(a) = \int_a^b \chi_{[a,s]} \, d\alpha_1 = \tilde{L}(\chi_{[a,s]})$$

$$= \int_a^b \chi_{[a,s]} \, d\alpha_2$$

$$= \alpha_2(s) - \alpha_2(a)$$

and so $\alpha_1(s) = \alpha_2(s) + \alpha_1(a) - \alpha_2(a)$ for all points s where both α_1 and α_2 are continuous. This implies, since the points of discontinuity are countable, that $\alpha_1(s) = \alpha_2(s) + \alpha_1(a) - \alpha_2(a)$ on a dense subset of the interval. From this $\alpha_1(s^+) = \alpha_2(s^+) + \alpha_1(a) - \alpha_2(a)$ and $\alpha_1(s^-) = \alpha_2(s^-) + \alpha_1(a) - \alpha_2(a)$. In particular

$$0 = \alpha_1(a^+) = \alpha_2(a^+) + \alpha_1(a) - \alpha_2(a)$$

implying that $\alpha_1(a) = \alpha_2(a)$. This implies that $\alpha_1(s^+) = \alpha_2(s^+)$ and $\alpha_1(s^-) = \alpha_2(s^-)$ for all s; since α_1 and α_2 are normalized, this shows that α_1 equals α_2.

PROBLEMS

6.17. Recall that the smallest σ-algebra containing all of the open sets consists of the *Borel sets*; a function f is *Borel measurable* if $f^{-1}(U)$ is a Borel set whenever U

is an open set. Show that a function f is Borel measurable if and only if there is a nondecreasing bounded sequence $\{f_n\}$ of continuous functions with the property that

$$\lim f_n(x) = f(x) \quad \text{for all} \quad x.$$

6.18. Let $\alpha \in BV[a, b]$ and let U be a (relatively) open subset of $[a, b]$. Show that $\int_a^b \chi_U \, d\alpha$ is well-defined.

6.19. *Measures induced by nondecreasing functions – see Problem 2.21.*

Definition. Let α be a nondecreasing function defined on $[a, b]$ and let $A \subseteq [a, b]$. We define the *outer α-measure of A* to be

$$\mu_\alpha^*(A) = \inf \left\{ \int_a^b \chi_U \, d\alpha : U \text{ is an open set and } A \subseteq U \right\}$$

where the value of the integral is given by Problem 6.18. We will say that E is α-measurable if

$$\mu_\alpha^*(A) = \mu_\alpha^*(A \cap E) + \mu_\alpha^*(A \cup E^C)$$

for all sets $A \subseteq [a, b]$. In this case, we define the *α-measure of E* to be

$$\mu_\alpha(E) = \mu_\alpha^*(E).$$

(a) Show that if G is a Borel set then G is α-measurable and that

$$\mu_\alpha^*(G) = \int_a^b \chi_G \, d\alpha.$$

(b) If $a \leq x \leq b$, show that $\{x\}$ is α-measurable and that $\mu_\alpha(\{x\}) = 0$ if and only if α is continuous at x.

(c) Show that if A is α-measurable, then so is A^C.

(d) Suppose that $\{A_n\}$ is a family of disjoint α-measurable subsets of $[a, b]$. Show that

$$\mu_\alpha \left(\bigcup A_n \right) = \sum \mu_\alpha(A_n).$$

(e) Show that α is absolutely continuous if and only if

$$\lambda(E) = 0 \quad \text{implies that } E \text{ is } \alpha\text{-measurable and} \quad \mu_\alpha(E) = 0.$$

(f) Let $a < b$; construct a set $B \subseteq [a, b]$ with the property that B fails to be α-measurable *for all* continuous, nondecreasing functions α.

6.20. Let $a < b$, let $\{x_n\} \subseteq [a, b]$, and let $\{a_n\}$ be a sequence of nonnegative real numbers for which

$$\sum a_n < \infty.$$

Show that there is a nondecreasing function α such that

$$\int_a^b f \, d\alpha = \sum a_n f(x_n)$$

for all $f \in C[a, b]$.

6.21. Show that the decomposition in 6.3.2 need not be unique.

Chapter 7

Integration over Abstract Spaces

7.1. ABSTRACT MEASURE SPACES. In the preceding chapters we have studied a variety of integrals and deduced many similar properties for them. The purpose of the current chapter is to try to unify these seemingly disparate integrals (including sums as "integrals" in ℓ_1) into a single coherent theory. Abstract measure spaces, introduced in this section, are one of the basic concepts of this theory; they turn out to be a special subclass of vector lattices, introduced in Section 7.4, which are the setting for the Daniell Integral, the unified theory which we will develop in this chapter.

7.1.1. Definition. Let X be an abstract set and let M be a collection of subsets of X; then M is a *σ-algebra* if
   (a) $X \in M$;
   (b) $A \in M$ implies $(X \backslash A) \in M$; and
   (c) If $\{A_n\}$ is a countable subfamily of M then $\cup A_n \in M$.
The elements of M are called *measurable sets* and the pair (X, M) is a *measurable space*.

7.1.2. Definition. Let X be a set, let M be a σ-algebra of subsets of X, and let μ be an (extended) real-valued function defined on M; then μ is a *measure* if
   (a) $\mu(A) \geq 0$ for all $A \in M$;
   (b) $\mu(\emptyset) = 0$; and
   (c) if $\{A_n\}$ is a disjoint countable family of measurable sets, then

$$\mu(\cup A_n) = \sum \mu(A_n).$$

We refer to the triple (X, M, μ) as a *measure space*. If $\mu(X) < \infty$, then we say that μ is a *finite measure*.

7.1.3. Examples. (i) As we saw in Section 2.4, if $X = \Re$, M is the class of Lebesgue measurable sets, and μ is Lebesgue measure, then (X, M, μ) is a measure space.

   (ii) Let X be an interval $[a, b] \subset \Re$ and let $\alpha \in \mathrm{BV}[a, b]$. Denote by M the Borel subsets of $[a, b]$ and, for $A \in M$, set

$$\mu_\alpha(A) = \int_A 1 \, d\alpha$$

(applying 6.3.4 to define μ_α). Then $([a, b], M, \mu_\alpha)$ is a measure space. [Only 7.1.2

(c) is not obvious, but can be deduced—see Problem 7.1].

(iii) Let X be any set and let $M = P(X) = \{A : A \subset X\}$. For $A \in M$, define

$$\nu(a) = \begin{cases} \infty & \text{if } A \text{ is infinite} \\ \text{number of elements in A} & \text{if } A \text{ is finite.} \end{cases}$$

Then $(X, P(X), \nu)$ is a measure space; ν is called the *counting measure*.

(v) Fix $f \in \mathcal{L}_1(\Re)$; for a Lebesgue measurable set E, take

$$\mu_f(E) = \int_E f \, d\lambda.$$

Then $(\Re, M, \mu_f)$ is a measure space.

Many of the properties of Lebesgue measure λ which we deduced in Section 2.4 carry over to abstract measures in general. In the next proposition we summarize some of these which are more or less immediate from the definitions.

7.1.4. Proposition. *Let (X, M, μ) be a measure space. Then:*
(a) *If $\{A_n\}$ is a countable subfamily of M, then $(\cap A_n) \in M$.*
(b) *If $\{A_1, \ldots, A_n\}$ is a finite family of disjoint measurable sets, then* $\mu(\cup_1^n A_j) = \sum_1^n \mu(A_j)$.
(c) *If A and B are measurable sets and $A \subset B$, then $\mu(A) \leq \mu(B)$.*
(d) *If A and B are measurable, then $\mu(A \cup B) = \mu(A) + \mu(B) - \mu(A \cap B)$.*

Proof. Part (a) is immediate from DeMorgan's laws; part (b) follows from 7.1.2 (c) and (b), upon defining $A_k = \emptyset$ for $k \geq n + 1$. For (c), observe that A and $(B \backslash A) = B \cap (X \backslash A)$ are disjoint measurable sets. Thus by part (b),

$$\mu(B) = \mu(A) + \mu(B \backslash A) \geq \mu(A)$$

by 7.1.2 (a). For (d), first observe that

$$\mu(B) = \mu(B \cap A) + \mu(B \backslash A)$$

and that

$$\mu(A \cup B) = \mu(A) + \mu(B \backslash A).$$

Combining these identities gives (d).

The "continuity" property of measures embodied in the following is crucial:

7.1.5. Theorem. *Let (X, M, μ) be a measure space and let $\{A_n\}$ be a sequence of measurable sets. Suppose that $\mu(A_1) < \infty$ and, for each n, that $A_{n+1} \subset A_n$. Then*

$$\mu(\cap A_n) = \lim \mu(A_n).$$

Proof. Set $B_0 = \cap_{n=1}^\infty A_n$ and set $B_n = A_n \backslash A_{n+1}$. Now

$$\cup_{n=0}^\infty B_n = A_1$$

and $B_i \cap B_k = \emptyset$ if $i \neq k$. Thus

$$\mu(A_1) = \sum_0^\infty \mu(B_n).$$

But $A_n = A_{n+1} \cup B_n$ and so, since A_{n+1} and B_n are disjoint,

$$\mu(A_n) = \mu(A_{n+1}) + \mu(B_n).$$

Since $\mu(A_{n+1}) \leq \mu(A_1) < \infty$, it follows that if $n \geq 1$

$$\mu(B_n) = \mu(A_n) - \mu(A_{n+1}).$$

Thus

$$\mu(A_1) = \sum_0^\infty \mu(B_n)$$

$$= \mu(B_0) + \sum_1^\infty \mu(A_n) - \mu(A_{n+1})$$

$$= \mu(B_0) + \lim_{m \to \infty} \sum_1^m \mu(A_n) - \mu(A_{n+1})$$

$$= \mu(B_0) + \lim_{m \to \infty} \mu(A_1) - \mu(A_{m+1})$$

$$= \mu(B_0) + \mu(A_1) - \lim_{m \to \infty} \mu(A_n).$$

Rearranging gives the conclusions.

We have already seen that the following property for Lebesgue measure was useful; it is likewise useful for measures in general.

7.1.6. Proposition. *Let $(X, \mathcal{M}, \mu)$ be a measure space and let $\{E_n\}$ be a countable family of measurable sets. Then $\mu(\cup_1^\infty E_n) \leq \sum_1^\infty \mu(E_n)$.*

Proof. Take $\{A_n\}$ to be the family of disjoint measurable sets given by $A_1 = E_1$ and

$$A_n = E_n \backslash (\cup_1^{n-1} E_k) \quad (k > 1).$$

Then

$$\mu(\cup_1^\infty E_n) \;=\; \mu(\cup_1^\infty A_n) \;=\; \sum_1^\infty \mu(A_n) \le \sum_1^\infty \mu(E_n).$$

Measure spaces can be classified in a variety of ways; we list in the next definition some of the various classes which will play a part in the sequel.

7.1.7. Definition. Let (X, M, μ) be a measure space. Then we say:

(i) μ is *finite* if $\mu(X) < \infty$.

(ii) μ is *σ-finite* if there is a countable family $\{E_n\}$ of measurable sets for which

$$X \;=\; \cup_1^\infty E_n \quad \text{and} \quad \mu(E_n) < \infty \quad \text{for each } \; n.$$

(iii) (X, M, μ) is *complete* if whenever $B \in M$, $\mu(B) = 0$ and $A \subseteq B$, then $A \in M$, i.e., every subset of a set of measure zero is measurable.

Note that Lebesgue measure on $\Re$ is both complete and σ-finite, whereas Lebesgue measure on the Borel sets fails to be complete.

PROBLEMS

7.1 Verify that 7.1.3(iii) is a measure space.

7.2. Let M be a σ-algebra of subsets from some set X. Show that M cannot be countably infinite (i.e., if M is infinite, then M necessarily has an uncountable subset).

7.3 Let (X, M, μ) be a measure space. Show that there is a σ-algebra $M' \supseteq M$ and a measure μ' defined on M' such that
(a) (X, M', μ') is complete;
(b) $\mu'(E) = \mu(E)$ for all $E \in M$;
(c) If $E \in M'$, then there are sets $A, B \in M$ and $C \in M'$ such that $E \subseteq A \cup C$, $C \subseteq B$, and $\mu(B) = 0$.

7.4. Let X be a set, let M be a σ-algebra of subsets of X, and let μ_1 and μ_2 be measures defined on M. Set $\nu = \mu_1 + \mu_2$, and show that (X, M, ν) is a measure space.

7.5. Let X be any uncountable set. Let

$$M = \{E \subseteq X : \quad \text{either } E \text{ is uncountable or } E^c \text{ is uncountable } \}.$$

(a) Show that M is a σ-algebra.
(b) For $E \in M$, set

$$\mu(E) = \begin{cases} 0 & \text{if } E \text{ is countable} \\ \infty & \text{if } E^c \text{ is countable.} \end{cases}$$

Show that (X, M, μ) is a measure space.

7.6. Let $\{E_n\}$ be a sequence of disjoint measurable sets. Show that

$$\mu(\cup E_n) = \lim \mu(\cup_{k=1}^n E_k).$$

7.7.   Let (X, M) be a measurable space and let μ_1 and μ_2 be measures defined on M; suppose that $\mu_1 \geq \mu_2$.

(a) Show that there is a measure μ_3 so that $\mu_2 = \mu_1 + \mu_3$.

(b) If μ_1 is σ-finite, show that the measure μ_3 in part (a) is unique.

(c) Show that, in general, μ_3 need not be unique.

7.2.   Measurable Functions and Integration.   Given an abstract measure space, we can define measurable functions in the conventional manner. As much of our development in sections 2.4—2.6 is independent of Lebesgue measure on $\Re$, we can define integration as in 2.6.2. Using 7.1.5, we can then deduce a special case of the Dominated Convergence Theorem (7.2.9) which is central to our unified theory.

7.2.1.   Definition.   Let (X, M, μ) be a measure space and let f be an extended real-valued function defined on X; we say that f is *measurable* if, for every open set $U \subset \Re$,

$$f^{-1}(U) \in M.$$

The analogs of 2.4.2—2.4.4 are now easily verified.

7.2.2.   Proposition.   *Let (X, M, μ) be a measure space and let f be an extended real-valued function defined on X; then f is measurable if and only if, for each $r \in \Re$, any (and hence all) of the following sets are measurable:*

(a) $\{x : g(x) < r\}$ *(b)* $\{x : g(x) \leq r\}$
(c) $\{x : g(x) > r\}$ *(d)* $\{x : g(x) \geq r\}$

7.2.3.   Proposition.   *If f and g are measurable functions defined on the measure space (X, M, μ) and if c is any real number, then each of the following functions are measurable:*

$$cf, \; f + g, \; f \vee g, \; f \wedge g, \; fg.$$

7.2.4.   Proposition.   *Let $\{f_n\}$ be a sequence of extended real-valued functions defined on the measure space (X, M, μ). Then each of the following functions are measurable:*

(a) $\sup f_n$ *(b)* $\inf f_n$
(c) $\overline{\lim} f_n$ *(d)* $\underline{\lim} f_n$.

If, in addition, $\lim f_n$ exists everywhere, then $\lim f_n$ is also measurable.

7.2.5.   Definition.   Let (X, M, μ) be a measure space, let $\{\alpha_1, \ldots, \alpha_n\}$ be real numbers and let $\{E_n, \ldots, E_n\}$ be measurable sets. A **simple function** is a function σ of the form

$$\sigma = \sum_1^n \alpha_i \chi_{E_i}.$$

(In contrast to 2.4.10, we now allow the possibility that $\mu(E_i) = \infty$.)

The following proposition is fundamental; as the proof is (essentially) identical with 2.4.11, it is omitted.

7.2.6. Proposition. *Let f be a non-negative measurable function defined on a measure space $(X, \mathcal{M}, \mu)$. Then there is a non-decreasing sequence $\{\sigma_n\}$ of simple functions such that $f = \lim \sigma_n$. If X is σ-finite then we may in addition choose sets $E_n \in \mathcal{M}$ so that $\sigma_n = 0$ on $X \backslash E_n$.*

We now define integration as in 2.6.2.

7.2.7. Definition. Let $(X, \mathcal{M}, \mu)$ be a measure space and let σ be a non-negative simple function defined on X; then the integral of σ is

$$\int_X \sigma \, d\mu = \sum_1^n c_i \mu(E_i)$$

where $\sigma = \sum_1^n c_i \chi_{E_i}$.

It is clear that this definition of $\int_X \sigma \, d\mu$ is independent of the representation of σ. Also if c is any non-negative real number and if σ_1 and σ_2 are any non-negative simple functions then

$$\int_X c\sigma_1 + \sigma_2 \, d\mu = c \int_X \sigma_1 \, d\mu + \int_X \sigma_2 \, d\mu, \quad \text{and}$$

$$\sigma_1 \leq \sigma_2 \text{ implies } \int_X \sigma_1 \, d\mu \leq \int_X \sigma_2 \, d\mu. \tag{1}$$

Finally observe that if $\int_X \sigma \, d\mu < \infty$, then there is a measurable set E and a number M such that:

$$\sigma \equiv 0 \text{ on } X \backslash E;$$
$$\mu(E) < \infty; \quad \text{and}$$
$$\sigma \leq M.$$

7.2.8. Definition. Let $(X, \mathcal{M}, \mu)$ be a measure space and let f be a non-negative measurable function defined on X. Then we define the integral of f to be

$$\int_X f \, d\mu \; = \; \sup\{\int_X \sigma \, d\mu : \sigma \text{ is a non-negative}$$

simple function and $\sigma \leq f\}$.

From this definition, it is clear that if f and g are non-negative measurable functions and α is a non-negative scalar, then

(i) $\int_X \alpha f \, d\mu \; = \; \alpha \int_X f \, d\mu$;

(ii) $\int_X f + g \, d\mu \; \geq \; \int_X f \, d\mu + \int_X g \, d\mu$; and

(iii) if $f \leq g$ then $\int_X f \, d\mu \; \leq \; \int_X g \, d\mu$.

While equality **does** hold in (ii), this is by no means obvious. Before establishing equality in (ii), we verify the following lemma; note the striking similarity with Dini's Theorem.

7.2.9. Lemma. *Let (X, M, μ) be a measure space and let $\{f_n\}$ be a non-increasing sequence of non-negative measurable functions defined on X. Suppose*

(a) $\lim f_n(x) = 0$ *for each x; and*

(b) $\int_X f_1 \, d\mu < \infty$.

Then $\lim \int_X f_n \, d\mu = 0$.

Proof. Set $g_n = f_1 - f_n$ and let φ be any non-negative simple function satisfying $0 \leq \varphi \leq f_1$. It suffices to show

$$\lim \int_X g_n \, d\mu \; \geq \; \int_X \varphi \, d\mu, \tag{2}$$

for then

$$\lim \int_X f_1 \, d\mu - \int_X f_n \, d\mu \; \geq \; \lim \int_X g_n \, d\mu \; \geq \; \int_X \varphi \, d\mu$$

and the conclusion follows upon taking the supremum over all $\varphi \leq f_1$.

To establish (2), observe that since $\varphi \leq f_1$ and $\int_X f_1 \, d\mu < \infty$, it follows that there is a measurable set E and a real number M such that

$$\varphi \equiv 0 \text{ on } X \backslash E;$$

$$0 \leq \varphi \leq M; \text{ and}$$

$$\mu(E) < \infty.$$

Now fix $\epsilon > 0$ and for each n set

$$E_n \; = \; \{x \in E : g_k(x) > (1 - \epsilon)\varphi(x) \text{ for all } k \geq n\}.$$

Then each E_n is measurable and, since $\lim g_n(x) = f_1(x) \geq \varphi(x)$, it follows that $\cup_1^\infty E_n = E$. As $E_{n+1} \supset E_n$ for all n, Theorem 29.5 implies

$$0 \; = \; \mu(\emptyset) \; = \; \lim \mu(E \backslash E_n).$$

Consequently

$$\lim \int_X g_n \, d\mu \geq \lim \int_{E_n} g_n \, d\mu$$

$$\geq \lim(1 - \epsilon) \int_{E_n} \varphi \, d\mu$$

$$= \lim(1 - \epsilon)\left(\int_E \varphi \, d\mu - \int_{E \backslash E_n} \varphi \, d\mu \right)$$

$$\geq \lim(1 - \epsilon) \int_X \varphi \, d\mu - M\mu(E \backslash E_n)$$

$$= (1 - \epsilon) \int_X \varphi \, d\mu.$$

Since $\epsilon > 0$ was arbitrary, this establishes (2), and hence the Lemma.

7.2.10. Corollary. *Let f be a non-negative measurable function and let $\{\sigma_n\}$ be a non-decreasing sequence of non-negative simple functions such that $\lim \sigma_n = f$. Then*

$$\int_X f \, d\mu = \lim \int_X \sigma_n \, d\mu.$$

Proof. Clearly $\overline{\lim} \int_X \sigma_n \, d\mu \leq \int_X f \, d\mu$ so we need only establish $\underline{\lim} \int_X \sigma_n \, d\mu \geq \int_X f \, d\mu$.

Let φ be any simple function satisfying $\varphi \leq f$; it suffices to show $\int_X \varphi \, d\mu \leq \underline{\lim} \int_X \sigma_n \, d\mu$. To do this we distinguish two cases.

I. $\int_X \varphi \, d\mu = \infty$.

In this case there is a measurable set E and a scalar $\xi > 0$ such that $\xi \chi_E \leq \varphi$ and $\mu(E) = +\infty$. If we then set

$$E_n = \{x \in E : \sigma_n \geq \xi/2\}$$

it follows that $\cup_1^\infty E_n = E$ and $\lim \mu(E_n) = +\infty$. But

$$\int_X \varphi \, d\mu \geq \int_{E_n} \sigma_n \, d\mu \geq \xi \mu(E_n)/2$$

and thus $\underline{\lim} \int_X \sigma_n \, d\mu = +\infty = \int_X \varphi \, d\mu$.

II. $\int_X \varphi \, d\mu < \infty$.

In this case, take

$$\psi_n = \sigma_n \wedge \varphi \quad \text{and} \quad \tilde{\psi}_n = \varphi - \sigma_n \wedge \varphi$$

so that each of ψ_n and $\tilde{\psi}$ are non-negative simple functions. Also, since $\{\sigma_n\}$ converges to f, it follows that $\{\tilde{\psi}_n\}$ is a non-increasing sequence of non-negative simple functions converging to zero. Then by 7.2.8

$$\int_X \varphi \, d\mu = \int_X \psi_n + \tilde{\psi}_n \, d\mu = \int_X \tilde{\psi}_n \, d\mu + \int_X \psi_n \, d\mu$$

$$\leq \varliminf \int_X \tilde{\psi}_n \, d\mu + \int_X \sigma_n \, d\mu$$

$$= \varliminf \int_X \sigma_n \, d\mu$$

and the corollary follows.

7.2.11. Corollary. *Let f and g be non-negative measurable functions. Then $\int_X f + g \, d\mu = \int_X f \, d\mu + \int_X g \, d\mu$.*

Proof. Select non-decreasing sequences $\{\varphi_n\}$ and $\{\psi_n\}$ of simple functions such that

$$f = \lim \varphi_n \quad \text{and} \quad g = \lim \psi_n$$

Then applying 7.2.10,

$$\int_X f + g \, d\mu = \lim \int_X \varphi_n + \psi_n \, d\mu$$

$$= \lim \int_X \varphi_n \, d\mu + \int_X \psi_n \, d\mu$$

$$= \int_X f \, d\mu + \int_X g \, d\mu.$$

We are now in a position to readily verify the Monotone Convergence Theorem.

7.2.12. Corollary (Monotone Convergence Theorem). *Let $\{f_n\}$ be a non-decreasing sequence of non-negative measurable functions and let $f = \lim f_n$. Then*

$$\int_X f \, d\mu = \lim \int_X f_n \, d\mu.$$

Proof. First suppose $\int_X f \, d\mu < \infty$. Then if $g_n = f - f_n$, it follows that $\{g_n\}$ is a non-increasing sequence of non-negative functions with $\lim g_n = 0$. Thus by 7.2.9,

$$\int_X f \, d\mu \;=\; \lim \int_X g_n + f_n \, d\mu$$

$$\leq \; \underline{\lim} \int_X f_n \, d\mu.$$

As $f_n \leq f$ for all n, it follows that

$$\overline{\lim} \int_X f_n \, d\mu \;\leq\; \int_X f \, d\mu \;\leq\; \underline{\lim} \int_X f_n \, d\mu,$$

and so the conclusion follows.

For the general case, let φ be a non-negative simple function satisfying $\varphi \leq f$. It suffices to argue that

$$\int_X \varphi \, d\mu \;\leq\; \underline{\lim} \int_X f_n \, d\mu.$$

If $\int_X \varphi \, d\mu < \infty$, then

$$\int_X \varphi \, d\mu \;=\; \lim \int_X \varphi \wedge f_n \, d\mu \;\leq\; \underline{\lim} \int_X f_n \, d\mu$$

and the inequality follows.

If $\int_X \varphi \, d\mu = +\infty$, then there is a scalar $\lambda > 0$ and a measurable set E satisfying $\mu(E) = +\infty$ and $\varphi \geq \lambda \chi_E$. Proceeding exactly as with 7.2.10, we set

$$E_n \;=\; \{x : f_n(x) \geq \lambda\}$$

so $\lim \mu(E_n) = +\infty$ and

$$\lim \int_X f_n \, d\mu \;\geq\; \underline{\lim} \int_{E_n} f_n \, d\mu \;\geq\; \underline{\lim} \lambda \mu(E_n) \;=\; \int_X \varphi \, d\mu.$$

We can now define $\mathcal{L}_1$ as in Chapter II.

7.2.13. Definition. Let $(X, \mathcal{M}, \mu)$ be a measure space; $\mathcal{L}_1(X\mathcal{M}, \mu)$ is the class of all measurable functions f defined on X corresponding to which there exist non-negative measurable functions φ and ψ for which

(a) $f = \varphi - \psi$
(b) Both $\int_X \varphi \, d\mu < \infty$ and $\int_X \psi \, d\mu < \infty$.

For $f \in \mathcal{L}_1(X, \mathcal{M}, \mu)$ we define the integral of f to be

$$\int_X f \, d\mu \;=\; \int_X \varphi \, d\mu - \int_X \psi \, d\mu$$

where $f = \varphi - \psi$ is any decomposition for f satisfying (a) and (b).

It is readily verified that this definition of $\int_X f \, d\mu$ is independent of the decomposition. This, and the properties listed below, are all immediate from the corresponding properties for non-negative measurable functions.

7.2.14. Theorem. *Let $(X, \mathcal{M}, \mu)$ be a measure space, let $f, g \in \mathcal{L}_1(X, \mathcal{M}, \mu)$ and let c be a real number. Then*

(a) $c \int_X f \, d\mu = \int_X cf \, d\mu$
(b) $f \leq g$ *implies* $\int_X f \, d\mu \leq \int_X g \, d\mu$
(c) $\int_X f + g \, d\mu = \int_X f \, d\mu + \int_X g \, d\mu$.

It is possible to deduce Fatou's Lemma (and thus the other convergence theorems as well) directly from Lemma 7.2.9. If the measure space is complete, we can also replace convergence with convergence a.e. (if the space is not complete, limits of sequences which only converge a.e. may not be measurable). We leave this direct development to the exercises, and instead place the above ideas in a slightly more general setting before proceeding.

We will need to know that $\mathcal{L}_1(X, \mathcal{M}, \mu)$ is a vector space, under the convention that f and g are the "same" vector if they agree a.e. (as in 9.3 (v)). It is routine to check that, if $f, g \in \mathcal{L}_1(X, \mathcal{M}, \mu)$ and $f \leq g$ a.e. then

$$\int_X f \, d\mu \leq \int_X g \, d\mu.$$

7.2.15. Proposition. *If $f \in \mathcal{L}_1(X, \mathcal{M}, \mu)$, then the set*

$$E = \{x : |f(x)| = +\infty\}$$

has measure zero.

Proof. Without loss of generality, suppose $f \geq 0$ and set

$$E_n = \{x \in X : f(x) \geq n\},$$

so that $E = \cap E_n$. For each n

$$n \chi_{E_n} \leq f$$

and thus

$$n \mu(E_n) = \int_X n \chi_{E_n} \, d\mu \leq \int_X f \, d\mu.$$

Since $f \in \mathcal{L}_1(X, \mathcal{M}, \mu)$, this implies $\lim \mu(E_n) = 0$. Since $E_{n+1} \subset E_n$ for each n, it follows that $\mu(E) = \lim \mu(E_n)$.

PROBLEMS

7.9. Prove 7.2.2.

7.10. Prove 7.2.3.

7.11. Prove 7.2.6.

7.12. Deduce Fatou's Lemma from 7.2.12.

7.13. Deduce the Dominated Convergence Theorem.

7.14. Prove 7.2.14.

7.15. Let (X, M, μ) be a measure space and let $f \in \mathcal{L}_1(X, M, \mu)$ be a non-negative function. Define ν on M by

$$\nu(E) = \int_X f \cdot \chi_E \, d\mu.$$

(i) Show that (X, M, ν) is a measure space.

(ii) Let h be a non-negative measurable function. Show that

$$\int_X h \, d\nu = \int_X hf \, d\mu.$$

7.16. Let (X, M, μ) be a complete measure space. Show that we may replace "convergence" with "convergenc a.e." in 7.2.12.

7.3. DANIELL FUNCTIONALS. In this section we will reproduce the results of Sections 2.1 and 2.2 in a more abstract setting. The essential features of the Riemann integral and Dini's Theorem are abstracted to the context of vector lattices and Daniell functionals; in the next section, the equivalence between these ideas and the measure-theoretic ideas of the preceding two sections is established. We will then apply these general results in a variety of settings, including, for example, product measures and signed measures.

7.3.1. Definition. Let X be an abstract set; a family of functions $V \equiv V(X)$ mapping X to $\Re$ is a *vector lattice* if V is a vector space and if for each $f, g \in V$ the functions $f \bigwedge g$ and $f \bigvee g$ are again in V.

A function $f \in V$ is *positive* if $f(x) \geq 0$ for all x.

A linear functional T mapping V to the real numbers is *positive* if $T(f) \geq 0$ whenever f is positive.

A positive linear functional T is a *Daniell functional* if whenever $\{f_n\}$ is a nonincreasing sequence in V satisfying $\lim f_n(x) = 0$ for each x, then

$$\lim T(f_n) = 0.$$

7.3.2. Examples. (i) If $X = \Re$, $V = \mathcal{C}_C(\Re)$, and T is the Riemann integral, then by Dini's Theorem, T is a Daniell functional.

(ii) Let X be a compact metric space, let $V = \mathcal{C}(M)$, and let T be any continuous linear functional defined on $\mathcal{C}(M)$. By the Jordan Decomposition Theorem

(6.3.2) $T = P - N$, where P and N are positive linear functionals. Dini's Theorem again implies that each of P and N is a Daniell functional. We will eventually use this example to give a general representation theorem for $(C(M))^*$.

(iii) Let $(X, \mathcal{M}, \mu)$ be a measure space; let $V = \{f \in \mathcal{L}_1(X, \mathcal{M}, \mu) : |f(x)| < \infty$ for all $x\}$, and define T on V by

$$T(f) = \int_X f \, d\mu.$$

Then, via 7.2.9, T is a Daniell functional. Observe that if $g \in \mathcal{L}_1(X, \mathcal{M}, \mu)$, then there exists $f \in V$ such that $f = g$ a.e.; i.e., if we adopt the convention that two functions $f, g \in \mathcal{L}_1(X, \mathcal{M}, \mu)$ are the "same" if they agree a.e., then $\mathcal{L}_1(X, \mathcal{M}, \mu)$ is a vector lattice and integration is a Daniell functional.

(iv) Let $X = [a, b]$ and let $\alpha \in \text{BV}[a, b]$; if $V = C[a, b]$ and

$$T(f) = \int_a^b f \, d\alpha$$

Dini's Theorem again implies that T is a Daniell functional.

(v) If $X = \Re^2$, $V = C_C(\Re^2)$, and T is the Riemann integral, then as a special case of (ii), T is a Daniell functional. We will pursue this example in much greater detail in Section 7.5.

(vi) If $X = \mathcal{N}$, $\mathcal{M} = \mathcal{P}(\mathcal{N})$, μ is the counting measure, and T is defined by $T(x_n) = \sum_1^\infty x_n$ for vectors $(x_n) \in \ell_1$, then T is a Daniell functional as a special case of (iii).

We now see that the setting of vector lattices and Daniell functionals is general enough to subsume the many special cases we have considered so far. Within this context we wish to imitate, insofar as possible, our development of the Lebesgue integral in Chapter 2. Throughout the remainder of this section X will be an arbitrary set, V will be a vector lattice of functions defined from X to $\Re$, and T will be a Daniell functional defined on X.

7.3.3. Definition. The class V^+ is the collection of functions f defined from X to $[0, \infty]$ for which there is a nondecreasing sequence $\{\varphi_n\} \subset V$ satisfying

$$\lim \varphi_n(x) = f(x) \quad \text{for all} \quad x \in X.$$

7.3.4. Lemma. *Let $f, g \in V^+$ and suppose that $f \le g$. If $\{\varphi_n\}$ and $\{\psi_n\}$ are nondecreasing sequences of functions in V converging, respectively, to f and g, then*

$$\lim T(\varphi_n) \le \lim T(\psi_n).$$

Proof. Fix k and set $h_n = (\varphi_k - \psi_n) \vee 0$. Then $\{h_n\}$ is a nonincreasing sequence which converges to zero and hence $\lim T(h_n) = 0$. Since $\varphi_k - \psi_n \le h_n$, it follows that $\lim T(\varphi_k - \psi_n) \leq 0$, i.e., that

$$T(\varphi_k) \leq \lim T(\psi_n).$$

The result follows upon letting $k \to \infty$.

7.3.5. Definition. If $f \in V^+$, we define $T(f)$ to be $\lim T(\varphi_n)$ where $\{\varphi_n\}$ is any nondecreasing sequence in V converging to f.

In light of 7.3.4, this extension of T to V^+ is well-defined, and if $f \leq g$ in V^+, then $T(f) \leq T(g)$.

The next lemma is analogous to 2.2.1, the fundamental tool needed to deduce the convergence theorems of Section 2.2. The proof is basically unchanged.

7.3.6. Lemma. *If $\{f_n\} \subset V^+$, then $\sum_1^\infty f_n \in V^+$ and $T(\sum_1^\infty f_n) = \sum_1^\infty T(f_n)$.*

Proof. Select for each n a nondecreasing, nonnegative sequence $\{\varphi_{n,k}\}_{k=1}^\infty \subset V$ such that $\lim_{k\to\infty} \varphi_{n,k} = f_n$. Set $\psi_n(x) = \sum_{j=1}^n \varphi_{j,n}(x)$; note that $\{\psi_n\}$ is nondecreasing since

$$\psi_{n+1}(x) = \sum_{j=1}^{n+1} \varphi_{j,n+1}(x) \geq \sum_{j=1}^n \varphi_{j,n+1}(x) \geq \sum_{j=1}^n \varphi_{j,n}(x) = \psi_n(x)$$

and thus $\psi_0 = \lim \psi_n$ exists and is an element of V^+; it is evident that $\psi_n \leq \sum_1^n f_j$ for each n and thus that $\psi_0 \leq \sum_1^\infty f_j$ and $T(\psi_0) \leq \sum_1^\infty T(f_j)$.

We next show that $\psi_0 = \sum_1^\infty f_j$; let us distinguish two cases.

Case 1. $\sum_1^\infty f_j = \infty$.

Fix $A > 0$ and select n so that $\sum_1^n f_j(x) \geq 2A$. Select $M > 0$ such that $m \geq M$ implies that

$$|f_j(x) - \varphi_{j,m}(x)| \leq A/n.$$

Then $m \geq \max\{M, n\}$ implies that

$$\psi_m(x) = \sum_1^m \varphi_{j,m}(x) \geq \sum_1^n \varphi_{j,m}(x) \geq \sum_1^n f_j(x) - A/n \geq A;$$

since A was arbitrary this shows that $\psi_0(x) = \infty$.

Case 2. $\sum_1^\infty f_j < \infty$.

Fix $\epsilon > 0$ and select n so large that $\sum_1^\infty f_j(x) - \epsilon \leq \sum_1^n f_j(x)$. Select $M > 0$ so that $m \geq M$ implies $|f_j(x) - \varphi_{j,m}(x)| \leq \epsilon/n$. Then $m \geq \max\{M, n\}$ implies that

$$\psi_m(x) = \sum_1^m \varphi_{j,m}(x) \geq \sum_1^n \varphi_{j,m}(x)$$

$$\geq \sum_1^n f_j(x) - \epsilon/n \geq \sum_1^\infty f_j(x) - 2\epsilon;$$

since ϵ was arbitrary, this shows that $\psi_0(x) \geq \sum_1^\infty f_j(x)$; the reverse inequality was noted earlier.

Thus $\sum_1^\infty f_j \in V^+$ and $T(\sum_1^\infty f_j) \leq \sum_1^\infty T(f_j)$. The reverse inequality follows from $\sum_1^\infty f_j \geq \sum_1^n f_j$, and thus $T(\sum_1^\infty f_j) \geq \sum_1^\infty T(f_j)$.

Note that, as a special case, if $f, g \in V^+$, then $f + g \in V^+$ and $T(f) + T(g) = T(f + g)$.

To pursue the ideas of Section 2.1, we first need to develop the notion of "null sets"; this is analogous to the extension in Section 2.1 of the Riemann integral to functions which are continuous almost everywhere.

7.3.7. Definition. A set $E \subset X$ is a *null set* if for each $\epsilon > 0$ there is a function $g \in V^+$ for which $g \geq \chi_E$ and $T(g) \leq \epsilon$.

It is evident that if $A \subset E$ and E is a null set, then A is also a null set. If $f \in V^+$ and $f = 0$ except on a null set, then it should happen that $T(f) = 0$; this is the content of the next lemma.

7.3.8. Lemma. *Let h be a mapping from X to $[0, \infty]$ and suppose that $E = \{x \in X : h(x) > 0\}$ is a null set. Then for each $\epsilon > 0$ there is a $g \in V^+$ for which $g \geq h$ and $T(g) \leq \epsilon$.*

Proof. For each n, take

$$E_n = \{x \in X : n < h(x) \leq n+1\},$$

so each E_n is a null set. Now fix $\epsilon > 0$ and select $g_n \in V^+$ so that $g_n \geq \chi_{E_n}$ and $T(g_n) \leq \epsilon\big((n+1)2^{n+1}\big)^{-1}$. Set

$$g = \sum_0^\infty (n+1)g_n;$$

by 7.3.6, $g \in V^+$ and, by construction, $g \geq h$. Also, 7.3.6 implies that

$$T(g) = \sum_0^\infty (n+1)T(g_n) \leq \epsilon,$$

completing the proof.

7.3.9. Lemma. *Let f and g be elements of V^+ and suppose that $f \leq g$ except on a null set. Then $T(f) \leq T(g)$. In particular, if $f = g$ except on a null set, then $T(f) = T(g)$.*

Proof. Set $h = (f - g) \wedge 0$, so $h \geq 0$ and $\{x : h(x) > 0\}$ is a null set E. Fix $\epsilon > 0$ and select, via 7.3.8, a function $\varphi \in V^+$ for which $\varphi \geq h$ and $T(\varphi) \leq \epsilon$. Note that if $x \in E$, then $f(x) - g(x) = h(x) \leq \varphi(x)$, and so $f(x) \leq g(x) + \varphi(x)$ for $x \in E$. On the other hand, if $x \in E^C$, then $f(x) \leq g(x) \leq g(x) + \varphi(x)$, so, in either case, $f \leq g + \varphi$. But then

$$T(f) \leq T(g) + T(\varphi) \leq T(g) + \epsilon;$$

since ϵ was arbitrary, this shows that $T(f) \leq T(g)$.

7.3.10. Definition. We denote by $\mathcal{L}^+$ the class of all functions f for which there is a function $g \in V^+$ with $f = g$ except on a null set. If $f \in \mathcal{L}^+$, we define $T(f)$ by

$$T(f) = T(g) \quad \text{where} \quad g \in V^+ \quad \text{and} \quad f = g$$

except on a null set.

In view of 7.3.9, the definition of $T(f)$ is independent of our choice of the function $g \in V^+$; also, if $\varphi, \psi \in \mathcal{L}^+$ and if $\varphi \leq \psi$ except on a null set, then $T(\varphi) \leq T(\psi)$. Note that $g \in \mathcal{L}^+$ if and only if there is a nondecreasing sequence $\{g_n\} \subset V$ such that $\lim g_n = g$ except on a null set, exactly as in Section 2.1. These and other readily deduced properties of $\mathcal{L}^+$ are summarized below.

7.3.11. Theorem. *Let $f, g \in \mathcal{L}^+$ and let $\alpha > 0$ Then*
　(a) $f + g \in \mathcal{L}^+$ *and* $T(f + g) = T(f) + T(g)$;
　(b) $\alpha f \in \mathcal{L}^+$ *and* $T(\alpha f) = \alpha T(f)$;
　(c) $f \bigwedge g$ *and* $f \bigvee g \in \mathcal{L}^+$; *and*
　(d) $f \leq g$ *except on a null set implies that* $T(f) \leq T(g)$.
　　If, in addition, $g \bigwedge 1 \in V$ *for all* $g \in V$, *then*
　(e) $f \bigwedge \alpha \in \mathcal{L}^+$;
　(f) *if* $E = \{x \in X : f(x) > \alpha\}$, *then* $\chi_E \in \mathcal{L}^+$; *and*
　(g) *if* $T(f) < \infty$ *then,* $\{x \in X : f(x) = +\infty\}$ *is a null set.*

Proof. Parts (a)—(d) follow exactly as the analogous results in Section 2.1, and part (e) is trivial; only (f) and (g) present any difficulty.

For (f), select $\varphi \in V^+$ so that $f = \varphi$ except on a null set and select a nondecreasing sequence $\{\varphi_n\}$ of nonnegative functions in V so that $\lim \varphi_n = \varphi$. Next define

$$\psi_n = \alpha^{-1}\varphi_n - [(\alpha^{-1}\varphi_n) \bigwedge 1]$$
$$\text{and} \quad \xi_n = 1 \bigwedge (n\psi_n).$$

It is easily checked that $\{\psi_n\}$ is a nondecreasing sequence and, by assumption, each $\psi_n \in V$; hence $\{\xi_n\}$ is a nondecreasing sequence of functions in V.

Now set $F = \{x \in X : \varphi(x) > \alpha\}$. If $x \in F$, then for n sufficiently large $\psi_n(x) > 0$ and so $\lim \xi_n(x) = 1$. If $x \notin F$, then $\psi_n(x) = 0$ for all x and so $\lim \xi_n(x) = 0$; in particular, $\lim \xi_n = \chi_F$, showing that $\chi_F \in V^+$. Since $\varphi = f$

except on a null set, it follows at once that $\chi_F = \chi_E$ except on a null set, and thus $\chi_E \in \mathcal{L}^+$.

For (g), let φ be a function in V^+ such that $\varphi = f$ except on a null set. Set $\alpha = T(\varphi)$ and fix $\epsilon > 0$; let

$$E = \{x \in X : \varphi(x) > \alpha\epsilon^{-1}\};$$

so that $\chi_E \in \mathcal{L}^+$ by (f). By definition, $\alpha\epsilon^{-1}\chi_E \le \varphi$ and thus by (d)

$$\alpha\epsilon^{-1}T(\chi_E) \le T(\varphi),$$

i.e., $T(\chi_E) \le \epsilon$. Since $\{x : \varphi(x) = +\infty\} \subset E$ and $\epsilon > 0$ was arbitrary, this verifies (g) for functions in V^+ and hence also for functions in $\mathcal{L}^+$.

7.3.12. Definition. Denote by $\mathcal{L}_1(T)$ the class of all functions f for which there are functions $g, h \in \mathcal{L}^+$ satisfying

(i) $T(g) < \infty$ and $T(h) < \infty$.

(ii) $f = g - h$ except on a null set.

If $f \in \mathcal{L}_1(T)$, we define $T(f) = T(g) - T(h)$; exactly as with Definition 2.1.11, this value is independent of the particular decomposition of f into functions g and $h \in \mathcal{L}^+$. [Notice that implicit in (ii) is the assumption that $g < \infty$ and $h < \infty$ except on a null set.]

Imitating proofs from Section 2.1, we can show that $\mathcal{L}_1(T)$ is a vector lattice:

7.3.13. Theorem. *Let $f, g \in \mathcal{L}_1(T)$ and let $\alpha \in \Re$. Then*

(a) $f + g \in \mathcal{L}_1(T)$ and $T(f + g) = T(f) + T(g)$;

(b) $\alpha f \in \mathcal{L}_1(T)$ and $T(\alpha f) = \alpha T(f)$;

(c) $f \wedge g$ and $f \vee g \in \mathcal{L}_1$;

(d) $|f| \in \mathcal{L}_1(T)$;

(e) $f \le g$ except on a null set implies that $T(f) \le T(g)$.

The convergence theorems of Section 2.2 also carry over (indeed, much of the work is already done in 7.3.6). Before turning to these results, we establish a technical lemma.

7.3.14. Lemma. *Let $\{E_n\}$ be a family of null sets; then $E = \cup E_n$ is again a null set.*

Proof. Fix $\epsilon > 0$ and select, for each n, a function $g_n \in V^+$ such that $g_n \ge \chi_{E_n}$ and $T(g_n) \le \epsilon/2^n$. Setting $g = \sum g_n$, it is evident that $g \ge \chi_E$; moreover, 7.3.6 implies that $T(g) = \sum T(g_n) \le \epsilon$ and so E is a null set.

7.3.15. Theorem. *Let $\{f_n\}$ be a collection of functions in $\mathcal{L}^+$ and set $f = \sum_1^\infty f_n$; then $f \in \mathcal{L}^+$ and $T(f) = \sum T(f_n)$.*

Proof. We reduce this to 7.3.6: Select $g_n \in V^+$ such that the set $E_n = \{x : g_n(x) \neq f_n(x)\}$ is a null set. Then $\sum g_n = \sum f_n$ except on the null set $E = \cup E_n$. Thus $f \in \mathcal{L}^+$ and

$$T(f) = T\left(\sum g_n\right) = \sum T(g_n) = \sum T(f_n).$$

7.3.16. Lemma. *Let $f \in \mathcal{L}_1(T)$; then for each $\epsilon > 0$ there are functions g and $h \in V^+$ such that $f = g - h$ except on a null set and $T(h) \leq \epsilon$.*

Proof. Select φ and $\psi \in V^+$ such that both $T(\varphi) < \infty$ and $T(\psi) < \infty$ and such that $f = \varphi - \psi$ except on a null set. Select nondecreasing sequences $\{\varphi_n\}$ and $\{\psi_n\}$ in V so that $\lim \varphi_n = \varphi$ and $\lim \psi_n = \psi$. Now choose N so large that $T(\psi_N) \geq T(\psi) - \epsilon$. Since $\{\varphi_n - \psi_N\}_{n=1}^{\infty}$ and $\{\psi_n + \psi_N\}_{n=1}^{\infty}$ are nondecreasing sequences in V converging to the nonnegative functions $\varphi + \psi_N$ and $\psi - \psi_N$, respectively, each $\varphi + \psi_N$ and $\psi - \psi_N$ are in V^+; set $g = \varphi + \psi_N$ and $h = \psi - \psi_N$. Clearly, $f = g - h$ except on a null set and, moreover,

$$T(h) = \lim_n T(\psi_n - \psi_N) = T(\psi) - T(\psi_N) \leq \epsilon,$$

so g and h are the required decomposition.

We can now establish the analogue of the Monotone Convergence Theorem:

7.3.17. Theorem. *Let $\{f_n\}$ be a nondecreasing sequence of nonnegative functions in $\mathcal{L}_1(T)$ and set $f = \lim f_n$. Then $f \in \mathcal{L}_1(T)$ if and only if $\lim T(f_n) < \infty$; if $f \in \mathcal{L}_1(T)$, then $\lim T(f_n) = T(f)$.*

Proof. First suppose that $\lim T(f_n) < \infty$. Set $g_0 = f_0$ and for $n \geq 1$, set $g_n = f_n - f_{n-1} \geq 0$. Since each $g_n \in \mathcal{L}_1(T)$, we may select $\varphi_n, \psi_n \in \mathcal{L}^+$ such that $g_n = \varphi_n - \psi_n$ except on a null set E_n and $T(\psi_n) \leq 2^{-n}$. Now

$$\sum_0^N \varphi_n - \psi_n = \sum_0^N g_n = f_N,$$

except possibly on the null set $E = \cup_1^{\infty} E_n$. By 7.3.15, $\sum_0^{\infty} \varphi_n \in \mathcal{L}^+$ and $\sum_0^{\infty} \psi_n \in \mathcal{L}^+$ and, moreover,

$$T\left(\sum_0^{\infty} \psi_n\right) = \sum_0^{\infty} T(\psi_n) \leq 1 \quad \text{by construction.}$$

A second application of 7.3.15 yields

$$T\left(\sum_0^{\infty} \varphi_n\right) = \sum_0^{\infty} T(g_n) + \sum_0^{\infty} T(\psi_n);$$

since $\sum_0^\infty T(g_n) = \lim T(f_n)$ by construction of the sequence $\{g_n\}$, $T(\sum_0^\infty \varphi_n) < \infty$. Taking $g = \sum_0^\infty \varphi_n$ and $h = \sum_0^\infty \psi_n$ is thus the desired decomposition of f into functions from $\mathcal{L}^+$.

If, on the other hand, $\lim T(f_n) = \infty$, then $f \notin \mathcal{L}_1(T)$, for if $f \in \mathcal{L}_1(T)$ then $f \geq f_n$ for all n implies that $T(f) \geq \lim T(f_n)$.

Next in order is Fatou's Lemma.

7.3.18. Theorem. *Let $\{f_n\}$ be a sequence of nonnegative functions in $\mathcal{L}_1(T)$ and suppose that there is a function f such that $\lim f_n = f$. Then $f \in \mathcal{L}_1(T)$ if $\underline{\lim} T(f_n) < \infty$. If $f \in \mathcal{L}_1(T)$, then $T(f) \leq \underline{\lim} T(f_n)$.*

Proof. Fix k and set for $n \geq k$, $g_n = (f_k \wedge f_{k+1} \wedge \cdots \wedge f_n)$. Then $\{g_k - g_n\}$ is a nondecreasing sequence (in n) of nonnegative functions in $\mathcal{L}^+$ such that $\lim_n T(g_k - g_n) \leq T(g_k) < \infty$. Thus if $\varphi_k = \inf\{g_n : n \geq k\}$, then $g_k - \varphi_k \in \mathcal{L}_1$ and so φ_k is also in $\mathcal{L}_1$ by 7.3.17; moreover, $T(\varphi_k) \leq \inf\{T(f_n) : n \geq k\}$.

Now $\{\varphi_k\}$ is a nondecreasing sequence in $\mathcal{L}_1(T)$, converging to f; by assumption $\lim T(\varphi_k) \leq \underline{\lim} T(f_n) < \infty$, and so $f \in \mathcal{L}_1(T)$. Moreover,

$$T(f) = \lim T(\varphi_k) \leq \underline{\lim} T(f_n).$$

Note in the course of proving 7.3.18 we have established the following.

7.3.19. Corollary. *Let $\{f_n\}$ be a sequence of nonnegative functions in $\mathcal{L}_1$; then $\inf\{f_n\}$ is in $\mathcal{L}_1(T)$.*

The Dominated Convergence Theorem follows exactly as before; we omit the proof.

7.3.20. Theorem. *Let $\{f_n\}$ be a sequence in $\mathcal{L}_1(T)$ converging to some function f. Suppose that there is a function $g \in \mathcal{L}_1(T)$ such that $|f_n| \leq g$ for all n. Then $f \in \mathcal{L}_1(T)$ and $T(f) = \lim T(f_n)$.*

The special case when V arises from $\mathcal{L}_1(X, \mathcal{M}, \mu)$ merits some special attention. The following theorems summarizing this situation are fairly routine, and so we leave their proofs to the exercises.

7.3.21. Theorem. *Let $(X, \mathcal{M}, \mu)$ be a measure space and define a functional T on $\mathcal{L}_1(X, \mathcal{M}, \mu)$ by $T(f) = \int_X f \, d\mu$. Then*
(a) $\mathcal{L}_1(X, \mathcal{M}, \mu)$ is a vector lattice.
(b) T is a Daniell functional.
(c) $g \in \mathcal{L}_1(T)$ if and only if there is a function $f \in \mathcal{L}_1(X, \mathcal{M}, \mu)$ and a set

$E \in \mathcal{M}$ such that

$$\{x \in X : g(x) \neq f(x)\} \subset E \quad \text{and}$$
$$\mu(E) = 0.$$

(d) If μ is a complete measure, then $\mathcal{L}_1(T) = \mathcal{L}_1(X, \mathcal{M}, \mu)$.

7.3.22. Monotone Convergence Theorem. *Let $(X, \mathcal{M}, \mu)$ be a measure space and let $\{f_n\}$ be a nondecreasing sequence of nonnegative functions in $\mathcal{L}_1(X, \mathcal{M}, \mu)$ with limit f. If μ is complete, then $f \in \mathcal{L}_1(X, \mathcal{M}, \mu)$ if and only if $\lim \int_X f_n \, d\mu < \infty$. If $f \in \mathcal{L}_1(X, \mathcal{M}, \mu)$, then*

$$\lim \int_X f_n \, d\mu = \int f \, d\mu.$$

7.3.23. Theorem. *Let $\{f_n\}$ be a sequence of nonnegative functions in $\mathcal{L}_1(X, \mathcal{M}, \mu)$ and suppose there is a function f such that $\lim f_n = f$. If μ is complete, then $f \in \mathcal{L}_1(X, \mathcal{M}, \mu)$ if $\varliminf \int f_n \, d\mu < \infty$. If $f \in \mathcal{L}_1(X, \mathcal{M}, \mu)$, then*

$$\int_X f \, d\mu \leq \varliminf \int_X f_n \, d\mu.$$

7.3.24. Theorem. *Let $\{f_n\}$ be a sequence in $\mathcal{L}_1(X, \mathcal{M}, \mu)$ with limit f. If there is a function $g \in \mathcal{L}(X, \mathcal{M}, \mu)$ such that $|f_n| \leq g$ for all n, then $f \in \mathcal{L}_1(X, \mathcal{M}, \mu)$ and*

$$\int_X f \, d\mu = \lim \int_X f_n \, d\mu.$$

PROBLEMS

7.17. Prove 7.3.13.

7.18. Prove 7.3.19.

7.19. Prove 7.3.21.

7.20. Let $\alpha \in AC[a, b]$ and suppose that α is nondecreasing. Take $V = C[a, b]$ and define T by $T(f) = \int_a^b f \, d\alpha$. Show that $f \in \mathcal{L}_1(T)$ if and only if f is measurable (in the sense of Chapter 2) and $\int_a^b |f\alpha'| \, d\lambda < \infty$.

7.21. Let $\alpha \in BV[a, b]$ and suppose that α is nondecreasing; let V and T be as in Problem 7.20. Show that if f is a bounded, Borel-measurable function, then $f \in \mathcal{L}_1(T)$.

7.4. THE DANIELL INTEGRAL. Given a vector lattice $V(X)$ and a Daniell functional T defined on $V(X)$, the preceding section extended T to a larger class of functions, $\mathcal{L}_1(T)$, and deduced the standard convergence theorems 7.3.17 to 7.3.20 for this class. Because of example 7.3.2(iii) this procedure applies to $\mathcal{L}_1(X, \mathcal{M}, \mu)$ whenever $(X, \mathcal{M}, \mu)$ is a measure space, so that integration can be regarded as a special case of the Daniell functional. In this section, we will show that the reverse is also true provided that $(1 \wedge f) \in V$ for all $f \in V$.

In particular, given a vector lattice $V(X)$ and a Daniell functional T defined on $V(X)$, we shall construct a σ-algebra $\mathcal{M}$ over X, and a measure μ on $\mathcal{M}$ in such a way that $\mathcal{L}_1(X, \mathcal{M}, \mu) = \mathcal{L}_1(T)$ and so that, for all $f \in \mathcal{L}_1$, $T(f) = \int_X f \, d\mu$. For this reason T is referred to as a Daniell integral; the development follows closely Sections 2.3 and 2.4.

Throughout this section, X is once again an arbitrary set, V is a vector lattice of functions defined on X, and T is a Daniell functional defined on V.

7.4.1. Definition. A nonnegative function f defined on X is (T)-*measurable* if $f \wedge g \in \mathcal{L}_1(T)$ for all $g \in \mathcal{L}_1(T)$.

7.4.2. Proposition. *Let f and g be nonnegative measurable functions and let $\alpha > 0$. Then*

    (a) $f \wedge g$ is measurable;

    (b) $f \vee g$ is measurable; and

    (c) αf is measurable.

    (d) If $\{f_n\}$ is a sequence of nonnegative measurable functions with $\lim f_n = f$, then f is measurable.

Proof. Let $h \in \mathcal{L}_1(T)$; since $\mathcal{L}_1(T)$ is a vector lattice and

$$h \wedge (f \wedge g) = (h \wedge f) \wedge (h \wedge g),$$
$$h \wedge (f \vee g) = (h \wedge f) \vee (h \wedge g), \text{ and}$$
$$h \wedge (\alpha f) = \alpha(\alpha^{-1} h \wedge f),$$

conclusions (a), (b), and (c) are immediate. Conclusion (d) follows from Theorem 7.3.20 and the inequality $|g \wedge f_n^+| \le |g|$ for all $g \in \mathcal{L}_1$.

7.4.3. Definition. A set $E \subset X$ is (T)-*measurable* if χ_E is a measurable function; if $\chi_E \in \mathcal{L}_1(T)$, we will say that E is *integrable*. Denote by $\mathcal{M}$ the class of all measurable sets.

Note that integrable sets are measurable since $\mathcal{L}_1(T)$ is a vector lattice. Also, if A and B are measurable, $A \subset B$, and B is integrable, then A is integrable. This follows from the observation that $\chi_A = \chi_A \wedge \chi_B \in \mathcal{L}_1(T)$ if $A \in \mathcal{M}$ and $\chi_B \in \mathcal{L}_1(T)$. We begin by showing that E is measurable if $\chi_E \wedge g \in \mathcal{L}_1(T)$ for all $g \in V$ [rather than for all $g \in \mathcal{L}_1(T)$].

7.4.4. Lemma. *A set $E \subset X$ is measurable if and only if $\chi_E \bigwedge g \in \mathcal{L}_1(T)$ for all* $g \in V$.

Proof. Necessity is obvious; for sufficiency fix $E \subset X$ and $f \in \mathcal{L}_1(T)$. Now both $f^+ = f \bigvee 0$ and $f^- = f \bigwedge 0$ are in $\mathcal{L}_1(T)$. Since at most one of $f^+(x)$ and $f^-(x)$ are nonzero for any x, it follows that

$$\chi_E \bigwedge f = \chi_E \bigwedge f^+ - f^-,$$

and hence it suffices to argue that $\chi_E \bigwedge f^+ \in \mathcal{L}_1(T)$. Select functions φ and ψ in V^+ so that $f^+ = \varphi - \psi$ except on a null set and both $T(\varphi) < \infty$ and $T(\psi) < \infty$. Select nondecreasing sequences $\{\varphi_n\}$ and $\{\psi_n\}$ in V so that $\lim \varphi_n = \varphi$ and $\lim \psi_n = \psi$. Now set $\overline{\varphi}_n = \varphi_n \bigvee \psi_n$ and $\overline{\psi}_n = \varphi_n \bigwedge \psi_n$, so $\{\overline{\varphi}_n\}$ and $\{\overline{\psi}_n\}$ are nondecreasing sequences in V converging to $\varphi \bigvee \psi$ and $\varphi \bigwedge \psi$, respectively. Since $0 \le f^+ = \varphi - \psi$ except on a null set, it follows that $\varphi \bigvee \psi = \varphi$ and $\varphi \bigwedge \psi = \psi$ except on a null set, i.e.,

$$\lim \overline{\varphi}_n - \lim \overline{\psi} = f^+$$

except on a null set. Set $\overline{\varphi} = \lim \overline{\varphi}_n$ and $\overline{\psi} = \lim \overline{\psi}_n$.
    Now fix k and set

$$h_n = (\overline{\varphi}_{n+k} - \overline{\psi}_k) \bigwedge \chi_E.$$

Since $\overline{\varphi} \ge \overline{\varphi}_{n+k} \ge \overline{\varphi}_k \ge \overline{\psi}_k$, it follows that $|h_n| \le \overline{\varphi} - \overline{\psi}_k$ for all n. Thus, applying 7.3.20 (Dominated Convergence) we see that $\lim h_n = (\overline{\varphi} - \overline{\psi}_k) \bigwedge \chi_E$ is in $\mathcal{L}_1(T)$ for each k. Since $0 \le (\overline{\varphi} - \overline{\psi}_k) \bigwedge \chi_E \le \overline{\varphi}$, a second application of 7.3.20 implies that $(\overline{\varphi} - \overline{\psi}) \bigwedge \chi_E$ is in $\mathcal{L}_1(T)$. Since $(\overline{\varphi} - \overline{\psi}) \bigwedge \chi_E = f^+ \bigwedge \chi_E$ except on a null set, $f^+ \bigwedge \chi_E$ is in $\mathcal{L}_1(T)$, and the proof is complete.

7.4.5. Theorem. *Let V be a vector lattice satisfying*

$$g \bigwedge 1 \in V \quad \text{for all } g \in V. \tag{$*$}$$

Then the collection of all measurable sets $\mathcal{M}$ is a σ-algebra.

Proof. By 7.4.4 and $(*)$ X is measurable. If $\{A_n\}$ is a countable collection of sets in $\mathcal{M}$, set $g_n = \max\{\chi_{A_j} : 1 \le j \le n\}$; then g_n is measurable. If $A = \cup A_n$, $\chi_A = \lim g_n$, and hence χ_A is measurable. Finally, if $A \in \mathcal{M}$ and $g \in \mathcal{L}_1(T)$,

$$\chi_{A^c} \bigwedge g = g^+ \wedge 1 - \chi_A \wedge g^+ - g^-$$

and thus $\chi_{A^c} \bigwedge g \in \mathcal{L}_1(T)$ if $(g \bigwedge \chi_A) \in \mathcal{L}_1(T)$. This implies that $A^C \in \mathcal{M}$ whenever $A \in \mathcal{M}$, and the proof is complete.

7.4.6. Definition. We define a set function μ on $\mathcal{M}$ by

$$\mu(E) = \begin{cases} T(\chi_E) & \text{if } \chi_E \in \mathcal{L}_1(T) \\ \infty & \text{otherwise.} \end{cases}$$

It is now routine to show, under the assumption (∗), that μ is a measure.

7.4.7. Proposition. *If* (∗) *holds, then the set function μ is a measure defined on the σ-algebra $\mathcal{M}$ of measurable sets.*

Proof. By definition, $\mu(\emptyset) = T(0) = 0$. Let $\{E_n\}$ be a family of disjoint measurable sets. If any one of the sets $\{E_n\}$ is not integrable, then $\cup_1^\infty E_n$ is not integrable and thus

$$\mu\left(\cup_1^\infty E_n\right) = \infty = \sum_1^\infty (E_n).$$

On the other hand, if each set E_n is integrable, then $\cup_1^\infty E_n$ is integrable if and only if

$$\sum_1^\infty T(\chi_{E_n}) \equiv \sum_1^\infty \mu(E_n) < \infty,$$

by 7.3.17. In any case, 7.3.17 implies that

$$\mu(\cup_1^\infty E_n) = T(\chi_{\cup E_n})$$

$$= T\left(\sum \chi_{E_n}\right)$$

$$= \sum T(\chi_{E_n})$$

$$= \sum \mu(E_n),$$

and so μ is a measure.

Thus, if (∗) holds, then $\mathcal{L}_1(X, \mathcal{M}, \mu)$ is well-defined, and because of 7.3.2(iii), the Monotone Convergence Theorem, Dominated Convergence Theorem and, Fatou's Lemma all hold for sequences in $\mathcal{L}_1(X, \mathcal{M}, \mu)$.

7.4.8. Theorem (Stone). *Let X be an arbitrary set, let V be a vector lattice of functions defined on X, and let T be a Daniell functional defined on V. Suppose that* (∗) *holds:*

$$1 \wedge f \in V \quad \text{for all } f \in V. \tag{∗}$$

Then there is a σ-algebra $\mathcal{M}$ of subsets of X and a measure μ defined on $\mathcal{M}$ such that $\mathcal{L}_1(X, \mathcal{M}, \mu) = \mathcal{L}_1(T)$. Moreover, for all $f \in \mathcal{L}_1$,

$$T(f) = \int_X f \, d\mu.$$

Proof. We take $\mathcal{M}$ to be the collection of all T-measurable subsets of X and μ to be the set function given by 7.4.6. Then in view of 7.4.5 and 7.2.2, the triple $(X, \mathcal{M}, \mu)$ is a measure space. By 7.2.13, $\mathcal{L}_1(T)$ is a vector lattice, and thus if $f \in \mathcal{L}_1(T)$, then f is measurable with respect to $\mathcal{M}$.

Now fix $f \in \mathcal{L}^+$ so that $T(f) < \infty$ and define, for integers n, and m, $E(n, m) = \{x \in X : f(x) > n2^{-m}\}$. Note that $E(n, m)$ is measurable [via 7.3.11(f)] and, since $\chi_{E(n,m)} \le \frac{2^m f}{n}$, it follows that $T(\chi_{E(n,m)}) < \infty$, and hence that $\mu(E(n, m)) < \infty$. Next set

$$\varphi_m = 2^{-m} \sum_{k=1}^{2^{2m}} \chi_{E(n,m)}$$

so $\varphi_m \in \mathcal{L}_1(T)$ and φ_m is a simple function in $\mathcal{L}_1(X, \mathcal{M}, \mu)$. As in the proof of 2.4.11, the sequence $\{\varphi_m\}$ is a nondecreasing sequence with limit f. Thus applying the Monotone Convergence Theorem twice yields

$$T(f) = \lim T(\varphi_m) = \lim 2^{-m} \sum_{k=1}^{2^{2m}} T(\chi_{E(k,m)})$$

$$= \lim 2^{-m} \sum_{k=1}^{2^{2m}} \mu(E(k, m))$$

$$= \lim \int \varphi_m \, d\mu$$

$$= \int f \, d\mu$$

and thus $f \in \mathcal{L}_1(X, \mathcal{M}, \mu)$. Since every $g \in \mathcal{L}_1(T)$ is the difference of two such functions in $\mathcal{L}^+$, it follows that if $g \in \mathcal{L}_1(T)$, then $g \in \mathcal{L}_1(X, \mathcal{M}, \mu)$ and

$$T(g) = \int_X g \, d\mu.$$

For the converse, let $f \in \mathcal{L}_1(X, \mathcal{M}, \mu)$ and suppose that $f > 0$. Define $E(n, m)$ and $\{\varphi_m\}$ as above; since $\int f \, d\mu < \infty$, each set $E(n, m)$ has finite measure, and thus $\varphi_n \in \mathcal{L}_1(T)$. Since $\{\varphi_m\}$ is a nondecreasing sequence converging to f, the Monotone Convergence Theorem again applies.

As we shall see shortly, 7.4.8 permits us to deduce a representation theorem for the dual space of $C(M)$ where M is a compact metric space. The approach is entirely analogous to 6.3.5.

PROBLEMS

7.22. Let α be a nondecreasing function defined on $[a, b]$ and define T on $C[a, b]$ by $T(f) = \int_a^b f \, d\alpha$. Let μ be the measure associated with T. Show that for any closed subinterval $[c, d] \subseteq [a, b]$

$$\mu([a, b]) = \int_c^d 1 \, d\alpha = \alpha(d-) - \alpha(c+).$$

Under what conditions is this true for open intervals?

7.23. Let $X = \{1, 2, \ldots\}$ and let

$$V = \{f : X \to \Re : f(n) \neq 0 \quad \text{for at most finitely many} \quad n \, \}.$$

Define $T : V \to \Re$ by

$$T(f) = \sum_{n=1}^{\infty} f(n).$$

(a) Show that T is a Daniell functional.
(b) Show that $\mathcal{L}_1(T) = \ell_1$.

7.24. Suppose that $V(X)$ is a vector lattice and suppose that there is a function $e \in V(X)$ such that $e(x) > 0$ for all x. Show that the conclusions of 7.4.8 remain valid.

7.5. THE PRODUCT OF TWO MEASURE SPACES.
Suppose that $(X, \mathcal{M}_1, \mu)$ and $(Y, \mathcal{M}_2, \nu)$ are two measure spaces; our goal in this section is to construct a σ-algebra $\mathcal{A}$ on $X \times Y$ and a measure $(\mu \times \nu)$ on $\mathcal{A}$ with the following properties:

   (i) If $f \in \mathcal{L}_1(X, \mathcal{M}_1, \mu)$ and $g \in \mathcal{L}_1(Y, \mathcal{M}_2, \nu)$, then $fg \in \mathcal{L}_1(X \times Y, \mathcal{A}, \mu \times \nu)$; and

   (ii) the integral in $\mathcal{L}_1(X \times Y, \mathcal{A}, \mu \times \nu)$ can be realized as an "iterated integral" over the factor spaces.

[We will make (ii) more precise when we formulate Fubini's Theorem below.]

7.5.1. Definition. A set $R \subset X \times Y$ is a *measurable rectangle* if there are measurable sets $A \in \mathcal{M}_1$ and $B \in \mathcal{M}_2$ such that $R = A \times B$. If, in addition, $\mu(A) < \infty$ and $\nu(B) < \infty$, then we will call R an *integrable rectangle*.

7.5.2. Definition. We will denote by $V \equiv V(X \times Y)$ the class of all real-valued functions σ defined on $X \times Y$ of the form

$$\sigma = \sum_1^n \lambda_k \chi_{S_k},$$

where $\{\lambda_i\} \subset \Re$ and $\{S_1, \ldots, S_n\}$ are integrable rectangles.

It is clear from 7.5.2 that V is a vector space; it is less clear that V is in fact a vector lattice over $X \times Y$. We can deduce the latter fact from the following theorem:

7.5.3. Theorem. *If $\sigma \in V(X \times Y)$, then there are disjoint integrable rectangles* $\{R_1, \ldots, R_N\}$ *and scalars* $\{\lambda_1, \ldots, \lambda_N\}$ *such that*

$$\sigma = \sum_{i=1}^{N} \lambda_i \chi_{R_i}.$$

A few pictures in the plane should convince you that this theorem is reasonable. We leave the formal proof, which is rather technical and set-theoretic in character, to the problems.

With Theorem 7.5.3 in hand, it is now relatively easy to show that $V(X \times Y)$ is a vector lattice to which the results of Section 7.4 apply.

7.5.4. Corollary. *The set $V(X, Y)$ is a vector lattice satisfying $(1 \wedge \sigma) \in V$ for all $\sigma \in V$.*

Proof. Clearly, $V(X, Y)$ is a vector space. Let $\sigma = \sum_{i=1}^{n} \lambda_i \chi_{R_i}$ where $R_1, \ldots, R_n$ are disjoint integrable rectangles. Observe that if α is any scalar and $x \in X \times Y$, then

$$(\alpha \wedge \sigma)(x) = \sum_{1}^{n} (\lambda_i \wedge \alpha) \chi_{R_i}(x) \quad \text{and}$$

$$(\alpha \vee \sigma)(x) = \sum_{1}^{n} (\lambda_i \vee \alpha) \chi_{R_i}(x)$$

since x is in at most one of the sets R_i, showing that $\alpha \wedge \sigma$ and $\alpha \vee \sigma$ are in V for all $\sigma \in V$ and scalars α.

In particular, $\sigma^+ = \sigma \vee 0$ and $\sigma^- = (-\sigma) \vee 0$ are members of V, and thus $|\sigma| = \sigma^+ + \sigma^-$ is also a member of V. But this is sufficient to guarantee that V is a vector lattice (see 4.3.8).

Our next step is to define a Daniell functional on V. If $\sigma = \sum_{1}^{n} \lambda_i \chi_{A_i \times B_i}$, we will define

$$T(\sigma) = \sum_{1}^{n} \lambda_i \mu(A_i) \nu(B_i);$$

however, as usual, it is first necessary to guarantee that this value is independent of the representation of σ.

7.5.5. Definition. Let $\sigma \in V(X \times Y)$; for fixed $x \in X$ define $\sigma_x : Y \to \Re$ by

$\sigma_x(y) = \sigma(x, y)$ and for fixed $y \in Y$ define $\sigma_y : X \to \Re$ by $\sigma_y(x) = \sigma(x, y)$.

7.5.6. Proposition. *Let* $\sigma \in V(X \times Y)$. *Then*

(a) $\sigma_x \in \mathcal{L}_1(Y, \mathcal{M}_2, \nu)$ *and* $\sigma_y \in \mathcal{L}_1(X, \mathcal{M}_1, \mu)$;

(b) $\int_Y \sigma_x \, d\nu \in \mathcal{L}_1(X, \mathcal{M}_1, \mu)$ *and* $\int_X \sigma_y \, d\mu \in \mathcal{L}_1(Y, \mathcal{M}_2, \nu)$; *and*

(c) $\int_X \int_Y \sigma_x \, d\nu \, d\mu = \int_Y \int_X \sigma_y \, d\mu \, d\nu.$

Proof. By symmetry it suffices to prove only the first half of each of parts (a) and (b). Suppose that $\sigma = \sum_1^n \lambda_i \chi_{R_i}$ and $R_i = A_i \times B_i$; note that $\chi_{R_i} = \chi_{A_i} \chi_{B_i}$.

For (a), fix $x \in X$, so that σ_x is the simple function

$$\sigma_x = \sum_1^n \lambda_i \chi_{A_i}(x) \chi_{B_i}$$

and is thus in $\mathcal{L}_1(Y, \mathcal{M}_2, \nu)$ and

$$\int_Y \sigma_x \, d\nu = \sum_1^n \lambda_i \nu(B_i) \chi_{A_i}(x).$$

Thus $\int_Y \sigma_x \, d\nu$ is a simple function in $\mathcal{L}_1(X, \mathcal{M}_1, \mu)$ and

$$\int_X \int_Y \sigma_x \, d\nu \, d\mu = \int_X \sum_1^n \lambda_i \nu(B_i) \chi_{A_i} \, d\mu$$

$$= \sum_1^n \lambda_i \nu(B_i) \mu(A_i).$$

A similar computation shows that

$$\int_Y \int_X \sigma_y \, d\mu \, d\nu = \sum_1^n \lambda_i \mu(A_i) \nu(B_i)$$

and thus (c) follows.

Since the value of the integral of a simple function is independent of the representation, we may now define our functional on $V(X \times Y)$:

7.5.7. Definition. Define a functional T on $V(X \times Y)$ by

$$T(\sigma) = \int_X \int_Y \sigma_x \, d\nu \, d\mu = \int_Y \int_X \sigma_y \, d\mu \, d\nu.$$

The stage is now nearly set to invoke the results of Sections 7.3 and 7.4; the following theorem summarizes the foregoing and completes the process.

7.5.8. Theorem. *Let* (X, M_1, μ) *and* (Y, M_2, ν) *be measure spaces and define* $V(X \times Y)$ *as in 7.5.2. Then*

(a) $V(X \times Y)$ *is a vector lattice;*

(b) *the functional* T *given by 7.5.7 is a Daniell functional defined on* $V(X \times Y)$; *and*

(c) *for all* $\sigma \in V(X \times Y)$, $1 \wedge \sigma \in V(X \times Y)$.

Proof. Parts (a) and (c) are 7.5.4. For part (b), first observe that it is clear from the definitions that T is a linear functional and that, if $\sigma \geq 0$, then

$$T(\sigma) = \int_X \int_Y \sigma \, d\nu \, d\mu \geq 0,$$

i.e., that T is a positive functional. Thus to complete the proof, we need only show that if $\{\sigma_n\}$ is a nonincreasing sequence of nonnegative functions in $V(X \times Y)$ which converge to zero, then $\lim T(\sigma_n) = 0$.

To see this, observe for each fixed x, that $\{(\sigma_n)_x\}$ is a nonincreasing sequence of nonnegative functions in $\mathcal{L}_1(Y, M_2, \nu)$, and thus (by Dominated Convergence)

$$\lim \int_Y (\sigma_n)_x \, d\nu = 0.$$

Since $\{(\sigma_n)_x\}$ is nonincreasing, so is the sequence $\{\int_Y (\sigma_n)_x \, d\nu\}$. Thus if

$$f_n(x) = \int_Y (\sigma_n)_x \, d\nu$$

then $\{f_n\}$ is a nonincreasing sequence of nonnegative functions in $\mathcal{L}_1(X, M_1, \mu)$ and hence

$$0 = \lim \int_X f_n \, d\mu = \lim \int_X \int_Y \sigma_n \, d\nu \, d\mu = \lim T(\sigma_n),$$

completing the proof.

7.5.9. Corollary. *Let* (X, M_1, μ), (Y, M_2, ν), $V(X \times Y)$ *and* T *be as in 7.5.8. Then there is a σ-algebra* $(M_1 \times M_2)$ *over* $X \times Y$ *and a complete measure* $(\mu \times \nu)$ *defined on* $(M_1 \times M_2)$ *with the property that*

$$\mathcal{L}_1(T) = \mathcal{L}_1(X \times Y, M_1 \times M_2, \mu \times \nu).$$

Moreover,

$$T(\sigma) = \int_{X \times Y} \sigma \, d(\mu \times \nu)$$

for all functions $\sigma \in \mathcal{L}_1$.

Proof. This is immediate from 7.5.8 and 7.4.8.

Note that if $f \in V(X \times Y)$, then the integral of f, $\int_{X \times Y} f \, d(\mu \times \nu)$, is realized as either of the iterated integrals $\int_X \int_Y f \, d\nu \, d\mu$ or $\int_Y \int_X f \, d\mu \, d\nu$. The remainder of this section is devoted to showing that this is true for arbitrary $f \in \mathcal{L}_1(X \times Y, M_1 \times M_2, \mu \times \nu)$. We begin by observing that it holds for $f \in V^+$.

7.5.10. Lemma. *Let (X, M_1, μ) and (Y, M_2, ν) be measure spaces and let $f \in V^+(T)$ where $V(X \times Y)$ and T are as in 7.5.8. Then*

    (a) for all x, the function $f_x(y) \equiv f(x, y)$ is in $\mathcal{L}_1(Y, M_2, \nu)$ and for all y, the function $f_y(x) \equiv f(x, y)$ is in $\mathcal{L}_1(X, M_1, \mu)$.

    (b) The function $\int_Y f_x \, d\nu$ is a member of $\mathcal{L}_1(X, M_1, \mu)$ and the function $\int_X f_y \, d\mu$ is a member of $\mathcal{L}_1(Y, M_2, \nu)$.

    (c) $\int_X \int_Y f \, d\nu \, d\mu = \int_{X \times Y} f \, d(\mu \times \nu) = \int_Y \int_X f \, d\mu \, d\nu$.

Proof. Select a nondecreasing sequence $\{\sigma_n\}$ of nonnegative functions in V for which $\lim \sigma_n = f$.

Now for each $x \in X$ $\{(\sigma_n)_x\}$ is a nondecreasing sequence of simple functions in $\mathcal{L}_1(Y, M_2, \nu)$ and thus, by the Monotone Convergence Theorem (7.2.12), $\lim(\sigma_n)_x = f_x$ is ν-measurable and

$$\lim \int_Y (\sigma_n)_x \, d\nu \;=\; \int_Y (f)_x \, d\nu.$$

Once again, the sequence $\{\int_Y (\sigma_n)_x \, d\nu\}$ is a nondecreasing sequence of simple functions in $\mathcal{L}_1(X, M_1, \mu)$ and thus

$$\int_Y f_x \, d\nu = \lim \int_Y (\sigma_n)_x \, d\nu$$

is μ-measurable and the Monotone Convergence Theorem again implies that

$$\lim \int_X \int_Y (\sigma_n)_x \, d\nu \, d\mu \;=\; \int_X \int_Y f \, d\nu \, d\mu.$$

But because $f \in V^+$, $\int_{X \times Y} f \, d(\mu \times \nu) = \lim \int_{X \times Y} \sigma_n \, d(\mu \times \nu)$ and so 7.5.7 implies that

$$\int_{X \times Y} f \, d(\mu \times \nu) \;=\; \int_X \int_Y f \, d\nu \, d\mu.$$

This proves the first half of each of the assertions (a)—(c); the second halves follow by a symmetrical argument.

Our second lemma deals with the structure of null sets in $\mathcal{L}_1(X \times Y, M_1 \times M_2, \mu \times \nu)$.

7.5.11. Lemma. *Let (X, M_1, μ) and (Y, M_2, ν) be complete measure spaces. Let E be a null set in $M_1 \times M_2$. Then for μ-almost all x, the set $E_x = \{y :$*

$(x, y) \in E\}$ *is measurable and* $\nu(E_x) = 0$. *Similarly, for* ν-*almost all* y, *the set* $E_y = \{x : (x, y) \in E\}$ *is measurable and* $\mu(E_y) = 0$.

Proof. Since $(\mu \times \nu)(E) = 0$, for each n there is a function $\varphi_n \in V^+$ such that $\varphi_n \geq \chi_E$ and $T(\varphi_n) \leq \frac{1}{n}$. Setting

$$\psi_n = \bigwedge_1^n (\varphi_i \wedge 1)$$

we obtain a nonincreasing sequence $\{\psi_n\} \subset V^+$ such that $1 \geq \psi_n \geq \chi_E$ and $T(\psi_n) \leq 1/n$. Set $\psi_\infty = \lim \psi_n$, so $\psi_\infty \in \mathcal{L}_1(X \times Y, \mathcal{M}_1 \times \mathcal{M}_2, \mu \times \nu)$ by the Dominated Convergence Theorem, and

$$0 = \int_{X \times Y} \psi_\infty \, d(\mu \times \nu) = \lim \int_{X \times Y} \psi_n \, d(\mu \times \nu)$$

$$= \lim \int_X \int_Y (\psi_n)_x \, d\nu \, d\mu, \tag{3}$$

applying 7.5.10. Applying the Dominated Convergence Theorem in $\mathcal{L}_1(Y, \mathcal{M}_2, \nu)$ to $(\psi_n)_x$ gives that $(\psi_\infty)_x \in \mathcal{L}_1(Y, \mathcal{M}_2, \nu)$ and

$$\int_Y (\psi_\infty)_x \, d\nu = \lim \int_Y (\psi_n)_x \, d\nu.$$

A third application of Dominated Convergence, this time in $\mathcal{L}_1(X, \mathcal{M}_1, \mu)$ to $\{\int_Y (\psi_n)_x \, d\nu\}$, gives $\int_Y (\psi_\infty)_x \, d\nu \in \mathcal{L}_1(X, \mathcal{M}_1, \nu)$ and

$$\int_X \int_Y \psi_\infty \, d\nu \, d\mu = \lim \int_X \int_Y \psi_n \, d\nu \, d\mu.$$

Combining this with (3) gives

$$\int_X \int_Y \psi_\infty \, d\nu \, d\mu = 0.$$

Since $\psi_\infty \geq \chi_E \geq 0$, it follows that for μ-almost all x,

$$\int_Y (\psi_\infty)_x \, d\nu = 0.$$

This implies that, for μ-almost all x, E_x is contained in the ν-null set

$$E_x \subset \{y : (\psi_\infty)_x(y) \geq 1\}.$$

By completeness of ν, this implies, for μ-almost all x, that E_x is ν-measurable and $\nu(E_x) = 0$.

A symmetric argument gives the same conclusions for E_y.

7.5.12. Fubini's Theorem. *Let* $(X, \mathcal{M}_1, \mu)$ *and* $(Y, \mathcal{M}_2, \nu)$ *be two complete measure spaces and let* $f \in \mathcal{L}_1(X \times Y, \mathcal{M}_1 \times \mathcal{M}_2, \mu_1 \times \mu_2)$. *Then*

(a) For μ-almost all x, the function $f_x(y) \equiv f(x, y)$ is in $\mathcal{L}_1(Y, M_2, \nu)$ and for ν-almost all y, the function $f_y(x) \equiv f(x, y)$ is in $\mathcal{L}_1(X, M_1, \mu)$.

(b) The function $\int_Y f_x \, d\nu$ is a member of $\mathcal{L}_1(X, M_1, \mu)$ and the function $\int_X f_y \, d\mu$ is a member of $\mathcal{L}_1(Y, M_2, \nu)$.

(c) $\int_X \int_Y f \, d\nu \, d\mu = \int_{X \times Y} f \, d(\mu \times \nu) = \int_Y \int_X d \, d\mu \, d\nu$.

Proof. As with 7.5.10, it suffices to verify only the first half of (a)—(c). If $f \in \mathcal{L}_1(X \times Y, M_1 \times M_2, \mu \times \nu)$, then there are functions $\varphi, \psi \in V^+$ such that

$$f = \varphi - \psi \qquad (\mu \times \nu) - \text{almost everywhere.}$$

Because of 7.5.11, for μ-almost all x

$$f_x = (\varphi - \psi)_x \qquad \nu - \text{almost everywhere,}$$

and so, since (Y, M_2, ν) is complete, (a) holds.

The function $g(x) = \int_Y f_x \, d\nu$ is thus defined μ-almost everywhere and where it is defined, agrees with $\int_Y (\varphi - \psi)_x \, d\nu$. Thus, via completeness of (X, M_1, μ), (b) holds and

$$\int_X \int_Y f \, d\nu \, d\mu = \int_X \int_Y \varphi - \psi \, d\nu \, d\mu.$$

But now by 7.5.10 and the fact that $f = \varphi - \psi$ $(\mu \times \nu)$ a.e.

$$\int_X \int_Y (\varphi - \psi) \, d\nu \, d\mu = \int_{X \times Y} (\varphi - \psi) \, d(\mu \times \nu) = \int_{X \times Y} f \, d(\mu \times \nu),$$

showing that (c) holds as well.

Note that in order to apply Fubini's Theorem to a function f defined on $X \times Y$, we must show both that f is $(\mu \times \nu)$ measurable and that $\int |f| \, d(\mu \times \nu) < \infty$. In many applications, the former can be established using topological considerations; if μ and ν are σ-finite, the integrability of f can be established using the following.

7.5.13. Tonelli-Hobson Theorem. *Let (X, M_1, μ) and (Y, M_2, ν) be σ-finite complete measure spaces and let f be a nonnegative measurable function defined on $X \times Y$. Then*

(a) *For μ-almost all x, the function $f_x(y) \equiv f(x, y)$ is ν-measurable and for ν-almost all y, the function $f_y(x) \equiv f(x, y)$ is ν-measurable.*

(b) *The function $\int_Y f_x \, d\nu$ is a μ-measurable function and the function $\int_X f_y \, d\mu$ is a ν-measurable function.*

(c) $\int_X \int_Y f \, d\nu \, d\mu = \int_{X \times Y} f \, d(\mu \times \nu) = \int_Y \int_X f \, d\mu \, d\nu$.

Proof. If (X, M_1, μ) and (Y, M_2, ν) are σ-finite, then so is $(X \times Y, M_1 \times M_2, \mu \times \nu)$ and thus there is a nondecreasing sequence $\{\sigma_n\}$ of simple functions defined on $X \times Y$ such that $\int_{X \times Y} \sigma_n \, d(\mu \times \nu) < \infty$ for each n, and $f = \lim \sigma_i$. By 7.5.12, the conclusions all hold for σ_n, and thus hold for f as well by the Monotone Convergence Theorem.

PROBLEMS

7.25. Verify 7.5.3.

7.26. Let $p > 1$. Suppose that
(a) $f \in \mathcal{L}_p[1, \infty)$;
(b) $f \geq 0$ a.e.

Define $g(x) = \int_1^\infty f(t)e^{-tx}\, dt$. Show that $g \in \mathcal{L}_1[1, \infty)$ and

$$\|g\|_1 \leq \frac{\|f\|_p}{(q-1)^{1/q}}$$

where $\frac{1}{p} + \frac{1}{q} = 1$.

7.27. Let $K \in C[0,1]$. For $f \in \mathcal{L}_1[0,1]$ set

$$\hat{f}(x) = \int_0^1 K(xt)f(t)\, dt \quad (0 \leq x \leq 1).$$

Show that for all $f, g \in \mathcal{L}_1[0,1]$,

$$\int_0^1 f\hat{g}\, d\lambda = \int_0^1 \hat{f}g\, d\lambda.$$

7.28. Suppose that f and g are nonnegative measurable functions defined on $[0, \infty)$. Suppose that

$$\int_0^\infty f(t)t^{-\frac{1}{2}}\, dt < \infty \quad \text{and}$$

$$\int_0^\infty g^2(t)\, dt < \infty$$

Show that

$$\int_0^\infty \int_0^x \frac{g(x)}{x} f(t)\, dt\, dx < \infty.$$

7.29. Let $X = Y = \{1, 2, \ldots\}$ and let $\mu = \nu$ be the counting measure. Set

$$f(x,y) = \begin{cases} 2 - 2^{-x} & \text{if } x = y \\ -2 + 2^{-x} & \text{if } x = y + 1 \\ 0 & \text{otherwise.} \end{cases}$$

Show that 7.5.12 and 7.5.13 fail for this f, and thus the assumption that $f \geq 0$ in 7.5.13 and $f \in \mathcal{L}_1$ in 7.5.12 cannot be omitted.

7.5.F. Evaluate, for $a > 0$

$$\sum_{k=0}^\infty \sum_{n=0}^\infty \frac{(-an)^k}{k!}.$$

7.30.   Use the formula

$$\frac{1}{x} = \int_0^\infty e^{-xt}\,dt$$

to deduce

$$\lim_{N\to\infty}\int_0^N \frac{\sin(x)}{x}\,dx = \frac{\pi}{2}.$$

7.6. SIGNED MEASURES.   As we have remarked previously, if f is a non-negative integrable function defined on a measure space (X, M, μ), then f induces a measure μ_f on M by the formula $\mu_f(E) = \int_E f\,d\mu$. If f is an arbitrary function in $\mathcal{L}_1(X, M, \mu)$, then f induces two measures,

$$\mu_{f+}(E) = \int_E f^+\,d\mu \quad \text{and} \quad \mu_{f-}(E) = \int_E f^-\,d\mu;$$

it is sometimes useful to consider the set function $\mu_f = \mu_{f+} - \mu_{f-}$. Since, in general, μ_f may assume negative values, μ_f need not be a measure in the sense of Section 7.1. However, μ_f does have many of the same properties as a measure, and set functions of this type turn out to be of independent interest.

7.6.1. Definition.   Let X be a set, let M be a σ-algebra of subsets of X, and let ν be an extended-real valued function defined on M. Then ν is a *signed measure* if

(i) ν assumes at most one of the values $+\infty$ and $-\infty$.
(ii) $\nu(\emptyset) = 0$.
(iii)   If $\{E_n\}$ is a sequence of disjoint measurable sets, then

$$\nu(\cup_1^\infty E_i) = \sum_1^\infty \nu(E_i)$$

[where $\sum_1^\infty \nu(E_i)$ converges absolutely if $\nu(\cup E_i) < \infty$ and otherwise properly diverges].

7.6.2. Example.   Let T be a bounded linear functional defined on $C[a, b]$; by 6.3.2 T can be decomposed as $T = P - N$, the difference of two positive linear functionals defined on $C[a, b]$. Each of P and N generate σ-algebras M_P and M_N and measures μ_P and μ_N so that for $f \in C[a, b]$,

$$P(f) = \int_a^b f\,d\mu_P \quad \text{and} \quad N(f) = \int_a^b f\,d\mu_N.$$

It can be argued (see Problem 7.22) that the Borel sets $\mathcal{B}$ are contained in $M_P \cap M_N$. Thus, if one defines

$$\nu(E) = \mu_P(E) - \mu_N(E)$$

for E in $\mathcal{B}$, then ν is a signed measure.

Note that, by the Riesz Representation Theorem, for $\mathcal{C}[a, b]$

$$T(f) = \int_a^b f \, d\alpha$$

for some $\alpha \in \mathrm{BV}[a, b]$. By the Jordan Decomposition Theorem, $\alpha = \alpha_1 - \alpha_2$ where α_1 and α_2 are nondecreasing functions defined on $[a, b]$. By 7.1.3(ii) each of α_1 and α_2 induces measures μ_{α_1} and μ_{α_2} on $\mathcal{B}$. Problem 7.32 investigates the relationship between μ_{α_1}, μ_{α_2} and μ_P, μ_N.

As we shall soon see, the above is a special example of an important way in which signed measures arise.

7.6.3. Definition. Let $(X, \mathcal{M})$ be a measurable space and let ν be a signed measure defined on $\mathcal{M}$. Let A be a measurable set with the property that $\nu(E) \geq 0$ for all $E \subset A$ and $E \in \mathcal{M}$; then we call A a *positive set*. Similarly A is a *negative set* if $\nu(E) \leq 0$ for all $E \subset A$ and $E \in \mathcal{M}$.

Our first goal in this section will be to show that every signed measure can be written as the difference of two measures (analogous to the Jordan Decomposition Theorem). Preliminary to this, we establish some lemmas.

7.6.4. Lemma. *Let $(X, \mathcal{M})$ be a measurable space and let ν be a signed measure defined on $\mathcal{M}$. If $\{A_n\}$ is a sequence of positive subsets of $\mathcal{M}$, then $A = \cup A_n$ is a positive subset of $\mathcal{M}$.*

Proof. Let $E \subset A$ and, for each n, set

$$E_n = (E \cap A_n) \backslash (\cup_1^{n-1} A_i).$$

Then $\nu(E_n) \geq 0$ for each n, $E = \cup_1^\infty E_n$ and $E_i \cap E_j = \emptyset$ if $i \neq j$. Thus

$$\nu(E) = \nu(\cup_1^\infty E_n) = \sum_1^\infty \nu(E_n) \geq 0.$$

7.6.5. Lemma. *Let $(X, \mathcal{M})$ be a measurable space and let ν be a signed measure defined on $\mathcal{M}$. Let E be a measurable set and suppose that $0 < \nu(E) < \infty$. Then there is a positive set $E_0 \subset E$ with $\nu(E_0) > 0$.*

Proof. We suppose for contradiction that E contains no positive sets (so, in particular, E itself is not positive). Take

$$n_1 = \min \left\{ n : \exists \text{ a measurable set } E_1 \subset E \text{ with } \nu(E_1) < -\frac{1}{n} \right\}.$$

Proceeding inductively, we may take

$$n_k = \min\{n : \exists \text{ a measurable set } E_k \subset E\backslash(\cup_1^{k-1}E_j)$$
$$\text{and } \nu(E_k) < -1/n\}.$$

[Note that for each k, $E\backslash(\cup_1^{k-1}E_j)$ is not a positive set by assumption.] Now set $E_0 = E\backslash \cup_1^\infty E_k$.

Then E can be written as the union of the disjoint sets $\{E_j\}_{j=0}^\infty$ and thus

$$\nu(E) = \sum_0^\infty \nu(E_j);$$

since $\nu(E) < \infty$, the series converges absolutely. But for $k \geq 1$, $\nu(E_k) < -\frac{1}{n_k}$, and thus $\sum_1^\infty \frac{1}{n_k} < \infty$ (so $n_k \to \infty$). Since $\nu(E_j) < 0$ for $j \geq 1$ and $\nu(E) > 0$, it also follows that $\nu(E_0) > 0$.

Now let $A \subset E_0$ and suppose that $\nu(A) < 0$. Then, for k sufficiently large, $\nu(A) < -\frac{1}{n_k}$. But observe that

$$A\cup E_k \subset E\backslash(\cup_1^{k-1}E_k)$$

and

$$\nu(A\cup E_k) = \nu(A) + \nu(E_k) < -\frac{1}{n_k} - \frac{1}{n_k} = -\frac{2}{n_k} = -\frac{1}{(n_k/2)},$$

which contradicts the minimality of n_k. Thus E_0 is a positive set.

7.6.6. Hahn Decomposition Theorem. *Let $(X, \mathcal{M})$ be a measurable space and let ν be a signed measure defined on $\mathcal{M}$. Then there is a positive set $A \in \mathcal{M}$ and a negative set $B \in \mathcal{M}$ such that $A \cap B = \emptyset$ and $A \cup B = X$.*

Proof. Since ν takes on at most one of the values $\pm\infty$, we may assume without loss of generality that $\nu(E) < \infty$ for all $E \in \mathcal{M}$.

Set $\varrho = \sup\{\nu(A) : A \text{ is a positive subset of } X\}$ (note $\varrho < \infty$). Select a sequence $\{A_k\}$ of positive subsets of X such that $\lim \nu(A_k) = \varrho$, and set $A = \cup_1^\infty A_k$. Then, by 7.6.4, A is a positive set, so $\nu(A) \leq \varrho$. On the other hand, $A_k \subset A$ for each k, and thus $\nu(A_k) \leq \nu(A)$. Letting $k \to \infty$ gives $\nu(A) = \varrho$.

Now suppose that $B = X\backslash A$ is not a negative set; then there is a set $C \subset B$ for which $\nu(C) > 0$. Since $\nu(C) < \infty$, there is a positive set $P \subseteq C$ for which $\nu(P) > 0$.

But then $P \cup A$ is a positive set and $\nu(P \cup A) = \nu(P) + \nu(A) > \varrho$, a contradiction.

7.6.7 Jordan Decomposition Theorem. *Let ν be a signed measure on the measurable space $(X, \mathcal{M})$. Then there are measures ν^+ and ν^- defined on $\mathcal{M}$ such that $\nu = \nu^+ - \nu^-$.*

Proof. Let A and B be a Hahn decomposition for $(X, \mathcal{M}, \nu)$. Then setting

$$\nu^+(E) = \nu(E \cap A) \quad \text{and} \quad \nu^-(E) = \nu(E \cap B)$$

the result follows at once.

      With the material in hand, we can now deduce a more general form of the Riesz Representation Theorem for $\mathcal{C}[a, b]$. First recall

7.6.8. Definition. Let (M, d) be a metric space. The Borel sets $\mathcal{B}$ are the smallest σ-algebra which contain the open sets. A signed measure μ defined on the Borel sets is a *Borel signed measure.*

7.6.9. Theorem. *Let (M, d) be a compact metric space and let T be a bounded linear functional defined on $\mathcal{C}(M)$. Then there is a Borel signed measure ν such that*

$$T(f) = \int_M f \, d\nu \quad \text{for all} \ f \in \mathcal{C}(M).$$

Proof. We may write $T = P - N$ as the difference of two Daniell functionals on $\mathcal{C}(M)$. There are measures μ_P and μ_N defined on $\mathcal{B}$ such that

$$P(f) = \int_M f \, d\mu_P \quad \text{and} \quad N(f) = \int_M f \, d\mu_N$$

for all $f \in \mathcal{C}(M)$. Setting $\nu = \mu_P - \mu_N$ gives the conclusion.

PROBLEMS

7.32. Let T be a bounded linear functional defined on $\mathcal{C}[a, b]$; let ν and α be as in 7.6.2. Show that

$$\mu_P(E) = \int_E 1 \, d\alpha_1$$

and

$$\mu_N(E) = \int_E 1 \, d\alpha_2.$$

7.33. Show that the decomposition given in 7.6.6 is unique up to sets of measure zero; i.e., if (A', B') is any other such decomposition, then

$$\nu(A' \cap B) = 0 = \nu(B' \cap A).$$

7.7. THE RADON-NIKODYM THEOREM. If α is an absolutely continuous function and if f is continuous, then the integration-by-parts formula tells us that

$$\int_a^b f \, d\alpha = \int_a^b f\alpha' \, d\lambda,$$

the latter integral being in the sense of Lebesgue. In particular, if α is absolutely continuous and nondecreasing, then the measure μ_α

$$\mu_\alpha(E) = \int_E 1 \, d\alpha$$

induced by α on the Borel sets can be realized as the integral of α', a member of $\mathcal{L}_1([a,b], \mathcal{M}, \lambda)$. The purpose of the present section is to deduce when a similar result can be obtained for measures μ, ν defined on the same measurable space $(X, \mathcal{M})$. Surprisingly, the Riesz Representation Theorem for Hilbert Space (3.6.3) will play a crucial role.

7.7.1.   Definition. Let $(X, \mathcal{M}, \mu)$ be a measure space and let $p > 0$. Denote by $\mathcal{L}_p(X, \mathcal{M}, \mu) = \mathcal{L}_p(\mu)$ the class of all functions f for which $|f|^p \in \mathcal{L}_1(X, \mathcal{M}, \mu)$.

7.7.2.   Theorem. *Let $(X, \mathcal{M}, \mu)$ be a measure space and let $p \geq 1$. For $f \in \mathcal{L}_p(X, \mathcal{M}, \mu)$ define $\|f\|_p = (\int_X |f|^p \, d\mu)^{\frac{1}{p}}$. Then $\| \cdot \|$ is a norm, making $\mathcal{L}_p(X, \mathcal{M}, \mu)$ into a Banach space.*

Proof. As the proof given in Section 3.2 for Lebesgue measure on $\mathfrak{R}$ relies only on pointwise inequalities, it carries over without change to this more general setting. The only tools required for completeness in Section 3.3 were the Monotone Convergence Theorem and the Dominated Convergence Theorem, and thus the proof of 3.3.1 also carries over without change.

7.7.3.   Corollary. *If $(X, \mathcal{M}, \mu)$ is a measure space, then $\mathcal{L}_2(X, \mathcal{M}, \mu)$ is a Hilbert space. In particular, $\left(\mathcal{L}_2(\mu)\right)^* = \mathcal{L}_2(\mu)$.*

Proof. If $f, g \in \mathcal{L}_2(\mu)$, then

$$\|f+g\|_2^2 + \|f-g\|_2^2 = 2\int_X |f|^2 + |g|^2 \, d\mu = 2\|f\|_2^2 + 2\|g\|_2^2,$$

verifying the parallelogram law.

We remark in passing that the proof that $\left(\mathcal{L}_p(\lambda)\right)^* = \mathcal{L}_q(\lambda)$ given in Sections 3.7 and 3.8 is likewise independent of Lebesgue measure and so carries over to this generality as well. For our immediate purposes, 7.7.3 suffices.

7.7.4.   Lemma. *Let $(X, \mathcal{M})$ be a measurable space and let μ and ν be measures* *defined on* M. *Suppose for all* $E \in M$ *that* $\nu(E) \leq \mu(E)$. *If* $f \in \mathcal{L}_2(\mu)$, *then* $f \in \mathcal{L}_2(\nu)$ *and* $\int_X |f|^2 \, d\nu \leq \int_X |f|^2 \, d\mu$.

Proof. Select a nondecreasing sequence $\{\sigma_n\}$ of simple functions such that $\lim \sigma_n = |f|^2$, so

$$\lim \int_X \sigma_n \, d\mu \;=\; \int_X |f|^2 \, d\mu \quad \text{and}$$

$$\lim \int_X \sigma_n \, d\nu \;=\; \int_X |f|^2 \, d\nu.$$

But it is clear for simple functions that $\int_X \sigma_n \, d\nu \leq \int_X \sigma_n \, d\mu$, and so the conclusion follows.

7.7.5. Definition. Let (X, M) be a measure space and let μ and ν be measures defined on M. We will say that ν is *absolutely continuous* with respect to μ (written $\nu \ll \mu$) if $\nu(E) = 0$ whenever $\mu(E) = 0$.

Note that if α is an absolutely continuous nondecreasing function and if $\mu_\alpha(E) = \int_E 1 \, d\alpha$ is the measure associated with α, the integration by parts formula implies that $\mu_\alpha \ll \lambda$. (Also compare 7.7.5 with 5.5.4.)

While the next lemma may seem rather peculiar, it is actually the central step in the proof which we present of the Radon-Nikodym Theorem.

7.7.6. Lemma. *Let* (X, M) *be a measurable space and let* μ *and* ν *be finite measures defined on* M *with* $\nu \ll \mu$. *Then there is an* M-*measurable function* φ *defined on* X *satisfying*
   (a) $\varphi(X) \subset [0, 1]$; *and*
   (b) $\int_X f(1 - \varphi) \, d\nu = \int_X f\varphi \, d\mu$ *for all* $f \in \mathcal{L}_2(\mu + \nu)$.

Proof. Observe that $(\mu + \nu)$ is a finite measure, and so $1 \in \mathcal{L}_2(\mu + \nu)$. Thus if $f \in \mathcal{L}_2(\mu + \nu)$, Hölder's inequality implies that

$$\int_X |f| \, d\nu \;\leq\; \int_X |f| \cdot 1 \, d(\mu + \nu) \;\leq\; \|f\|_2 \left((\mu + \nu)(X) \right)^{\frac{1}{2}}.$$

Thus we may define a linear functional T on $\mathcal{L}_2(\mu + \nu)$ by

$$T(f) \;=\; \int_X f \, d\nu \qquad (f \in \mathcal{L}_2(\mu + \nu)).$$

Moreover, again applying Hölder's inequality,

$$|T(f)| \leq \int_X |f| \, d\nu \leq \left(\int_X |f|^2 \, d\nu \right)^{\frac{1}{2}} (\nu(X))^{\frac{1}{2}}$$

$$\leq \|f\|_2 (\nu(X))^{\frac{1}{2}},$$

and to T is a bounded linear functional. Consequently, there is a function $h \in \mathcal{L}_2(\mu + \nu)$ such that

$$T(f) = \int_X f h \, d(\mu + \nu).$$

Next we verify that $h \geq 0$ $(\mu + \nu)$-almost everywhere. If $A = \{x : h(x) < 0\}$, then $A \in M$ and since μ and ν are finite, $\chi_A \in \mathcal{L}_1(\mu + \nu)$. Then

$$0 \leq \int_X \chi_A \, d\nu = T(\chi_A)$$

$$= \int_X \chi_A h \, d(\mu + \nu).$$

If $(\mu + \nu)(A) > 0$, then $\int_X \chi_A h \, d(\mu + \nu) < 0$, a contradiction; thus $h \geq 0$ $(\mu + \nu)$-almost everywhere.

Since for all $g \in \mathcal{L}_1(\mu + \nu)$, $\int_X g \, d(\mu + \nu) = \int_X g \, d\mu + \int_X g \, d\nu$, we know that

$$\int_X f(1 - h) \, d\nu = \int_X f h \, d\mu. \tag{1}$$

Now let $B = \{x : h(x) \geq 1\}$. Then taking $f = \chi_B$ gives

$$0 \leq \mu(B) = \int_X \chi_B \, d\mu \leq \int_X \chi_B h \, d\mu = \int_X \chi_B (1 - h) \, d\nu \leq 0,$$

implying that $\mu(B) = 0$. Since $\nu \ll \mu$, it follows that $\nu(B) = 0$ as well.

Now set $\varphi = h \chi_{A^c} \chi_{B^c}$, so $0 \leq \varphi(x) < 1$ for all x and $\varphi = h$ both ν-almost everywhere and μ-almost everywhere. Thus by (1)

$$\int_X f(1 - \varphi) \, d\nu = \int_X f \varphi \, d\mu$$

for all $f \in \mathcal{L}_2(\mu + \nu)$, completing the proof.

7.7.7. Radon-Nikodym Theorem. *Let* (X, M) *be a measurable space and let* μ *and* ν *be finite measures defined on* M *with* $\nu \ll \mu$. *Then there is a nonnegative function* $\left[\frac{d\nu}{d\mu} \right]$ *in* $\mathcal{L}_1(\mu)$ *such that*

(a) $\int_X f \, d\nu = \int_X f \left[\frac{d\nu}{d\mu} \right] d\mu$

for all nonnegative M-measurable functions f defined on X. If $f \in \mathcal{L}_1(\nu)$, then $f\left[\frac{d\nu}{d\mu}\right] \in \mathcal{L}_1(\mu)$ and (a) holds. In particular,

(b) $\nu(A) = \int_A \left[\frac{d\nu}{d\mu}\right] d\mu$ for all $A \in M$.

Moreover, the function $\left[\frac{d\nu}{d\mu}\right]$ is unique in the sense that if g is any other function in $\mathcal{L}(\mu)$ satisfying (b), then $g = \left[\frac{d\nu}{d\mu}\right]$ μ-a.e.

Proof. Denote by φ the function of 7.7.6, and let $f \geq 0$ be any bounded M-measurable function. Then for each n the function

$$\psi_n = f\left(\sum_0^{n-1} \varphi^k\right)$$

is bounded and measurable, and hence $\psi_n \in \mathcal{L}_2(\mu + \nu)$ since $\mu + \nu$ is a finite measure. By 7.7.6(b)

$$\int_X (1 + \cdots + \varphi^{n-1}) f(1 - \varphi)\, d\nu = \int_X (1 + \cdots + \varphi^{n-1}) f\varphi\, d\mu;$$

since $0 \leq \varphi(x) < 1$ for all x, we may rewrite this as

$$\int_X (1 - \varphi^n) f\, d\nu = \int_X \frac{\varphi(1 - \varphi^n)}{1 - \varphi} f\, d\mu. \tag{2}$$

Now $\{(1 - \varphi^n)f\}$ is a nondecreasing sequence, and thus letting $n \to \infty$ in (2) gives, via Monotone Convergence,

$$\int_X f\, d\nu = \int_X \frac{\varphi}{1 - \varphi} f\, d\mu \tag{3}$$

for all bounded measurable functions $f \geq 0$.

Set $\left[\frac{d\nu}{d\mu}\right] = \frac{\varphi}{1-\varphi}$; taking $f = \chi_A$ in (3) gives (b). If f is an arbitrary nonnegative M-measurable function, then $\{f \wedge n\}$ is a nondecreasing sequence of bounded functions converging to f, and so (a) follows from (3) and the Monotone Convergence Theorem. The conclusions regarding $f \in \mathcal{L}_1$ are now obvious consequences. We leave the proof of uniqueness to the problems.

As with the Hahn Decomposition, it is possible to extend this to σ-finite measure spaces, as well as to signed measures. We leave both extensions to the exercises.

Recall that if $\alpha \in BV[a, b]$, then $\alpha = g - h$ where $g \in AC[a, b]$, $h \in BV[a, b]$ and $h' = 0$ λ-a.e. (Theorem 5.5.8). There is a similar decomposition for measures, and this is the final topic of this section.

7.7.8.   Definition. Let (X, M) be a measurable space and let μ and ν be measures defined on M. Suppose that there is a measurable set A such that

$$\mu(A) = 0 = \nu(A^C).$$

Then we say that μ and ν are *mutually singular* and write $\mu \perp \nu$.

7.7.9. Lebesgue Decomposition Theorem. *Let (X, M) be a measurable space and let μ and ν be finite measures defined on M. Then there are unique measures ν_1 and ν_2 satisfying:*

      (a) $\nu = \nu_1 + \nu_2$.

      (b) $\nu_1 \ll \mu$.

      (c) $\nu_2 \perp \mu$.

Proof. First observe that $\nu \ll (\mu + \nu)$, and thus there is a measurable function $\varphi \equiv \left[\frac{d\nu}{d(\mu+\nu)} \right]$ such that

$$\int_X f \, d\nu = \int_X f\varphi \, d(\mu + \nu)$$

for all nonnegative M-measurable functions f. We first show that $\varphi \leq 1$ $(\mu + \nu)$-almost everywhere.

Set $E = \{x : \varphi(x) > 1\}$ and suppose that $(\mu + \nu)(E) > 0$. Then there is a set $F \subset E$ and a number $\alpha > 1$ such that $(\mu + \nu)(F) > 0$ and $\varphi \geq \alpha$ on F. Then

$$\nu(F) = \int_X \chi_F \, d\nu = \int_X \chi_F \varphi \, d(\mu + \nu)$$

$$\geq \alpha(\mu + \nu)(F),$$

and so $(1 - \alpha)\nu(F) \geq \alpha\mu(F)$. Since $\alpha \geq 1$, this implies that $\alpha\mu(F) \leq 0$, i.e., that $\mu(F) = 0$. If $\nu(F) > 0$, then $\alpha\mu(F) < 0$, again a contradiction, so $\nu(F) = 0$ as well. Thus $(\mu + \nu)(E) = 0$, as claimed.

Now set $\psi = 1 \wedge \varphi$, so $0 \leq \psi \leq 1$ and

$$\int_X f \, d\nu = \int_X f\psi \, d(\mu + \nu)$$

for all nonnegative M-measurable functions f.

Next, let A be the set

$$A = \{x \in X : \psi(x) = 1\}.$$

Then

$$\nu(A) = \int_X \chi_A \, d\nu = \int_X \chi_A \psi \, d(\mu + \nu)$$

$$= \mu(A) + \nu(A),$$

and thus $\mu(A) = 0$. Define the measures ν_1, ν_2 on $\mathcal{M}$ by

$$\nu_2(C) = \nu(C\cap A)$$

$$\nu_1(C) = \nu(C\cap A^C).$$

Then $\nu = \nu_1 + \nu_2$ and $\nu_2\perp\mu$. We next show that $\nu_1 \ll \mu$.
Let $D \in \mathcal{M}$ and suppose that $\mu(D) = 0$. Then

$$\nu_1(D) = \nu(D\cap A^C) = \int_{D\cap A^C} 1\, d\nu$$

$$= \int_{D\cap A^C} \psi\, d\mu + \int_{D\cap A^C} \psi\, d\nu$$

$$= \int_{D\cap A^C} \psi\, d\nu,$$

and thus $\int_{D\cap A^C}(1-\psi)\, d\nu = 0$. Since $(1-\psi)>0$ on A^C, this implies that $\chi_{D\cap A^C}(1-\psi) = 0$ ν-almost everywhere, i.e., that $\nu_1(D) = 0$. Consequently, $\nu_1 \ll \mu$.

For uniqueness, suppose that $\nu = \nu_1+\nu_2 = \tilde{\nu}_1+\tilde{\nu}_2$ are two decompositions for ν satisfying (a)—(c). Since both $\nu_2 \perp \mu$ and $\tilde{\nu}_2 \perp \mu$, there are measurable sets A and $\tilde{A}$ such that

$$\mu(A) = \mu(\tilde{A}) = 0 \text{ and } \nu_2(A^C) = \nu_2(\tilde{A}^C) = 0.$$

Now for $E \in \mathcal{M}$, we may write

$$E = [E\cap(A\cup\tilde{A})] \cup [E\cap(A\cup\tilde{A})^C].$$

Now $\mu(E \cap (A \cup \tilde{A})) = 0$; since $\nu_1 \ll \mu$ and $\tilde{\nu}_1 \ll \mu$, this implies that $\nu_1(E \cap (A \cup \tilde{A})) = \tilde{\nu}_1(E \cap (A \cup \tilde{A})) = 0$. Since

$$\nu(E\cap(A\cup\tilde{A})) = \tilde{\nu}_1(E\cap(A\cup\tilde{A}))$$

$$= \nu_1(E\cap(A\cup\tilde{A})) + \nu_2(E\cap(A\cup\tilde{A})),$$

this implies that

$$\nu_2(E\cap(A\cup\tilde{A})) = \tilde{\nu}_2(E\cap(A\cup\tilde{A})).$$

Thus

$$\nu_2(E) = \nu_2\big(E\cap(A\cup\tilde{A})\big) + \nu_2\big(E\cap(A\cup\tilde{A})^C\big)$$

$$= \nu_2\big(E\cap(A\cup\tilde{A})\big) + 0$$

$$= \tilde{\nu}_2\big(E\cap(A\cup\tilde{A})\big) + 0$$

$$= \tilde{\nu}_2(E).$$

Thus $\nu_2 = \tilde{\nu}_2$. Since $\nu_1 + \nu_2 = \tilde{\nu}_1 + \tilde{\nu}_2$, this implies that $\nu_1 = \tilde{\nu}_1$ as well.

PROBLEMS

7.34. Prove the uniqueness assertion of the Radon-Nikodym Theorem.

7.35. Let $X = [0,1]$ and M be the Borel subsets of X. Let $\alpha(t) = t^2$ and $\beta(t) = t^3$ and define measures μ and ν on M by

$$\mu(E) = \int_E 1\, d\alpha \quad \text{and}$$

$$\nu(E) = \int_E 1\, d\beta.$$

Only *one* of

$$\left[\frac{d\mu}{d\nu}\right] \quad \text{and} \quad \left[\frac{d\nu}{d\mu}\right]$$

exists. Decide which one and compute its value.

7.36. Let (X, M) be a measurable space and let μ and ν be σ-finite measures defined on M. If $\nu \ll \mu$, show that there is a nonnegative function $\left[\frac{d\nu}{d\mu}\right]$ in $\mathcal{L}_1(\mu)$ such that

(a)

$$\int_X f\, d\nu = \int_X f\left[\frac{d\nu}{d\mu}\right] d\mu$$

for all nonnegative M-measurable functions f defined on X. If $f \in \mathcal{L}_1(\nu)$, then

$$f\left[\frac{d\nu}{d\mu}\right] \in \mathcal{L}_1(\mu)$$

and (a) holds. In particular,

(b)

$$\nu(A) = \int_A \left[\frac{d\nu}{d\mu}\right] d\mu$$

for all $A \in M$.

7.37. Let (X, M) be a measurable space and let μ, ν and ξ be finite measures defined on M. If $\mu \ll \nu$ and $\nu \ll \xi$, show that $\mu \ll \xi$ and that

$$\left[\frac{d\mu}{d\xi}\right] = \left[\frac{d\mu}{d\nu}\right] \cdot \left[\frac{d\nu}{d\xi}\right].$$

7.8. THE FOURIER TRANSFORM. The results in the foregoing sections have many profound applications. Some of the most important, especially for probability and differential equations, involve the Fourier transform. As this transform is a complex-valued function, we digress briefly to discuss some of the foregoing results for complex-valued functions.

Throughout, C will denote the complex number field; a complex number $z \in C$ can be uniquely written as $z = a + bi$, where a and b are real numbers, $a = Re(z)$ (the *real part* of z) and $b = Im(z)$ (the *imaginary part* of z). The *modulus* of a complex number z is the real number $|z|$ given by

$$|z| = \sqrt{Re(z)^2 + Im(z)^2}.$$

Notice that if $z_1, z_2 \in C$, then

$$|z_1 \cdot z_2| = |z_1| \cdot |z_2|.$$

Further, if $z \in C$, then we may write

$$z = \varrho e^{i\theta}$$

where

$$\varrho = |z| \quad \text{and}$$

$$\theta = \begin{cases} \arctan\left(\frac{Im(z)}{Re(z)}\right) & \left(-\frac{\pi}{2} < \theta < \frac{\pi}{2}\right) & Re(z) \neq 0 \\ \text{sgn}(Im(z)) \cdot \pi/2 & & Re(z) = 0. \end{cases}$$

For a complex number z,

$$e^z = e^{Re(z)}\left(\cos(Im(z)) + i\sin(Im(z))\right).$$

Notice that $|\cdot|$ is a *norm* over C and that the functions $Im(\cdot)$ and $Re(\cdot)$ are continuous, real-valued mappings.

7.8.1. Definition. Let (X, M, μ) be a measure space, and let f be a mapping from X to C; then we say that f is *measurable* if

$$f^{-1}(U) \in M \quad \text{for all open sets } U \subset C.$$

7.8.2. Proposition. *Let (X, M, μ) be a measure space and let f be a mapping from X to C. Then f is measurable if and only if each of the function $Re(f)$ and $Im(f)$ are measurable functions from X to $\mathfrak{R}$.*

Proof. This is immediate from the continuity of $Re(\cdot)$ and $Im(\cdot)$.

7.8.3. Corollary. *If f is a complex-valued measurable function, then the modulus of f, $|f|$, is a real-valued measurable function.*

7.8.4. Definition. A complex-valued measurable function f defined on a measure space (X, M, μ) is *integrable* if $|f| \in \mathcal{L}_1(X, M, \mu)$.

7.8.5. Proposition. *Let (X, M, μ) be a measure space and let f be a complex-valued integrable function defined on X. Then both $Re(f)$ and $Im(f)$ are members of $\mathcal{L}_1(X, M, \mu)$.*

Proof. Both $|Re(f)|$ and $|Im(f)|$ are nonnegative measurable functions by 7.8.2; since

$$|Re(f)| \leq |f| \quad \text{and} \quad |Im(f)| \leq |f|$$

it follows that $\int_X |Re(f)| \, d\mu < \infty$ and $\int_X |Im(f)| \, d\mu < \infty$.

In view of 7.8.5, the following definition is now possible.

7.8.6. Definition. *Let f be an integrable, complex-valued function defined on the measure space (X, M, μ). Then we define the integral of f to be the complex number*

$$\int_X f \, d\mu \;=\; \int_X Re(f) \, d\mu + i \int_X Im(f) \, d\mu.$$

7.8.7. Proposition. *Let $f, g \in \mathcal{L}_1(X, M, \mu)$ be a complex-valued functions and let $\alpha \in C$. Then:*
(i) Both $f + g$ and αf are in $\mathcal{L}_1(X, M, \mu)$; moreover,

$$\int_X f + g \, d\mu = \int_X f \, d\mu + \int_X g \, d\mu \quad \text{and}$$

$$\int_X \alpha f \, d\mu = \alpha \int_X f \, d\mu.$$

(ii) $\left| \int_X f \, d\mu \right| \leq \int_X |f| \, d\mu$.

Proof. The assertions in (i) are obvious, and so we will prove only (ii). Write the complex number $\int_X f \, d\mu$ as $\varrho e^{i\theta}$; set $e^{-i\theta} f \equiv g = g_1 + i g_2$. Then

$$\int_X g \, d\mu = \int_X g_1 \, d\mu + i \int_X g_2 \, d\mu$$
$$= e^{-i\theta} \int_X f \, d\mu$$
$$= e^{-i\theta} \varrho e^{i\theta}$$
$$= \varrho.$$

In particular, $\int_X g \, d\mu$ is real, implying that $\int_X g_2 \, d\mu = 0$ and $\int_X g_1 \, d\mu = |\int_X f \, d\mu|$. Now

$$g_1 \leq |g_1|$$
$$\leq |e^{-i\theta} f|$$
$$= |f|,$$

and so

$$|\int_X f \, d\mu| = \int_X g_1 \, d\mu$$
$$\leq \int_X |f| \, d\mu,$$

establishing (ii).

7.8.8. Dominated Convergence Theorem. *Let $(X, \mathcal{M}, \mu)$ be a measure space and let $\{f_n\}$ be a sequence of complex-valued integrable functions defined on X such that $f(x) = \lim f_n(x)$ exists a.e. Suppose there is a function $g \in \mathcal{L}_1(X, \mathcal{M} \, \mu)$ such that $|f_n| \leq g$ a.e. Then f is integrable and*

$$\int_X f \, d\mu = \lim \int_X f_n \, d\mu.$$

Proof. Let $\varphi_n = Re(f_n)$ and $\psi_n = Im(f_n)$; then $\{\varphi_n\} \subset \mathcal{L}_1(X, \mathcal{M}, \mu)$, $\{\psi_n\} \subset \mathcal{L}_1(X, \mathcal{M}, \mu)$, $|\psi_n| \leq |f_n| \leq g$ and $|\varphi_n| \leq |f_n| \leq g$. If $\varphi = Re(f)$ and $\psi = Im(f)$, then $\lim \psi_n = \psi$ a.e. and $\lim \varphi_n = \varphi$ a.e. Thus, applying the real-valued dominated convergence theorem to the sequences $\{\varphi_n\}$ and $\{\psi_n\}$, gives $\varphi, \psi \in \mathcal{L}_1(X, \mathcal{M}, \mu)$ and

$$\int_X \varphi \, d\mu = \lim \int_X \varphi_n \, d\mu, \quad \int_X \psi \, d\mu = \lim \int_X \psi_n \, d\mu.$$

Since $|f| = (\varphi^2 + \psi^2)^{\frac{1}{2}} \leq |\varphi| + |\psi|$, it follows at once that $|f| \in \mathcal{L}_1(X, \mathcal{M}, \mu)$, so f is integrable. Finally,

$$\int_X f \, d\mu = \int_X \varphi \, d\mu + i \int_X \psi \, d\mu$$

$$= \lim \int_X \varphi_n \, d\mu + i \int_X \psi_n \, d\mu$$

$$= \lim \int_X f_n \, d\mu,$$

completing the proof.

Before proceeding with the Fourier transform, we need the following formula for integration by substitution. (Much sharper formulas than the one below are available—this special case suffices for our purposes.)

7.8.9. Lemma. *Let $f \in \mathcal{L}_1(\Re)$ and let φ be a continuously differentiable function defined on $\Re$. If $\varphi' > 0$, then*

$$\int_\Re f(\varphi(t)) \, dt = \int_\Re (\varphi'(u))^{-1} f(u) \, d\mu,$$

while if $\varphi' < 0$, then

$$\int_\Re f(\varphi(t)) \, dt = - \int_\Re (\varphi'(u))^{-1} f(u) \, d\mu.$$

Proof. If $f \in C_C(\Re)$, each formula reduces to the conventional substitution formula from calculus. Consequently, the former formula holds for all $f \in \mathcal{L}^+$, such f being the monotonic limit a.e. of functions from $C_C(\Re)$, and thus holds for all $f \in \mathcal{L}_1(\Re)$. The latter formula is an obvious consequence of the former.

7.8.10. Corollary. *Let f be an integrable, complex-valued function defined on $\Re$ with Lebesgue measure, and let φ be as in 7.8.9. If $\varphi' > 0$, then*

$$\int_\Re f(\varphi(t)) \, dt = \int_\Re (\varphi'(u))^{-1} f(u) \, d\mu,$$

while if $\varphi' < 0$, then

$$\int_\Re f(\varphi(t)) \, dt = - \int_\Re (\varphi'(u))^{-1} f(u) \, d\mu.$$

Proof. Apply 7.8.9 to each of $Re(f)$ and $Im(f)$.

7.8.11. Proposition. *Let $f \in \mathcal{L}_1(\Re)$. Then for each $x \in \Re$ the function $e^{-itx}f(t)$ is an integrable function in t. If*

$$\hat{f}(x) = \int_{\Re} e^{-itx} f(t)\, dt,$$

then $\hat{f}$ is continuous and $\lim_{|x| \to \infty} |\hat{f}(x)| = 0$.

Proof. Since $|e^{i\theta}| = |\cos(\theta) + i\sin(\theta)| = 1$, it follows that $e^{-itx} f(t)$ is integrable as a function of t. Moreover,

$$|e^{-i(x+h)t} f(t)| = |f(t)|$$

and so the Dominated Convergence Theorem implies that

$$\lim_{h \to 0} \hat{f}(x+h) = \lim_{h \to 0} \int_{\Re} e^{-i(x+h)t} f(t)\, dt = \int_{\Re} e^{-ixt} f(t)\, dt = \hat{f}(x)$$

and thus $\hat{f}$ is continuous.

We will verify the final conclusion first for $f \in \mathcal{C}_c(\Re)$. Note that

$$\hat{f}(x) = \int_{\Re} f(t) e^{-itx}\, dt$$

$$= -e^{-i\pi} \int_{\Re} f(t) e^{-itx}\, dt$$

$$= -\int_{\Re} f(t) \exp\left(-ix(t + \frac{\pi}{x})\right) dt$$

$$= -\int_{\Re} f(t - \frac{\pi}{x}) e^{-itx}\, dt.$$

Thus

$$2|\hat{f}(x)| = \left| \int_{\Re} f(t) e^{-itx}\, dt - \int_{\Re} f\left(t - \frac{\pi}{x}\right) e^{-itx}\, dt \right|$$

$$\leq \int_{\Re} \left| f(t) - f\left(t - \frac{\pi}{x}\right) \right| dt.$$

Since $f \in \mathcal{C}_c(\Re)$ and $\lim_{x \to \infty} |f(t) - f(t - \frac{\pi}{x})| = 0$, it follows that $\lim_{x \to \infty} \hat{f}(x) = 0$.

For the general case, let $f \in \mathcal{L}_1(\Re)$ and let $\epsilon > 0$ be arbitrary. Choose $\varphi \in \mathcal{C}_c(\Re)$ so that $\|f - \varphi\|_1 \leq \epsilon$. Then for all x,

$$|\hat{f}(x) - \hat{\varphi}(x)| = \left| \int_{\Re} (f(t) - \varphi(t)) e^{-itx}\, dt \right|$$

$$\leq \int_{\Re} |f - \varphi|\, d\lambda$$

$$\leq \epsilon.$$

Since $\lim_{x\to\infty} \hat{\varphi}(x) = 0$ and ϵ was arbitrary, this shows that $\lim_{x\to\infty} \hat{f}(x) = 0$, as desired.

The reader should compare this result with problem 4.C.

7.8.12. Definition. If $f \in \mathcal{L}_1(\Re)$, then the *Fourier transform* of f is the function $\hat{f}(x) = \int_{\Re} e^{-itx} f(t) \, dt$.

In practice it is often difficult to compute directly the Fourier transform of a given $f \in \mathcal{L}_1(\Re)$; we conclude this section by computing the Fourier transform for a particular function which plays a central role in probability theory. The example will also be used in Section 7.9.

7.8.13. Example. For $c > 0$ set $\varphi_c(t) = \frac{1}{c\sqrt{2\pi}} \exp(-\frac{t^2}{2c^2})$. Then $\varphi_c \in \mathcal{L}_1(\Re)$, $\|\varphi_c\| = 1$, and $\hat{\varphi}(x) = \exp(-\frac{x^2 c^2}{2})$.

The fact that $\varphi_c \in \mathcal{L}_1(\Re)$ and $\|\varphi_c\| = 1$ is a famous problem (which involves Fubini's Theorem!). We leave this result to the problems and prove only the assertion regarding $\hat{\varphi}$.

7.8.14. Lemma. *Let f be a nonnegative function in $\mathcal{L}_1(\Re)$ and suppose that $\|f\| = 1$ while $\int_{\Re} |t| f(t) \, dt < \infty$. Then $\hat{f}$ is everywhere differentiable and*

$$\hat{f}'(x) \;=\; -i \int_{\Re} t f(t) e^{-itx} \, dt.$$

Proof. Observe that

$$\left| \frac{e^{i(x+h)t} - e^{ixt}}{h} \right| \;=\; \left| \frac{\cos(ht) - 1 + i\sin(ht)}{h} \right|$$

$$= \; \left| \frac{2 - 2\cos(ht)}{h^2} \right|^{\frac{1}{2}}$$

$$\leq \; 2|t|.$$

Thus, since $\int_{\Re} |t| f(t)\, dt < \infty$, we may apply Dominated Convergence as follows:

$$-i \int_{\Re} t f(t) e^{-itx}\, dt \;=\; \int_{\Re} \frac{d}{dx}\left(e^{-itx}\right) f(t)\, dt$$

$$=\; \int_{\Re} \lim_{h \to 0} \frac{e^{-i(x+h)t} - e^{-itx}}{h} f(t)\, dt$$

$$=\; \lim_{h \to 0} \frac{\hat{f}(x+h) - \hat{f}(x)}{h}.$$

Proof of 7.8.13. A simple integration verifies

$$\int_0^\infty t e^{-\frac{t^2}{2c^2}}\, dt < \infty,$$

and thus, by symmetry of φ_c about 0, $\int_{\Re} |t| \varphi_c(t)\, dt < \infty$. Hence 7.8.14 implies that $\hat{\varphi}_c$ is differentiable and

$$c\sqrt{2\pi}\, \hat{\varphi}_c'(x) \;=\; -i \int_{\Re} t e^{-itx} e^{-\frac{t^2}{2c^2}}\, dt.$$

Now fix $a < b$ and integrate by parts:

$$\int_a^b -it e^{-\frac{t^2}{2c^2}} e^{-itx}\, dt$$

$$=\; c^2 i e^{-\frac{t^2}{2c^2}} e^{-itx} \Big|_{t=a}^b \;-\; \int_a^b c^2 x e^{-itx} e^{-\frac{t^2}{2c^2}}\, dt.$$

Letting $a \to -\infty$ and $b \to +\infty$ therefore gives

$$\hat{\varphi}_c'(x) \;=\; -c^2 x \hat{\varphi}_c(x).$$

Since $\hat{\varphi}_c(0) = 1$, solving the differential equation above gives

$$\hat{\varphi}_c(x) \;=\; \exp\left(-\frac{x^2 c^2}{2}\right).$$

PROBLEMS

7.38. Show that $\|\varphi_c\|_1 = 1$.

7.39. For each n let

$$k_n(t) = e^{(-|t|/n)}.$$

Find $\hat{k}_n$ and $\|k_n\|_1$. (The sequence $\{k_n\}$ is known as *Fejer's Kernel.*)

7.40. For each n let
$$f_n(t) = \begin{cases} \left(1 - \frac{|t|}{n}\right) & \text{if } -n \leq t \leq n \\ 0 & \text{otherwise.} \end{cases}$$

Find $\hat{f}_n$ and $\|f_n\|_1$. (The sequence $\{f_n\}$ is known as *Abel's Kernel*.)

7.41. Suppose that the function f is continuously differentiable and
   (a) $\lim_{|x| \to 0} f(x) = 0$;
   (b) $f \in \mathcal{L}_1(\mathfrak{R})$; and
   (c) $f' \in \mathcal{L}_1(\mathfrak{R})$.

Show that $ix\hat{f}(x) = \hat{f'}(x)$ for each x.

7.9. THE BANACH ALGEBRA $\mathcal{L}_1(\mathfrak{R})$. In the next section we will address when and how a function $f \in \mathcal{L}_1(\mathfrak{R})$ can be recovered from its Fourier transform. The proof of the inversion formula will rely heavily on some special properties of $\mathcal{L}_1(\mathfrak{R})$ which we discuss in this section.

Students may recall from a course in elementary differential equations or probability that the convolution of two functions f and g is the function $f * g$ given by
$$(f * g)(x) \;=\; \int_{\mathfrak{R}} f(x - t)g(t)\, dt,$$

provided that the integral exists and is finite. We begin this section by showing that if f and g are members of $\mathcal{L}_1(\mathfrak{R})$, then $f * g$ is almost everywhere defined and moreover is again a member of $\mathcal{L}_1(\mathfrak{R})$. The latter fact entails examining iterated integrals of the form
$$\int_{\mathfrak{R}} \int_{\mathfrak{R}} f(x - t)g(t)\, dt\, dx,$$

and, not unexpectedly, Fubini's Theorem will play a crucial role. Before we can appeal to Fubini's Theorem, we must establish that the function $f(x - t)g(t)$ is $(\lambda \times \lambda)$-measurable. The following lemma will start this investigation.

7.9.1. Lemma. *Let E be a subset of $\mathfrak{R}$ for which $\lambda(E) = 0$. Then $\{(x,y) : x - y \in E\}$ has $(\lambda \times \lambda)$ measure zero.*

Proof. For each n, set
$$E_n \;=\; \{(x,y) : x - y \in E \text{ and } -n \leq x \leq n\}.$$

It suffices to show $(\lambda \times \lambda)E_n = 0$ for each n. Thus fix n and fix $\epsilon > 0$. Select open intervals $(a_m, b_m) \subset \mathfrak{R}$ so that $E \subset \cup_1^\infty (a_m, b_m)$ and $\sum_1^\infty (b_m - a_m) \leq \frac{\epsilon}{2n}$. Now set
$$A(m,n) \;=\; \{(x,y) : -n \leq x \leq n \text{ and } x - y \in (a_m, b_m)\}.$$

Then $(\lambda \times \lambda)\big(A(m,n)\big) = 2(b_m - a_m)n$. [The set $A(m,n)$ is a parallelogram—draw a picture and compute the area!] Also,

$$E_n = \{(x,y) : x - y \in E \text{ and } -n \le x \le n\}$$

$$\subset \{(x,y) : x - y \in \cup_1^\infty (a_m, b_m) \text{ and } -n \le x \le n\}$$

$$= \cup_1^\infty A(m,n).$$

But

$$(\lambda \times \lambda)(\cup_1^\infty A(n,m)) \le \sum_{m=1}^{\infty} (\lambda \times \lambda)\big(A(m,n)\big)$$

$$= \sum_{m=1}^{\infty} 2(b_m - a_m)n \le \epsilon.$$

Since $\epsilon > 0$ was arbitrary, this implies that $(\lambda \times \lambda)E_n = 0$, as desired.

7.9.2. Lemma. *Let $f, g \in \mathcal{L}_1(\Re)$ and define*

$$h(x,y) = f(x-y)g(y).$$

Then h is a measurable function defined on $\Re \times \Re$.

Proof. Since we may write $f = f^+ - f^-$, it suffices to verify only the case when $f \ge 0$. Fix $\alpha \in \Re$ and note that

$$\{(x,y) : f(x-y)g(y) < \alpha\}$$

$$= \cup_{r \in Q}\{(x,y) : f(x-y)r < \alpha\} \cap \big(\Re \times \{y : g(y) < r\}\big).$$

Thus it suffices to show that sets of the form

$$U = \{(x,y) : f(x-y) < \beta\}$$

are measurable for $\beta \in \Re$.

If f is continuous, then U is an open subset of $\Re \times \Re$, and thus can be written as a countable union of open rectangles. Hence if $f \in \mathcal{C}_c(\Re)$, then U is $(\lambda \times \lambda)$-measurable.

Next let $\{f_n\}$ be a nondecreasing sequence of functions in $\mathcal{C}_c(\Re)$ and set $f = \lim f_n$. Then

$$U = \{(x,y) : f(x-y) < \beta\}$$

$$= \cup_{j=1}^{\infty} \cap_{n=1}^{\infty} \{(x,y) : f_n(x-y) \le \beta - \frac{1}{j}\}$$

and so U is once again measurable.

Now with f as in the preceding paragraph, if $\varphi = f$ ($\lambda \times \lambda$)-almost everywhere, then we may write $\{(x, y) : \varphi(x - y) < \beta\} = A \cup B$ where

$$A = \{(x, y) : f(x - y) < \beta \text{ and } \varphi(x - y) = f(x - y)\} \text{ and}$$

$$B = \{(x, y) : \varphi(x - y) < \beta \text{ and } \varphi(x - y) \neq f(x - y)\}.$$

By Lemma 7.9.2, the set $C = \{(x, y) : \varphi(x - y) \neq f(x - y)\}$ is a null set and so B, being a subset of C, is a null set and hence measurable. One the other hand, A is the intersection of the measurable sets C^c and U and so is also measurable. Thus φ is a measurable function. Since every measurable function defined on $\Re$ can be realized as the difference of two such functions φ, this completes the proof.

The needed facts about convolutions now follow readily.

7.9.3. Theorem. *Let* $f, g \in \mathcal{L}_1(\Re)$. *Then for almost all x the function $h_x(y) = f(x - y)g(y)$ is in $\mathcal{L}_1(\Re)$. For such x we define the convolution of f and g to be*

$$(f * g)(x) = \int_{\Re} f(x - y)g(y)\, dy.$$

Then $f * g \in \mathcal{L}_1(\Re)$ and $\|f * g\|_1 \leq \|f\|_1 \|g\|_1$.

Proof. The preceding lemmas show that $f(x - y)g(y)$ is a $(\lambda \times \lambda)$-measurable function. By the Tonelli-Hobson Theorem

$$\int_{\Re} \int_{\Re} |f(x - y)|\, |g(y)|\, dx\, dy = \int_{-\infty}^{\infty} |g(y)| \int_{-\infty}^{\infty} |f(x - y)|\, dx\, dy$$

$$= \int_{-\infty}^{\infty} |g(y)|\, \|f\|_1\, dy$$

$$= \|f\|_1\, \|g\|_1.$$

Thus the hypotheses of Fubini's Theorem are satisfied, so $h_x \in \mathcal{L}_1(\Re)$ for almost all x, and $f * g \in \mathcal{L}_1(\Re)$. Finally,

$$\|f * g\|_1 = \int_{\Re} \left| \int_{\Re} f(x - y)g(y)\, dy \right|\, dx$$

$$\leq \|f\|_1\, \|g\|_1.$$

It should be reasonably evident that $f * g = g * f$, $(\alpha f) * g = \alpha(f * g)$, $(f * g) * h = f * (g * h)$, and $f * (g + h) = f * g + f * h$. We leave the verification of these simple properties to the reader. Less obvious, but of crucial importance, is the following:

7.9.4. Theorem. *If* $f, g \in \mathcal{L}_1(\Re)$, *then* $\widehat{f * g} = \hat{f}\hat{g}$.

Proof. Applying Fubini's Theorem gives us

$$
\begin{aligned}
\widehat{f * g}(x) &= \int_\Re (f * g)(t) e^{-itx} \, dt \\
&= \int_\Re \int_\Re e^{-itx} f(t - u) g(u) \, du \, dt \\
&= \int_\Re \int_\Re e^{-itx} f(t - u) g(u) \, dt \, du \\
&= \int_\Re g(u) \int_\Re e^{-i(y+u)x} f(y) \, dy \, du \qquad (\text{set } y = t - u) \\
&= \int_\Re g(u) e^{-iux} \hat{f}(x) \, du \\
&= \hat{g}(x) \hat{f}(x).
\end{aligned}
$$

Before proceeding with the inversion theorem, there is another technical fact about functions in $\mathcal{L}_1(\Re)$ which we need to verify; 7.9.6, which is due to Lebesgue, can be regarded as a refinement of 5.5.7.

7.9.5. Lemma. *Let* $f \in \mathcal{L}_1[a, b]$. *Then there is a set* $E \subset (a, b)$ *such that* $\lambda([a, b] \backslash E) = 0$ *and*

$$
\lim_{h \to 0+} \frac{1}{h} \int_x^{x+h} |f(t) - \alpha| \, dt = \lim_{h \to 0+} \frac{1}{h} \int_{x-h}^x |f(t) - \alpha| \, dt = |f(x) - \alpha|
$$

for all scalars $\alpha \in \Re$ *and all* $x \in E$.

Proof. Let $\{r_n\}$ be an enumeration of the rational numbers. For each n define $\varphi_n(t) = |f(t) - r_n|$, so $\varphi_n \in \mathcal{L}_1([a, b])$. Thus by 5.5.7 there is a set $E_n \subset (a, b)$ such that $\lambda([a, b] \backslash E_n) = 0$ and

$$
\lim_{h \to 0+} \frac{1}{h} \int_x^{x+h} \varphi_n \, d\lambda = \lim_{h \to 0+} \frac{1}{h} \int_{x-h}^x \varphi_n \, d\lambda = \varphi_n(x)
$$

for all $x \in E_n$. If $E = \cap E_n$, then $\lambda([a, b] \backslash E) = 0$.

Now fix $\epsilon > 0$ and $\alpha \in \Re$. Select r_n so that $|r_n - \alpha| < \frac{\epsilon}{2}$, so that

$$
\big| |f(t) - \alpha| - |f(t) - r_n| \big| \leq |r_n - \alpha| < \frac{\epsilon}{2} \qquad \text{for all } t \in [a, b].
$$

This in turn implies that

$$\left| \frac{1}{h} \int_x^{x+h} |f(t) - \alpha| \, dt - \frac{1}{h} \int_x^{x+h} |f(t) - r_n| \, dt \right| \le \frac{\epsilon}{2}.$$

Consequently, if $x \in E$,

$$\overline{\lim}_{h \to 0+} \left| \frac{1}{h} \int_x^{x+h} |f(t) - \alpha| \, dt - |f(x) - \alpha| \right|$$

$$\le \overline{\lim}_{h \to 0+} \left| \frac{1}{h} \int_x^{x+h} |f(t) - \alpha| \, dt - \frac{1}{h} \int_x^{x+h} |f(t) - r_n| \, dt \right|$$

$$+ \left| \frac{1}{h} \int_x^{x+h} \varphi_n(t) \, dt - \varphi_n(x) \right| + |r_n - \alpha|$$

$$\le \epsilon/2 + \epsilon/2 \ = \ \epsilon.$$

Since ϵ was arbitrary, this shows that if $x \in E$, then

$$\lim_{h \to 0+} \frac{1}{h} \int_x^{x+h} |f(t) - \alpha| \, dt \ = \ |f(x) - \alpha|.$$

A similar argument shows that

$$\lim_{h \to 0+} \frac{1}{h} \int_{x-h}^{x} |f(t) - \alpha| \, dt \ = \ |f(x) - \alpha|$$

for all $x \in E$, completing the proof.

7.9.6. Theorem. *Let* $f \in \mathcal{L}_1[a, b]$. *Then*

$$\lim_{h \to 0+} \frac{1}{h} \int_0^h |f(x + t) + f(x - t) - 2f(x)| \, dt \ = \ 0$$

for almost all $x \in (a, b)$.

Proof. Let E be the set described in 7.9.5. Then for $x \in E$

$$\overline{\lim}_{h \to 0+} \frac{1}{h} \int_0^h |f(x+t) + f(x-t) - 2f(x)| \, dt$$

$$\leq \lim_{h \to 0+} \frac{1}{h} \int_0^h |f(x+t) - f(x)| \, dt + \frac{1}{h} \int_0^h |f(x-t) - f(x)| \, dt$$

$$= \lim_{h \to 0+} \frac{1}{h} \int_x^{x+h} |f(u) - f(x)| \, du + \frac{1}{h} \int_{x-h}^x |f(u) - f(x)| \, du = 0$$

upon taking $\alpha = f(x)$ in 7.9.5.

7.9.7. Corollary. *Let $f \in \mathcal{L}_1(\mathfrak{R})$. Then there is a set $E \subset \mathfrak{R}$ such that $\lambda(\mathfrak{R} \backslash E) = 0$ and for all $x \in E$*

$$\lim_{h \to 0+} \frac{1}{h} \int_0^h |f(x+t) + f(x-t) - 2f(x)| \, dt = 0.$$

*(The set E is called the **Lebesgue set** for f.)*

PROBLEMS

7.42. Let $f, g, h \in \mathcal{L}_1(\mathfrak{R})$ and let $\alpha \in \mathfrak{R}$. Verify
(a) $f * g = g * f$.
(b) $(\alpha f) * g = \alpha(f * g)$.
(c) $(f * g) * h = f * (g * h)$.
(d) $f * (g + h) = f * g + f * h$.

7.43. Show that there is no function $e \in \mathcal{L}_1(\mathfrak{R})$ with the property that $e * f = f$ for all $f \in \mathcal{L}_1(\mathfrak{R})$.

7.44. Let

$$e_n(t) = \begin{cases} \frac{n}{2} & -\frac{1}{n} \leq t \leq \frac{1}{n} \\ 0 & \text{elsewhere.} \end{cases}$$

Show that $\lim_{n \to \infty} \|e_n * f - f\|_1 = 0$. (The sequence $\{e_n\}$ is call an *approximate identity.*)

7.45. Fix $f \in \mathcal{L}_1(\mathfrak{R})$ and fix p so that $1 < p < \infty$; fix $g \in \mathcal{L}_p(\mathfrak{R})$. As usual q is the conjugate of p.
(a) Show that the functions $y \mapsto f(x-y)g(y)$ and $y \mapsto f(y)g(x-y)$ are members of $\mathcal{L}_1(\mathfrak{R})$ for almost all $x \in \mathfrak{R}$. For such x write

$$(f * g)(x) = \int_{\mathfrak{R}} f(x-y)g(y) \, dy \quad \text{and} \quad (g * f)(x) = \int_{\mathfrak{R}} f(y)g(x-y) \, dy.$$

[*Hint:* For arbitrary $h \in \mathcal{L}_q(\mathfrak{R})$, apply Hölder's inequality to

$$\int_{\mathfrak{R}} \int_{\mathfrak{R}} |f(x-y)g(y)h(x)| \, dx \, dy.]$$

(b) Define a mapping T from $\mathcal{L}_q(\Re)$ to $\Re$ by

$$T(h) = \int_{\Re} h(x)(f * g)(x)\, dx.$$

Show that T is a bounded linear functional defined on $\mathcal{L}_q(\Re)$.
(c) Show that $f * g \in \mathcal{L}_p(\Re)$ and $f * g = g * f$ a.e.

7.10. THE FOURIER INVERSION THEOREM.

The stage is now nearly set for this major result. We need two more brief preliminary lemmas.

7.10.1. Lemma. *For each n, set*

$$k_n(t) = \frac{1}{2\pi} \exp\left(-\frac{t^2}{2n^2}\right).$$

Then (a) $\hat{k}_n(x) = \frac{n}{\sqrt{2\pi}} \exp\left(-\frac{x^2 n^2}{2}\right)$.
(b) $\hat{k}_n \in \mathcal{L}_1(\Re)$ *and* $\|\hat{k}_n\|_1 = 1$.
(c) $\hat{k}_n(-x) = \hat{k}_n(x)$.

This first lemma is just a restatement of 7.8.13. The collection of functions $\{k_n\}$ is referred to as *Gauss's kernel.* Families of functions of this type are important because there is no "unit" in $\mathcal{L}_1(\Re)$ (see Problem 7.43). We will first recover f from $\{\hat{f}k_n\}$. The next lemma relates $\hat{f}k_n$ and $f * \hat{k}_n$.

7.10.2. Lemma. *Let* $f \in \mathcal{L}_1(\Re)$. *Then for each n and x*

$$\int_{\Re} \hat{f}(y) k_n(y) e^{ixy}\, dy = (f * \hat{k}_n)(x) = \int_{\Re} f(x - s)\hat{k}_n(s)\, ds.$$

Proof. Applying Fubini's Theorem gives

$$\int_{\Re} \hat{f}(y) k_n(y) e^{ixy}\, dy$$

$$= \int_{\Re} \int_{\Re} f(t) e^{-iyt} k_n(y) e^{ixy}\, dt\, dy$$

$$= \int_{\Re} \int_{\Re} f(t) e^{-iyt} k_n(y) e^{ixy}\, dy\, dt$$

$$= \int_{\Re} f(t) \int_{\Re} k_n(y) e^{-i(t-x)y}\, dy\, dt$$

$$= \int_{\Re} f(t) \hat{k}_n(t - x)\, dt$$

$$= f * \hat{k}_n(x).$$

[Note that $f(t)\hat{k}_n(y) \in \mathcal{L}_1(\Re \times \Re)$; we have then applied Fubini's Theorem separately to the imaginary and real parts of $f(t)e^{-iyt} k_n(y) e^{ixy}$.]

Finally, we are able to recover f from its transform via the following.

7.10.3. Theorem. *Let f be a member of $\mathcal{L}_1(\Re)$, let $\{k_n\}$ denote Gauss's kernel, and let E be the Lebesgue set of f. Then for all $x \in E$*

$$\lim_{n \to \infty} \int_{\Re} \hat{f}(y) k_n(y) \exp(ixy)\, dy = f(x).$$

Proof. Fix x in the Lebesgue set of f; in view of 7.10.2, it suffices to show that

$$\lim_{n \to \infty} f * \hat{k}_n(x) = f(x).$$

Computing

$$f * \hat{k}_n(x) - f(x) \;=\; \int_{\Re} f(x-t)\hat{k}_n(t)\,dt \;-\; \int_{\Re} f(x)\hat{k}_n(t)\,dt$$

$$=\; \int_0^\infty \big(f(x-t) - f(x)\big)\hat{k}_n(t)\,dt + \cdots$$

$$\cdots + \int_{-\infty}^0 \big(f(x-t) - f(x)\big)\hat{k}_n(t)\,dt$$

$$=\; \int_0^\infty \big(f(x+t) + f(x-t) - 2f(x)\big)\hat{k}_n(t)\,dt$$

since $\hat{k}_n(-t) = \hat{k}_n(t)$ for all t. (Note that the integrand above provides the connection with 7.9.7.)

Set $\varphi(t) = f(x+t) + f(x-t) - 2f(x)$ and $\psi(t) = \hat{k}_1(t)$. Then, since $\hat{k}_n(t) = n\psi(nt)$, the above yields

$$|f * \hat{k}_n(x) - f(x)| \;\leq\; \int_0^\infty |\varphi(t)| n\psi(nt)\,dt.$$

We will show that $\displaystyle\lim_{n\to\infty} \int_0^\infty |\varphi(t)| n\psi(nt)\,dt = 0$.

To this end, fix $\epsilon > 0$ and set

$$\alpha = \max\{3\psi(0), \epsilon, 6 + 3\psi(1), 6(\|f\|_1 + |f(x)|)\}.$$

Fix $h > 0$ and set $\Phi(h) = \int_0^h |\varphi(t)|\,dt$; since $x \in E$ there is a number $\beta \in (0, 1)$ such that

$$\frac{1}{h}\Phi(h) \;<\; \frac{\epsilon}{\alpha} \quad \text{for } h \in (0, \beta). \tag{1}$$

Now ψ is a nonincreasing, nonnegative function in $\mathcal{L}_1(\Re^+)$ and thus there is a number N with the following two properties:

$$\int_{n\beta}^\infty \psi(t)\,dt < \frac{\epsilon}{\alpha} \quad \text{(for } n \geq N)$$

and

$$\frac{1}{\beta}\int_{n\beta}^\infty \psi(t)\,dt < \frac{\epsilon}{\alpha} \quad \text{(for } n \geq N). \tag{2}$$

From this latter inequality

$$n\psi(n\beta) = \frac{1}{\beta}\frac{\beta n}{1}\psi(n\beta) \leq \frac{1}{\beta}\int_{n\beta}^{2n\beta} \psi(t)\, dt$$

$$\leq \frac{1}{\beta}\int_{n\beta}^{\infty} \psi(t)\, dt < \frac{\epsilon}{\alpha}$$

i.e.,

$$n\psi(n\beta) < \frac{\epsilon}{\alpha} \quad \text{(for } n \geq N\text{).} \tag{3}$$

Now let $n \geq \max\{N, \frac{1}{\beta}\}$. Then

$$\int_0^{\infty} |\varphi(t)|n\psi(nt)\, dt = \int_0^{\frac{1}{n}} |\varphi(t)|n\psi(nt)\, dt + \cdots$$

$$\cdots + \int_{\frac{1}{n}}^{\beta} |\varphi(t)|n\psi(nt)\, dt + \int_{\beta}^{\infty} |\varphi(t)|n\psi(nt)\, dt.$$

We will show that each of these three integrals is majorized by $\frac{\epsilon}{3}$, and thus $\lim \int_0^{\infty} |\varphi(t)|n\psi(nt)\, dt = 0$, as desired.

First observe, applying (1),

$$\int_0^{\frac{1}{n}} |\varphi(t)|n\psi(nt)\, dt \leq \int_0^{\frac{1}{n}} |\varphi(t)|n\psi(0)\, dt$$

$$= n\Phi(\frac{1}{n})\psi(0) < \psi(0)\frac{\epsilon}{\alpha} \leq \frac{\epsilon}{3},$$

showing that the first integral is majorized by $\frac{\epsilon}{3}$.

For the second integral, use (3) and integrate by parts:

$$\int_{\frac{1}{n}}^{\beta} |\varphi(t)|n\psi(nt)\, dt$$

$$= \Phi(\beta)n\psi(n\beta) - \Phi(\frac{1}{n})n\psi(1) - \int_{\frac{1}{n}}^{\beta} \Phi(t)n^2\psi'(nt)\, dt$$

$$\leq \frac{1}{\beta}\Phi(\beta)n\psi(n\beta) - \int_{\frac{1}{n}}^{\beta} \Phi(t)n^2\psi'(nt)\, dt$$

$$\leq \left(\frac{\epsilon}{\alpha}\right)^2 - \int_{\frac{1}{n}}^{\beta} t\frac{\epsilon}{\alpha}n^2\psi'(nt)\, dt \qquad \text{[by (1) and } \psi' \leq 0]$$

$$= \left(\frac{\epsilon}{\alpha}\right)^2 - \int_1^{n\beta} \frac{\epsilon u}{\alpha} \psi'(u) \, du$$

$$= \left(\frac{\epsilon}{\alpha}\right)^2 - \frac{\epsilon}{\alpha} \left[n\beta\psi(n\beta) - \psi(1) - \int_1^{n\beta} \psi(u) \, du \right]$$

$$\leq \left(\frac{\epsilon}{\alpha}\right)^2 + \frac{\epsilon}{\alpha}\psi(1) + \frac{\epsilon}{\alpha}\int_1^{n\beta} \psi(u) \, du$$

$$\leq \frac{\epsilon}{\alpha}\left(\frac{\epsilon}{\alpha} + \psi(1) + 1\right)$$

$$\leq \epsilon/3$$

majorizing the second integral.

The third integral is not as difficult as the second:

$$\int_\beta^\infty |\varphi(t)| n\psi(nt) \, dt$$

$$\leq \int_\beta^\infty \big(|f(x+t)| + |f(x-t)| \big) n\psi(nt) \, dt + 2|f(x)|n\int_\beta^\infty \psi(nt) \, dt$$

$$\leq n\psi(n\beta)2\|f\|_1 + 2|f(x)|\int_{n\beta}^\infty \psi(u) \, du$$

$$\leq \frac{2\epsilon}{\alpha}\|f\|_1 + 2|f(x)|\frac{\epsilon}{\alpha} \qquad\qquad \text{[by (2) and (3)]}$$

$$\leq \epsilon/3,$$

completing the proof.

Using the above we can now directly find f from $\hat{f}$ without passing through Gauss's kernel, provided that $\hat{f} \in \mathcal{L}_1$.

7.10.4. Corollary (Fourier Inversion Formula). *Let f be a member of $\mathcal{L}_1(\Re)$ and suppose that $\hat{f}$ is also in $\mathcal{L}_1(\Re)$. Then*

$$\frac{1}{2\pi}\int_\Re \hat{f}(t)\exp(ixt) \, dt = f(x)$$

for every point in the Lebesgue set of f.

Proof. Since $\hat{f} \in \mathcal{L}_1(\Re)$, we may apply Dominated Convergence to the conclusion of 7.10.3 and let $n \to \infty$. Since $\lim k_n(t) = \frac{1}{2\pi}$ for all t, the conclusion is immediate.

[The presence of the factor $\frac{1}{2\pi}$ in the inversion formula is sometimes a nuisance. For this reason many authors multiply Lebesgue measure by a factor of $(2\pi)^{-\frac{1}{2}}$ when discussing these results. This makes the theorems more elegant to state and prove but makes them more difficult to use, especially in probability, where this normalization is not standard. Because of this, we have not normalized Lebesgue measure in this fashion.] The restriction in 7.10.4 that $\hat{f} \in \mathcal{L}_1(\Re)$ appears to be a rather formidable one; this condition turns out to be less restrictive than might be expected with judicious use of Gauss's kernel (or some other suitable kernel). By way of example, we prove the following continuity theorem. This theorem appears — without proof — in every elementary probability text and is the crucial fact needed to verify such fundamental results as the Central Limit Theorem and the Weak Law of Large Numbers (see the problems).

7.10.5. Continuity Theorem. *Let $\{f_n\}$ be a sequence of functions in $\mathcal{L}_1(\Re)$ and let $f \in \mathcal{L}_1(\Re)$. Suppose that*

(a) *$f_n, f \geq 0$.*

(b) *$\int_{\Re} f_n \, d\lambda = \int_{\Re} f \, d\lambda = 1$.*

(c) *$\lim \hat{f}_n(x) = \hat{f}(x)$ for all x.*

Then

(d) *$\lim \int_{-\infty}^{x} f_n(t) \, dt = \int_{-\infty}^{x} f(t) \, dt$ for all x.*

Proof. For each n, set $F_n(x) = \int_{-\infty}^{x} f_n(t) \, dt$, so F_n is a nondecreasing function and $0 \leq F_n(x) \leq 1$ for all x. By Helly's Theorem (6.2.4), there is a nondecreasing function $\mu_1(x)$ defined on $[-1, 1]$ and a subsequence $\{F_{n(k,1)}\}$ of $\{F_n\}$ such that $\lim_k F_{n(k,1)}(x) = \mu_1(x)$ on $[-1, 1]$. A second application of Helly's Theorem yields a nondecreasing function $\mu_2(x)$ defined on $[-2, 2]$ and a subsequence $\{F_{n(k,2)}\}$ of $\{F_{n(k,1)}\}$ such that

$$\mu_1(x) = \mu_2(x) \text{ if } x \in [-1, 1]$$
$$\lim_k F_{n(k,2)}(x) = \mu_2(x).$$

Continuing in this fashion we obtain, for each m, a function $\mu_m(x)$ and a subsequence $\{F_{n(k,m)}\}$ of $\{F_{n(k,m-1)}\}$ satisfying

$$\mu_m(x) = \mu_{m-1}(x) \text{ if } x \in [-m+1, m-1]$$
$$\lim_k F_{n(k,m)}(x) = \mu_m(x).$$

Setting $\mu(x) = \lim_m \mu_m(x)$ and $F_{n_k}(x) = F_{n(k,k)}(x)$, we obtain a subsequence $\{F_{n_k}\}$ of $\{F_n\}$ such that

$$\lim_k F_{n_k}(x) = \mu(x). \qquad (x \in \Re)$$

Note that $0 \leq \mu(x) \leq 1$. We begin by showing that $\mu(x) = \int_{-\infty}^{x} f(t) \, dt$.

For $c > 0$, set $\varphi_c(t) = \frac{1}{c\sqrt{2\pi}} \exp\left(-\frac{t^2}{2c^2}\right)$, so $\varphi_c \geq 0$, $\int_{\Re} \varphi_c(t)\, dt = 1$ and $\hat{\varphi}_c(x) = \exp(-x^2 c^2/2)$. Set $\psi_{n_k}(t) = (f_{n_k} * \varphi_c)(t)$. Fix numbers $a < b$ in $\Re$.

Claim 1

$$\lim_k \int_a^b \psi_{n_k}(t)\, dt = \frac{1}{\sqrt{2\pi}} \int_{-\infty}^\infty e^{-u^2/2}\left[\mu(b - cu) - \mu(a - cu)\right] d\mu.$$

Proof. By definition,

$$\int_a^b \psi_{n_k}(t)\, dt = \int_a^b \int_{-\infty}^\infty f_{n_k}(x)\frac{1}{c\sqrt{2\pi}} \exp\left(-\frac{(t-x)^2}{2c^2}\right) dx\, dt$$

$$= \int_{-\infty}^\infty f_{n_k}(x) \int_a^b \frac{1}{c\sqrt{2\pi}} \exp\left(-\frac{(t-x)^2}{2c^2}\right) dt\, dx \qquad \text{(Fubini)}$$

$$\left(\text{change of variables}\quad u = \frac{t-x}{c}\right)$$

$$= \int_{-\infty}^\infty f_{n_k}(x) \int_{\frac{a-x}{c}}^{\frac{b-x}{c}} \frac{1}{\sqrt{2\pi}} \exp\left(-\frac{u^2}{2}\right) du\, dx$$

$$= \int_{-\infty}^\infty \frac{1}{\sqrt{2\pi}} e^{-\frac{u^2}{2}} \int_{a-cu}^{b-cu} f_{n_k}\, dx\, du \qquad \text{(Fubini)}$$

$$= \int_{-\infty}^\infty \frac{1}{\sqrt{2\pi}} e^{-\frac{u^2}{2}} \left[F_{n_k}(b - cu) - F_{n_k}(a - cu)\right] du.$$

Since $0 \leq F_{n_k} \leq 1$, we may apply the Dominated Convergence Theorem to obtain

$$\lim_k \int_a^b \psi_{n_k}(t)\, dt = \int_{-\infty}^\infty \frac{1}{\sqrt{2\pi}} e^{-\frac{u^2}{2}} \left[\mu(b - cu) - \mu(a - cu)\right] du.$$

Claim 2

$$\lim_k \int_a^b \psi_{n_k}(t)\, dt = \int_a^b (f * \varphi_c)(t)\, dt.$$

Proof. Note that

$$\hat{\psi}_{n_k}(x) = \hat{f}_{n_k}(x) \exp\left(-\frac{c^2 x^2}{2}\right).$$

Now $|\hat{f}_{n_k}(x)| \leq \|f_{n_k}\|_1 = 1$, and thus $\hat{\psi}_{n_k} \in \mathcal{L}_1(\Re)$. Consequently, we may apply the Fourier inversion formula:

$$\int_a^b \psi_{n_k}(t)\,dt = \frac{1}{2\pi} \int_a^b \int_{-\infty}^\infty \hat{\psi}_{n_k}(x) e^{itx}\,dx\,dt$$

$$= \frac{1}{2\pi} \int_a^b \int_{-\infty}^\infty \hat{f}_{n_k}(x) \exp\left(-\frac{c^2 x^2}{2}\right) e^{itx}\,dx\,dt$$

(Fubini)

$$= \frac{1}{2\pi} \int_{-\infty}^\infty \int_a^b \hat{f}_{n_k}(x) \exp\left(-\frac{c^2 x^2}{2}\right) e^{itx}\,dt\,dx$$

(calculus)

$$= \frac{1}{2\pi} \int_{-\infty}^\infty \hat{f}_{n_k}(x) \exp\left(-\frac{c^2 x^2}{2}\right) \frac{e^{ibx} - e^{iax}}{ix}\,dx.$$

Since $|\hat{f}_{n_k}(x) \frac{e^{ibx}-e^{iax}}{ix}| \le (b-a)\|f_{n_k}\|_1 = (b-a)$, we may again apply dominated convergence to obtain

$$\lim_k \int_a^b \psi_{n_k}(t)\,dt = \frac{1}{2\pi} \int_{-\infty}^\infty \hat{f}(x) \exp\left(-\frac{c^2 x^2}{2}\right) \frac{e^{ibx} - e^{iax}}{ix}\,dx.$$

Now reversing the steps in the computation for $\psi_{n_k} = f_{n_k} * \varphi_c$, with f replacing f_{n_k}, we see that

$$\lim_k \int_a^b \psi_{n_k}(t)\,dt = \int_a^b (f * \varphi_c)(t)\,dt.$$

Combining Claims 1 and 2 gives

$$\frac{1}{\sqrt{2\pi}} \int_{-\infty}^\infty e^{-\frac{u^2}{2}} \left(\mu(b - cu) - \mu(a - cu)\right)\,d\mu = \int_a^b (f * \varphi_c)(t)\,dt.$$

Now let c tend to zero. Since $|\mu(b - cu) - \mu(a - cu)| \le 2$, dominated convergence gives

$$\lim_{c \to 0} \frac{1}{\sqrt{2\pi}} \int_{-\infty}^\infty e^{-\frac{u^2}{2}} \left(\mu(b - cu) - \mu(a - cu)\right)\,d\mu$$

$$= \frac{1}{\sqrt{2\pi}} \int_{-\infty}^\infty e^{-\frac{u^2}{2}} [\mu(b) - \mu(a)]$$

$$= \mu(b) - \mu(a).$$

The right-hand side is only slightly more complicated (we essentially show that $\{\varphi_c\}$ is an "approximate unit" in $\mathcal{L}_1(\Re)$):

$$\int_a^b (f * \varphi_c)(t)\, dt \;=\; \int_a^b \int_{-\infty}^\infty f(y)\varphi_c(t-y)\, dy\, dt$$

(Fubini)

$$= \int_{-\infty}^\infty f(y) \int_a^b \frac{1}{c\sqrt{2\pi}} \exp\left(-\frac{(t-y)^2}{2c^2}\right) dt\, dy$$

(change of variables: $u = \dfrac{t-y}{c}$)

$$= \int_{-\infty}^\infty f(y) \int_{\frac{a-y}{c}}^{\frac{b-y}{c}} \frac{1}{\sqrt{2\pi}} \exp\left(-\frac{u^2}{2}\right) du\, dy$$

(Fubini)

$$= \int_{-\infty}^\infty \frac{1}{\sqrt{2\pi}} e^{-\frac{u^2}{2}} \int_{a-cu}^{b-cu} f(y)\, dy\, du.$$

Again, applying Dominated Convergence and $\int_{-\infty}^\infty \frac{1}{\sqrt{2\pi}} e^{-\frac{u^2}{2}}\, du = 1$ gives

$$\lim_{c\to 0} \int_a^b (f * \varphi_c)(t)\, dt \;=\; \int_a^b f(t)\, dt.$$

Combining these observations gives, for all $a < b$,

$$\mu(b) - \mu(a) \;=\; \int_a^b f(t)\, dt. \tag{$*$}$$

Now μ is nondecreasing so $\mu(-\infty) = \lim_{a\to -\infty} \mu(a)$ and $\mu(+\infty) = \lim_{b\to\infty} \mu(b)$ exists Since $0 \le \mu \le 1$, $0 \le \mu(+\infty) - \mu(-\infty) \le 1$. However, $\int_{-\infty}^\infty f(t)\, dt = 1$ by assumption, and hence $\mu(+\infty) - \mu(-\infty) = 1$ by ($*$). But then $\mu(+\infty) = 1$ and $\mu(-\infty) = 0$. In particular, letting $a \to -\infty$ in ($*$) gives

$$\mu(b) \;=\; \int_{-\infty}^b f(t)\, dt.$$

Finally, suppose that $\{F_n(x)\}$ does not converge to $\int_{-\infty}^x f(t)\, dt = \mu(x)$. Then there is an $x_0 \in \Re$ and a subsequence $\{f_{n_k}\}$ of $\{f_n\}$ for which $\lim F_{n_k}(x_0)$ exists, but

$$\lim \int_{-\infty}^{x_0} f_{n_k}(t)\, dt \;\ne\; \int_{-\infty}^{x_0} f(t)\, dt.$$

But the sequence $\{F_{n_k}\}$ satisfies the assumptions of the theorem and we have just shown that any such sequence has a subsequence $\{F_{n'_k}\}$ such that

$$\lim F_{n'_k}(x_0) \;=\; \int_{-\infty}^{x_0} f(t)\, dt,$$

a contradiction. This completes the proof.

PROBLEMS

7.46. Let $f \in \mathcal{L}_1[a, b]$ and let $\varphi \in AC[a, b]$. If $F(x) = \int_a^x f(t)\,dt$, show that

$$\int_a^b f(t)\varphi(t)\,dt \;=\; F(b)\varphi(b) - F(a)\varphi(a) - \int_a^b F(t)\varphi'(t)\,dt,$$

justifying the integration by parts in the proof of 7.10.3.

7.47. Let $\alpha > 0$ and β be real numbers and set

$$\delta_\alpha(t) = \begin{cases} \frac{1}{2\alpha} & \text{if } \beta - \alpha \le t \le \beta + \alpha \\ 0 & \text{otherwise.} \end{cases}$$

Show that

(a) $\int_{\Re} \delta_\alpha(t)\,dt = 1$.

(b) $\lim_{\alpha \to 0} \hat{\delta}(x) = e^{-ix\beta}$.

7.48. Let $\{f_n\}$ and $\{g_n\}$ be sequences in $\mathcal{L}_1(\Re)$ and suppose that

(a) both $f_n \ge 0$ and $g_n \ge 0$ for all n;

(b) $\int_{\Re} f_n\,d\lambda = \int_{\Re} g_n\,d\lambda = 1$;

(c) $\lim \hat{f}(x) = \lim \hat{g}(x)$ for all x.

Show that

$$\lim \int_{-\infty}^x f_n(t)\,dt = \lim \int_{-\infty}^x g_n(t)\,dt$$

for all x. (*Hint:* Apply Claim 1.)

7.49. *Weak Law of Large Numbers.* Let f be a nonnegative function in $\mathcal{L}_1(\Re)$ and suppose that $\int_{\Re} |t| f(t)\,dt < \infty$. Set $f_1 = f$ and, for $n > 1$, set $f_n = f * f_{n-1}$. Set $\mu = \int_{\Re} t f(t)\,dt$. Show that for every $\epsilon > 0$,

$$\lim_{n \to \infty} \int_{\mu - \epsilon}^{\mu + \epsilon} n f_n(nt)\,dt \;=\; 1.$$

7.50. *Central Limit Theorem.* Let f be a nonnegative function in $\mathcal{L}_1(\Re)$. Suppose that:

(a) $\int_{\Re} |t| f(t)\,dt < \infty$ and $\int_{\Re} t f(t)\,dt = 0$.

(b) $\int_{\Re} t^2 f(t)\,dt = 1$.

Set $f_1 = f$ and, for $n \ge 1$, set $f_{n+1} = f * f_n$. Show that for all x

$$\lim_{n \to \infty} \int_{-\infty}^x \sqrt{n} f_n(t\sqrt{n})\,dt \;=\; \int_{-\infty}^x \frac{1}{\sqrt{2\pi}} e^{-\frac{t^2}{2}}\,dt.$$

(You will need 7.8.13.)

7.51. Suppose that $f, g \in \mathcal{L}_1(\Re)$ and $\hat{f} = \hat{g}$. Show that $f = g$ a.e.

7.52. Suppose that $f \in \mathcal{L}_1(\Re) \cap \mathcal{L}_\infty(\Re)$ and that $\hat{f}$ is real-valued and nonnegative. Show that $\hat{f} \in \mathcal{L}_1(\Re)$.

7.53. Let $f \in \mathcal{L}_1(\Re) \cap \mathcal{L}_2(\Re)$. Show that $\hat{f} \in \mathcal{L}_2(\Re)$.

7.54. Let $f \in \mathcal{L}_1(\Re)$ and suppose that $f * f = f$ a.e. Show that $f = 0$ a.e.

7.55. Let $f \in \mathcal{L}_1(\Re)$ and suppose that $f * f = 0$ a.e. Show that $f = 0$ a.e.

7.56. Let
$$D = \left\{ \hat{f} : f \in \mathcal{L}_1(\Re) \right\}.$$
Show that D is dense in $C_0(\Re)$ (under the uniform norm).

Chapter 8

Functional Analysis

8.1. THE HAHN-BANACH THEOREM. This chapter is devoted to a brief introduction to some of the fundamental ideas of functional analysis. There are three basic principles of functional analysis, two of which, the Principle of Uniform Boundedness and the Open Mapping Theorem (Problem 4.33), derive from the Baire Category Theorem. (You may want to review the Principle of Uniform Boundedness before proceeding with this chapter; in addition, you should review the Axiom of Choice, if you have not already done so.) The third basic principle of functional analysis is the Hahn-Banach Theorem, the topic of the current section.

A major topic of this text has been the extension of linear functionals originally defined on a space X to a larger space $Y \supset X$. In Sections 2.1 and 2.2 we extended the Riemann integral from $C_c(\Re)$ to $\mathcal{L}_1(\Re)$; later we extended the Riemann-Stieltjes integral to the same class and, in the preceding chapter, Daniell functionals on $V(X)$ to $\mathcal{L}_1(X)$. In each case the extended functionals—the Lebesgue integral, the Lebesgue-Stieltjes integral, and the Daniell integral—retained the continuityproperties of the original functional. Extensions of this type are fundamental to many problems in functional analysis and were first given precise abstract formulation by Hahn in 1927 and by Banach in 1929 (although Helly anticipated their results as early as 1912). In this section we present this major theorem and discuss some of its consequences.

8.1.1. Hahn-Banach Theorem. *Let Y be a vector space and let p be a mapping from Y to $[0, \infty)$ satisfying for all $x, y \in Y$ and all scalars $\lambda > 0$*

(i) $p(\lambda x) = \lambda p(x)$; and

(ii) $p(x + y) \leq p(x) + p(y)$.

Let X be a subspace of Y and let f be a linear mapping from X to $\Re$ satisfying for all $x \in X$

(iii) $f(x) \leq p(x)$.

Then there is a linear mapping g from Y to $\Re$ satisfying

(iv) $f(x) = g(x)$ for all $x \in X$; and

(v) $g(x) \leq p(x)$ for all $x \in Y$.

In many applications, Y is a normed linear space and $p(x) = \|x\|$; a functional satisfying (i) and (ii) is called a *seminorm* or *ecart*. The functional g is an extension of f (iv) and preserves the boundedness property (v).

Proof of 8.1.1. Denote by P the class of all pairs (E, h) satisfying

$$E \text{ is a vector space and } X \subseteq E \subseteq Y$$

$$h \text{ is a linear mapping from } E \text{ to } Y$$

$$h(x) \leq p(x) \text{ for all } x \in E \text{ and}$$

$$h(x) = f(x) \text{ for all } x \in X.$$

Next define a partial ordering on P by saying $(E_1, h_1) \preceq (E_2, h_2)$ if and only if

$$E_1 \subseteq E_2; \text{ and}$$

$$h_1(x) = h_2(x) \text{ for all } x \in E_1.$$

(It is readily checked that $\preceq$ is a partial order.)

Now let $\{(E_\alpha, h_\alpha)\}$ be a totally ordered subset of $(P, \preceq)$. Set $E = \cup_\alpha E_\alpha$ and define $h : E \to \Re$ by $h(x) = h_\alpha(x)$, where α is any index with $x \in E_\alpha$.

As with our earlier arguments, we need to verify that this definition of h is independent of the particular choice of α. Thus suppose that $x \in E_{\alpha_1} \cap E_{\alpha_2}$; since $\{(E_\alpha, h_\alpha)\}$ is totally ordered, either $(E_{\alpha_1}, h_{\alpha_1}) \preceq (E_{\alpha_2}, h_{\alpha_2})$ or else $(E_{\alpha_2}, h_{\alpha_2}) \preceq (E_{\alpha_1}, h_{\alpha_1})$. In either case, it follows from the definition of $\preceq$ that $h_{\alpha_1}(x) = h_{\alpha_2}(x)$, as desired. A similar argument shows that h is linear.

Moreover, $h(x) = h_\alpha(x) \leq p(x)$ for all $x \in E$ and for all $x \in X$, $h(x) = h_\alpha(x) = f(x)$, and thus $(E, h) \in P$. Since $(E, h) \succeq (E_\alpha, h\alpha)$ for all α, it follows from Zorn's Lemma that $(P, \preceq)$ has a maximal element (E_0, h_0). We complete the proof by showing that $E_0 = Y$.

If $E_0 \neq Y$, we may select $y \in Y \backslash E_0$. Now observe for any $u, v \in E_0$ that

$$h_0(u) - h_0(v) = h_0(u - v)$$
$$\leq p(u - v)$$
$$\leq p(u + y) + p(-y - v).$$

This in turn implies that

$$-p(-y - v) - h_0(v) \leq p(u + y) - h_0(u)$$

and hence, for all $u \in E_0$, that

$$\sigma \equiv \sup_{v \in E_0} \{-p(-y - v) - h_0(v)\} \leq p(u + y) - h_0(u) < \infty. \qquad (*)$$

Now set $E^+ = \mathrm{sp}(E_0 \cup \{y\})$. Since $y \notin E_0$, for each vector $x \in E^+$ there is a unique vector $u \in E$ and a unique scalar $t \in \Re$ such that $x = u + ty$. We define the mapping h^+ on E^+ by

$$h^+(x) = h^+(u + ty) =_{def} h_0(u) + t\sigma.$$

It is clear that h^+ is a linear functional, that $h^+ = h_0$ on E_0, and hence that $h^+ = f$ on X. If h^+ satisfies $h^+(x) \leq p(x)$ for all $x \in E^+$ as well, then we will have constructed a pair $(E^+, h^+) \in P$ for which $(E^+, h^+) \succeq (E_0, h_0)$ and

$(E^+, h^+) \neq (E_0, h_0)$, contradicting the maximality of (E_0, h_0). This contradiction shows that $E_0 = Y$, as desired.

Fix $x = w + ty \in E^+$ and consider three cases.

Case 1: $t = 0$.

In this case $h^+(x) = h(w) \leq p(w) = p(x)$ by construction.

Case 2: $t > 0$.

In (∗) take $u = w/t$ to see that

$$\sigma \leq p(w/t + y) - h_0(w/t)$$

which implies, upon multiplying by t, that

$$t\sigma \leq p(w + ty) - h_0(w)$$

from which

$$\begin{aligned} h^+(x) &= h^+(w + ty) \\ &= h_0(w) + t\sigma \\ &\leq p(w + ty) \\ &= p(x). \end{aligned}$$

Case 3: $t < 0$.

Again from (∗)

$$-p(-y - w/t) - h_0(w/t) \leq \sigma;$$

multiplying by $-t > 0$ gives

$$-p(ty + w) + h_0(w) \leq -t\sigma,$$

which in turn implies that

$$\begin{aligned} h^+(x) &= h^+(w + ty) \\ &= h_0(w) + t\sigma \\ &\leq p(ty + w) \\ &= p(x). \end{aligned}$$

Thus in each case $h^+(x) \leq p(x)$, so $(E^+, h^+) \in \mathcal{P}$. This completes the proof.

The following corollary is sometimes referred to as the analytic version of the Hahn-Banach Theorem.

8.1.2. Corollary. *Let Y be a normed linear space and let X be a subspace of Y. For every $f \in X^*$ there is a $g \in Y^*$ such that $g(x) = f(x)$ for all $x \in X$ and $\|g\| = \|f\|$.*

Proof. Fix $f \in X^*$ and define $p(x) = \|f\| \cdot \|x\|$, so that p satisfies (i) and (ii) of 8.1.1. Clearly, $f(x) \leq p(x)$ for all $x \in X$, and thus there is a linear function g such that $g(x) = f(x)$ for all $x \in X$ and such that $g(x) \le p(x)$ for all $x \in Y$. The latter inequality implies that both

$$g(-x) \le \|f\| \cdot \|x\| \quad \text{and}$$
$$g(x) \le \|f\| \cdot \|x\|$$

for all $x \in Y$. From this, $g \in Y^*$ and $\|g\| \le \|f\|$. Since $g = f$ on X, $\|g\| \ge \|f\|$, and so $\|g\| = \|f\|$.

8.1.3. Corollary. *Let Y be a normed linear space and let $x \in Y$. Then there is an $x^* \in Y^*$ such that $\|x^*\| = 1$ and $x^*(x) = \|x\|$.*

Proof. Take $X = \mathrm{sp}\{x\}$ and define $f(tx) = t\|x\|$ for $tx \in X$.

8.1.4. Corollary. *Let X be a normed linear space. If $x, y \in X$ and $x \ne y$, then there is an $x^* \in X^*$ for which $x^*(x - y) = 1$.*

Proof. By 8.1.3 there is an $f \in X^*$ for which $f(x-y) = \|x-y\|$. Since $\|x-y\| \ne 0$, we may take $x^* = \|x - y\|^{-1} \cdot f$.

A somewhat deeper consequence is the following geometric form of the Hahn-Banach Theorem. Note that the functional p in the proof of this next result is *not* a norm, in contrast with the preceding corollaries. Recall that a set S contained in a vector space X is *convex* if the line segment

$$\{tx + (1 - t)y : 0 \le t \le 1\}$$

is contained in S whenever $x, y \in S$.

8.1.5. Theorem. *Let K be a closed and convex subset of a Banach space X and suppose that $x \notin K$. Then there is an $x^* \in X^*$ for which*

$$x^*(x) < \inf\{x^*(u) : u \in K\}.$$

Proof. Set $\hat{K} = K - x$; it suffices to show that there is an $x^* \in X^*$ such that

$$0 < \inf\left\{x^*(u) : u \in \hat{K}\right\},$$

for then

$$x^*(x) < \inf\left\{x^*(u) : u \in \hat{K}\right\} + x^*(x)$$
$$= \inf\{x^*(u - x) : u \in K\} + x^*(x)$$
$$= \inf\{x^*(u) : u \in K\}.$$

Since $\hat{K}$ is closed and $0 \notin \hat{K}$ there is an $\epsilon > 0$ such that $B(0; \epsilon) \cap \hat{K} = \emptyset$. Fix $z \in \hat{K}$ and set

$$C = B(0; \epsilon) - \hat{K} + z$$
$$\equiv \left\{ y : y = u - v + z \text{ where } \|u\| \leq \epsilon \text{ and } v \in \hat{K} \right\}.$$

(Note that C is convex.) Since $B(0, \epsilon) \cap \hat{K} = \emptyset$, $z \notin C$; taking $v = z$ shows that $B(0, \epsilon) \subseteq C$. Now define, for $y \in X$,

$$p(y) = \inf \left\{ \lambda > 0 : \lambda^{-1} y \in C \right\}.$$

Since $\epsilon \|y\|^{-1} y \in C$ for all $y \in X$, $p(y) \leq \epsilon^{-1} \|y\|$ for all y; clearly $p(y) \geq 0$ for all y.

We next show that p satisfies the assumptions (i) and (ii) of 8.1.2. If $\xi \geq 0$, then

$$p(\xi y) = \inf \left\{ \lambda > 0 : \lambda^{-1} \xi y \in C \right\}$$
$$= \inf \left\{ \lambda > 0 : (\lambda/\xi)^{-1} y \in C \right\}$$
$$= \xi \cdot p(y),$$

so that 8.1.2(i) holds. Now fix $u, v \in X$ and select sequences of positive real numbers $\{\mu_n\}$ and $\{\nu_n\}$ so that $\lim \mu_n = p(u)$, $\lim \nu_n = p(v)$, $\mu_n^{-1} u \in C$, and $\nu_n^{-1} v \in C$. Since C is convex,

$$\frac{u + v}{\mu_n + \nu_n} = \left(1 - \frac{\nu_n}{\mu_n + \nu_n} \right) \frac{u}{\mu_n} + \left(\frac{\nu_n}{\mu_n + \nu_n} \right) \frac{v}{\nu_n} \in C$$

and hence

$$p(u + v) \leq \lim(\mu_n + \nu_n) = p(u) + p(v),$$

establishing 8.1.1(ii).

Now take $E = \{tz : t \in \Re\}$ and define the linear functional f on E by $f(tz) = t \cdot p(z)$. Then $f(tz) \leq p(tz)$ for all t, for if $t \geq 0$, then $f(tz) = tp(z) = p(tz)$, while if $t \leq 0$, then $f(tz) = tp(z) \leq 0 \leq p(tz)$. Consequently, we may apply the Hahn-Banach Theorem to obtain a linear functional g defined on all of X for which $g = f$ on E and $g(y) \leq p(y)$ for all $y \in X$.

Since $p(y) \leq \epsilon^{-1} \|y\|$, it now follows that $g(-y) \leq \epsilon^{-1} \|y\|$ and that $g(y) \leq \epsilon^{-1} \|y\|$, from which $g \in X^*$ and $\|g\| \leq \epsilon^{-1}$.

Since C is convex, $0 \in C$ and $z \notin C$, it follows that $p(z) \geq 1$. Now fix $v \in \hat{K}$ and $u \in B(0; \epsilon)$, so that $(u - v + z) \in C$; then $g(u - v + z) \leq p(u - v + z) \leq 1$, and thus

$$g(u) + g(z) - 1 \leq g(v).$$

Since $g(z) = f(z) = p(z) \geq 1$, this in turn implies that

$$\sup \left\{ g(u) : \|u\| \leq \epsilon \right\} \leq \inf \left\{ g(v) : v \in \hat{K} \right\}.$$

Now select $y \in B(0; \epsilon)$ for which $g(y) > 0$ (for example, $y = \epsilon \|z\|^{-1}z$). Then

$$0 = g(0) < g(0) + g(y)$$
$$= g(y)$$
$$\leq \sup \{g(u) : \|u\| \leq \epsilon\}$$

and thus

$$0 < \inf \left\{ g(w) : w \in \hat{K} \right\}.$$

The following easy corollary will be of use in the next section.

8.1.6. Corollary. *Let K be a closed and convex subset of a Banach space X and suppose that $x \notin K$. Then there is an $x^* \in X^*$ for which $\|x^*\| = 1$ and $x^*(x) > \sup\{x^*(u) : u \in K\}$.*

Proof. By 8.1.5 there is a $g \in X^*$ for which $g(x) < \inf \{g(u) : u \in K\}$. Thus $\|g\| \neq 0$ and we may take $x^* = -\|g\|^{-1}g$.

PROBLEMS

8.1. *Prove or find a counterexample.* If X is a Banach space and if $x^* \in X^*$, then there is an $x \in X$ with $\|x\| = 1$ and $x^*(x) = \|x^*\|$. (This is the *dual conclusion* to 8.1.3.)

8.2. Show by example that the extension given by 8.1.2 is not unique.

8.3. Use the Hahn-Banach Theorem to prove the Riesz Representation Theorem for $C[0,1]$.

8.4. Let $X = C_c(\Re)$ endowed with the $\mathcal{L}_1$ norm and let Y be the completion of X (see Problem 4.10). Use the Hahn-Banach Theorem to extend the Reimann integral to Y and show that $Y = \mathcal{L}_1(\Re)$.

8.5. *Hausdorff Moment Problem.* A sequence $\{c_n\}$ of real numbers is said to be a *moment sequence* if there is a function $\alpha \in BV[0,1]$ such that

$$\int_0^1 t^n \, d\alpha(t) = c_n.$$

Show that $\{c_n\}$ is a moment sequence if and only if there exists an $M > 0$ such that whenever $\{a_n\}$ is a sequence with at most finitely many nonzero terms,

$$\left| \sum_{n=1}^{\infty} a_n c_n \right| \leq M \max \left\{ \left| \sum_{n=1}^{\infty} a_n t^n \right| : 0 \leq t \leq 1 \right\}.$$

8.6. Let Y be a Banach space, let X be a proper closed subspace of Y and let $z \in Y \backslash X$. Show that there is an $x^* \in Y^*$ with the following properties:
(a) $\|x^*\| = 1$;
(b) $x^*(z) = \|z\|$; and

(c) $x^*(X) \equiv 0$.

8.7. *Banach limits.*

(a) Let ℓ_∞ be the vector space of all bounded sequences of real numbers $\vec{x} = \{x_n\}$. Show that $\|\vec{x}\| = \sup x_n$ defines a complete norm on ℓ_∞.

(b) Let m_0 be the smallest closed subspace of ℓ_∞ containing subsequences of the form $\vec{y} = (x_1, x_2 - x_1, x_3 - x_2, \ldots)$ where $\{x_n\} \in \ell_\infty$. Show that the sequence $\vec{e} \equiv (1, 1, \ldots)$ is not in m_0.

(c) Show that there is a linear functional $x^* \in \ell_\infty^*$ with the following properties: $\|x^*\| = 1$; $x^*(\vec{e}) = 1$; and $x^*(m_0) \equiv 0$.

(d) For $\vec{x} \in \ell_\infty$, define the Banach limit of $\vec{x}$ to be

$$\text{LIM}\, x_n = x^*(\vec{x})$$

where x^* is as in part *(c)*. Show that LIM has the following properties:

(i) $\text{LIM}\, x_n = \text{LIM}\, x_{n+1}$.

(ii) $\text{LIM}\, \alpha x_n + \beta y_n = \alpha \, \text{LIM}\, x_n + \beta \, \text{LIM}\, y_n$.

(iii) $\text{LIM}\, x_n \geq 0$ if $x_n \geq 0$.

(iv) $\liminf x_n \leq \text{LIM}\, x_n \leq \limsup x_n$.

(v)   if $\{x_n\}$ converges, then $\lim x_n = \text{LIM}\, x_n$.

8.8. *Minimum norm estimates.* **Definition.** Let Y be a normed linear space and let X be a subspace of Y. The *annihilator* of X in Y^* is the set

$$X^\perp = \{x^* \in Y^* : x^*(x) = 0 \quad \text{for all} \quad x \in X\}.$$

Let X and Y be as above and let $y \in Y$. Show that

$$\inf \{\|y - x\| : x \in X\} = \max \{x^*(y) : x^* \in X^\perp \quad \text{and} \quad \|x^*\| \leq 1\},$$

where the maximum on the right is attained at some $x_0^* \in X^\perp$.

8.2. THE WEAK AND WEAK-∗ TOPOLOGIES.

Throughout this text we have discussed a number of different kinds of convergence: pointwise, in measure, in norm, and in some of the problems, "weak" convergence. In this section and in the following two sections we will concentrate on the notions of weak and weak-∗ convergence. These ideas are best understood from a topological point of view, so we begin with a brief review of some elementary topological notions.

8.2.1. Definition. Let X be an arbitrary set. A collection τ of subsets of X is a *topology* if

(a) both $\emptyset \in \tau$ and $X \in \tau$;

(b) whenever U_1 and U_2 are in τ, then $U_1 \cap U_2 \in \tau$;

(c) whenever $\{U_\alpha\}$ is any collection in τ, $\cup_\alpha U_\alpha \in \tau$.;

The elements of τ are said to be *open sets* and the pair (X, τ) is said to be a *topological space.* If $Y \subseteq X$, then Y *inherits* a topology from X, namely

$$\tau_Y = \{U \cap Y \,:\, U \in \tau\}.$$

One of the most obvious examples of a topological space is a metric space; here a set U is open if and only if U can be written as the union of open balls. [Proposition 4.1.4 shows that (M, d) is a topological space.] So far all of our examples of topological spaces have arisen from metrics in just this manner; however this is neither the only, nor the most powerful, way of defining open sets.

In the case of metric spaces, open sets are realized as unions of more elementary sets (open balls), and the properties of τ are deduced by examining these simpler sets. This is in general a very useful approach and is the motivation for the next definition.

8.2.2. Definition. Let (X, τ) be a topological space and let $\mathcal{B}$, be a subset of τ. We will say that $\mathcal{B}$ is a *basis for the topology* τ if

$$U \in \tau \;\leftrightarrow\; U = \bigcup \{S \in \mathcal{B} \,:\, S \subseteq U\}.$$

Not every subfamily of τ will be a basis for τ, nor is every family $\mathcal{B}$ a basis for some topology τ. The following proposition gives necessary and sufficient conditions for a collection $\mathcal{B}$ generate a topology.

8.2.3. Proposition. *Let X be a set and let $\mathcal{B}$ be a collection of subsets of X. Then $\mathcal{B}$ is a base for a topology on X if and only if*
(i) $X = \cup_{B \in \mathcal{B}} B$; and
(ii) whenever B_1 and $B_2 \in \mathcal{B}$ and $x \in B_1 \cap B_2$, there is a B_3 in $\mathcal{B}$ such that $x \in B_3 \subset B_1 \cap B_2$.

Proof. Problem 8.9.

In practice, one usually defines a basis for a topology rather than defining the topology explicitly; this is exactly the approach which we took earlier when we discussed metric spaces. We are now ready to describe the two main additional topologies with which we will be concerned.

8.2.4. Definition. Let X be a normed linear space with dual space X^*. The *weak topology on X* has basis sets in the form

$$B(x; x_1^*, \ldots, x_n^*; \epsilon) \equiv \{y \in X \,:\, |x_i^*(y - x)| < \epsilon, \, i = 1, \ldots, n\},$$

where $x \in X$, $\{x_1^*, \ldots, x_n^*\}$ is a finite collection from X^* and $\epsilon > 0$.

The *weak-* topology on X^** has basis sets of the form

$$B(x^*; x_1, \ldots, x_n; \epsilon) \equiv \{y^* \in X^* \,:\, |y^*(x_i) - x^*(x_i)| < \epsilon, \, i = 1, \ldots, n\},$$

where $x^* \in X^*$, $\{x_1, \ldots, x_n\}$ is a finite collection from X and $\epsilon > 0$.

Thus a set $U \subseteq X$ is weakly open if and only if for each $x \in U$ there is a finite collection $\{x_1^*, \ldots, x_n^*\} \subseteq X^*$ and an $\epsilon > 0$ such that

$$B(x; x_1^*, \ldots, x_n^*; \epsilon) \subseteq U.$$

Similarly, a set $U^* \subseteq X^*$ is weak-$*$ open if and only if for each $x^* \in U^*$ there is a finite collection $\{x_1, \ldots, x_n\} \subseteq X\}$ and an $\epsilon > 0$ such that

$$B(x^*; x_1, \ldots, x_n; \epsilon) \subseteq U^*.$$

The norm tolopolgy on either X or X^* is generally called the *strong topology*. Since the sets $B(x; x_1^*, \ldots, x_n^*, \epsilon)$ and $B(x^*; x_1, \ldots, x_n; \epsilon)$ are strongly open, the following observation is immediate.

8.2.5. Proposition. *Let X be a normed linear space with dual space X^*. Then every weakly open subset of X and every weak-$*$ open subset of X^* is strongly open.*

[The converse to 8.2.5 is true if and only if X is finite dimensional.]

It is a routine matter to verify that either of the definitions above indeed give rise to topologies (Problem 8.10) and that the vector space operations are continuous in both topologies (Problem 8.11). We turn our attention at this point to reformulating some of the standard notions of convergence and compactness in the context of these two topologies.

8.2.6. Definition. Let X be a normed linear space with dual space X^*. Let $\{x_n\}$ be a sequence in X and let $\{x_n^*\}$ be a sequence in X^*. Then $\{x_n\}$ *converges weakly to $x_\infty \in X$* if and only if for each weakly open set U about x_∞ there is an N such that $n \geq N$ implies that $x_n \in U$. Similarly, $\{x_n^*\}$ *converges weakly to $x_\infty^* \in X^*$* if and only if for each weak-$*$ open set U^* about x_∞^* there is an N such that $n \geq N$ implies implies $x_n^* \in U^*$.

These notions of weak and weak-$*$ convergence are phrased to correspond exactly with the "open set" formulations of convergence given in the earlier chapters. The following proposition rephrases these ideas in more intuitive form.

8.2.7. Proposition. *Let X be a normed linear space with dual space X^* and let $\{x_n\}$ and $\{x_n^*\}$ be sequences in X and X^*, respectively. Then $\{x_n\}$ converges weakly to $x_\infty \in X$ if and only if*

$$\lim_{n \to \infty} x^*(x_n) = x^*(x)$$

for each $x^* \in X^*$. Similarly, $\{x_n^*\}$ converges weak-$*$ to $x_\infty^* \in X^*$ if and only if

$$\lim_{n \to \infty} x_n^*(x) = x_\infty^*(x)$$

for each $x \in X$.

Proof. We will prove only the assertion for weakly convergent sequences and leave the similar proof for weak-$*$ convergent sequences to the reader.

Suppose that $\{x_n\}$ converges weakly to x_∞. Let $x^* \in X^*$ and let $\epsilon > 0$. Since $B(x_\infty; x^*; \epsilon)$ is a weakly open set, there is an N such that $j \geq N$ implies that $x_j \in B(x_\infty; x^*; \epsilon)$, i.e., $j \geq N$ implies that

$$|x^*(x_j) - x^*(x_\infty)| \leq \epsilon.$$

Since $\epsilon > 0$ was arbitrary, this shows that $\lim x^*(x_n) = x^*(x)$.

For the converse, suppose that $\lim x^*(x_n) = x^*(x_\infty)$ for each $x^* \in X^*$. Let U be a weakly open set containing x_∞ and select $\{x_1^*, \ldots, x_n^*\} \subseteq X^*$ and $\epsilon > 0$ so that

$$B(x_\infty; x_1^*, \ldots, x_n^*; \epsilon) \subseteq U.$$

Corresponding to each x_j^* we may select an N_j so that $m \geq N_j$ implies that

$$|x_j^*(x_m) - x_j^*(x_\infty)| \leq \epsilon.$$

If $N = \max\{N_1, \ldots, N_n\}$, then $m \geq N$ implies that

$$x_m \in B(x_\infty; x_1^*, \ldots, x_n^*; \epsilon) \subseteq U,$$

showing that $\{x_n\}$ converges weakly to x_∞.

8.2.8. Corollary. *Let X be a normed linear space with dual space X^*. If $\{x_n\}$ [respectively, $\{x_n^*\}$] converges strongly to $x_\infty \in X$ [resp., to $x_\infty^* \in X^*$], then $\{x_n\}$ [resp., $\{x_n^*\}$] converges weakly to x_∞ [resp. converges weak-$*$ to x_∞^*].*

The converse to 8.2.8 is not true if X is infinite-dimensional (Problem 8.12). In addition, it is *not* possible to describe either the weak or weak-$*$ topologies exclusively in terms of convergent sequences (see Problem 8.8.18). In a similar vein, it is not possible to describe continuity in these topologies with sequences; it is necessary, instead, to use open sets.

8.2.9. Definition. Let X be a normed linear space with dual space X^*. A mapping f from X [respectively, from X^*] to $\Re$ is said to be *weakly continuous* [resp., *weak-$*$ continuous*] if $f^{-1}(U)$ is weakly [resp., weak-$*$] open whenever U is an open subset of $\Re$.

One of the powerful properties of the weak topology is that it is the smallest toplogy which makes all of the functionals in X^* continuous (Problem 8.20). A similar statement holds for the weak-* toplogy (Problem 8.21). Notice that if $x \in X$, then x can be used to define a continuous linear functional $\varsigma(x)$ on X^* in the following way:

$$\varsigma(x)(x^*) = x^*(x).$$

Since

$$\varsigma(x)(\alpha x^* + \beta y^*) = (\alpha x^* + \beta y^*)(x)$$
$$= \alpha x^*(x) + \beta y^*(x)$$
$$\alpha\varsigma(x)(x^*) + \beta\varsigma(x)(x^*)$$

and

$$|\varsigma(x)(x^*)| = |x^*(x)| \le \|x^*\| \cdot \|x\|$$

it follows that $\varsigma(x)$ is a linear functional defined on X^* with norm at most $\|x\|$ for each $x \in X$. The weak-* topology is the smallest topology which makes the mappings $\{\varsigma(x) : x \in X\}$ continuous on X^*. The mapping $\varsigma : X \to X^*$ is called the *natural embedding* of X into X^{**}. Notice that this mapping need *not* be surjective.

8.2.10. Proposition. *Let X be a normed linear space and let $\varsigma : X \to X^{**}$ be the natural embedding. Then ς is a linear mapping and $\|\varsigma(x)\| = \|x\|$ for each $x \in X$. Moreover, if X is a Banach space, then $\varsigma(X)$ is a closed linear subspace of X^{**}.*

Proof. A routine argument show that ς is linear; the preceding remarks show that $\|\varsigma(x)\| \le \|x\|$. Thus it will suffice to show that $\|\varsigma(x)\| \ge \|x\|$.

If $x = 0$, there is nothing to prove and so we suppose that $x \ne 0$; by 8.1.3 there is an $x^* \in X^*$ with $\|x^*\| = 1$ and $x^*(x) = \|x\|$. Thus

$$\|x\| = x^*(x) = |\varsigma(x)(x^*)|$$
$$\le \sup\{|\varsigma(x)(z^*)| : z^* \in X^* \quad \text{and} \quad \|z^*\| = 1\}$$
$$= \|\varsigma(x)\|.$$

8.2.11. Theorem. *Let X be a normed linear space with dual space X^*. Then each $x^* \in X^*$ is weakly continuous and, for each $x \in X$, the mapping $\varsigma(x) \in X^{**}$ is weak-* continuous.*

Proof. We will prove only that each $x^* \in X^*$ is weakly continuous and leave the similar proof about weak-* continuity to the reader.

If $x^* \in X^*$ and if (a, b) is an open inverval of real numbers, it suffices to show that $(x^*)^{-1}((a,b))$ is a weakly open subset of X. However,

$$(x^*)^{-1}((a,b)) = \cup\Big\{ B(z;x^*;\epsilon) :$$

$$x^*(z) \in (a,b) \quad\text{and}\quad \epsilon = \min\{x^*(z) - a;\; b - x^*(z)\} \Big\}$$

from which the conclusion is immediate.

Since it is easier for sequences to converge weakly (or weak-∗) than strongly, there are more weakly (weak-∗) convergent sequences than there are strongly convergent sequences. This is another way of saying that there are fewer weakly (weak-∗) open sets that there are strongly open sets. These observations have a dramatic impact on compactness.

8.2.12. Definition. Let X be a normed linear space with dual space X^*. A subset $F \subseteq X$ is said to be *weakly closed* if $X\backslash F$ is weakly open. Similiarly, a subset $F^* \subseteq X^*$ is said to be *weak-∗ closed* if $X^*\backslash F^*$ is weak-∗ open.

8.2.13. Definition. If (X,τ) is a topological space and if $C \subseteq X$, then we say that C is *compact* if every cover of C by members of τ admits a finite subcover. If $C = X$, then we say that X is compact. If every sequence in C has a convergent subsequence, then we say that C is *sequentially compact.*

In particular, let X be a normed linear space with dual space X^*. A subset $K \subseteq X$ is said to be *weakly compact* if every cover of K by weakly open sets admits a finite subcover. Similarly, a subset $K^* \subseteq X^*$ is said to be *weak-∗ compact* if every cover of K^* by weak-∗ open sets admits a finite subcover. The set K is *weakly sequentially compact* if whenever $\{x_n\}$ is a sequence in K, there is a subsequence $\{x_{n'}\}$ which converges weakly to an element of K. The set K^* is *weak-∗ sequentially compact* if whenever $\{x_n^*\}$ is a sequence in K^*, there is a subsequence $\{x_{n'}^*\}$ which converges weak-∗ to an element of K^*.

In a metric space, the notions of sequential compactness and compactness are the same. This turns out to be true for the weak topology as well, but this is a very deep result which is beyond the scope of this text (the Eberlein-Smul'yan Theorem). The two notions of compactness need *not* be the same in the weak-∗ topology (see exercise 8.14).

Since there are fewer weakly open sets than strongly open sets, there are more weakly compact sets than there are strongly compact sets. It turns out in many applications that weak compactness is enough to obtain the desired results, and so using weakly compact sets can greatly increase the scope of results. In a similar way, weak-∗ compact sets often play a useful role in applications, although they are generally more difficult to work with than are weakly compact sets (in part because weak-∗ compactness and weak-∗ sequential compactness are different).

The following proposition lists some elementary facts about weakly (weak-∗) closed and compact sets.

8.2.14. Proposition. *Let X be a normed linear space with dual space X^*. Then*

(i). *every strongly closed convex subset of X is weakly closed;*

(ii) *every weakly compact subset of X is bounded;*

(iii) *every weak-∗ compact subset of X^* is bounded;*

(iv) *every strongly compact subset of X is weakly compact;*

(v) *every strongly compact subset of X^* is weak-∗ compact.*

Proof. (i) Let C be a strongly closed convex subset of X and fix $z \notin C$. By 8.1.5 there is an $x^* \in X^*$ with $\|x^*\| = 1$ and

$$x^*(z) > \sigma \equiv \sup\{x^*(x) : x \in C\}.$$

If $\epsilon = \frac{1}{2}(x^*(z) - \sigma)$, then

$$B(z; x^*; \epsilon) \subseteq X \backslash C,$$

showing that $X \backslash C$ is weakly open and hence that C is weakly closed.

(ii) Let K be a weakly compact subset of X. If K is not bounded then we may select a sequence $\{x_n\} \subseteq K$ with $\lim \|x_n\| = \infty$. Now each $x^* \in X^*$ is weakly continuous, and thus $x^*(K)$ is a compact set of real numbers. In particular, for each $x^* \in X^*$ there is a number $M_{x^*} > 0$ so that

$$\sup_n |\varsigma(x_n)(x^*)| = \sup_n |x^*(x_n)| \le M_{x^*}.$$

Since X^* is complete, we may apply the Principle of Uniform Boundedness to conclude that there is a number $M > 0$ so that

$$\sup_n \|\varsigma(x_n)\| = \sup_n \|x_n\| \le M,$$

an evident contradiction to our choice of $\{x_n\}$.

(iii) Exercise 8.22.

(iv) and (v) Both of these are immediate from 8.2.5.

PROBLEMS

8.9. Prove proposition 8.2.3.

8.10. Verify that Definition 8.2.4 does in fact give rise to topologies.

8.11. Let X be a normed linear space with dual space X^*. Verify each of the following assertions:

(a)   If $U \subseteq X$ is weakly open and if $x \in U$ and $\alpha \in \Re \backslash \{0\}$, then the sets αU and $x + U$ are weakly open.

(b)   If $U^* \subseteq X^*$ is weak-$*$ open and if $x^* \in U^*$ and $\alpha \in \Re \backslash \{0\}$, then the sets αU^* and $x^* + U^*$ are weak-$*$ open.

8.12.   Find examples of weakly convergent and weak-$*$ convergent sequences which fail to converge strongly.

8.13.   Show that ς need not be surjective (consider c_0).

8.14.   *Compactness and sequential compactness in the weak-$*$ topology.*
(a)   Let K denote the set $\{x^* \in \ell_\infty^* : \|x^*\| \leq 1\}$. Show that K is weak-$*$ compact.
(b)   For each n define a functional f_n on ℓ_∞ by setting

$$f_n(\{x_j\}) = x_n.$$

Show that $f_n \in \ell_\infty^*$ and that $\|f_n\| = 1$ for all n.
(c)   Show that $\{f_n\}$ has no weak-$*$ convergent subsequence.

8.15.   **Definition.** A sequence $\{x_n\}$ is *weakly Cauchy* if $\{x^*(x_n)\}$ is a Cauchy sequence of real numbers for each $x^* \in X^*$.
(a)   Let $\{f_n\}$ be a weakly Cauchy sequence in $\mathcal{L}_1[0,1]$. Show that there exists a function $f \in \mathcal{L}_1[0,1]$ such that

$$\lim \int_E f_n \, d\lambda = \int_E f \, d\lambda$$

for all measurable sets $E \subseteq [0,1]$.
(b)   Show that, with $\{f_n\}$ and f as in (a) that $\{f_n\}$ converges weakly to f (i.e., that $\mathcal{L}_1[0,1]$ is *weakly sequentially complete*).

8.16.   Show that c_0 is not weakly sequentially complete.

8.17.   Show that $C_0(\Re)$ is weak-$*$ dense in $\mathcal{L}_\infty(\Re)$.

8.18.   Let E be the subset of ℓ_2 consisting of the vectors x_{mn}, where the mth coordinate of x_{mn} is 1, the nth coordinate is n, and all other coordinates are zero. Show that the zero vector is in the weak closure of E but that no subsequence of elements in E converges weakly to the zero vector. (This example is due to von Neumann.)

8.19.   Show that the norm is a weakly lower semicontinuous functional.

8.20.   Let τ be a topology on the normed linear space X with the property that $f^{-1}(U) \in \tau$ whenever $f \in X^*$ and $U \subseteq \Re$ is an open set. Show that every weakly open set is a member of τ.

8.21.   Let X be a normed linear space and let τ be a topology on the dual space X^* with the property that $(\varsigma(x))^{-1}(U) \in \tau$ whenever $x \in X$ and $U \subseteq \Re$ is an open set. Show that every weak-$*$ open set is a member of τ.

8.22.   Prove 8.2.14(iii).

8.23.   Let X be a normed linear space and suppose that X^* is separable. Show that X is separable.

8.3. WEAK-* COMPACTNESS. While norm-compact subsets of X^* must be norm closed and totally bounded, weak-* compact subsets of X^* need only be weak-* closed and *bounded*. As it is generally both highly restrictive and difficult to verify that a family of functions is totally bounded, the fact that weak-* compact sets need only be (norm-) bounded is often of great value. Our main goal in this section is to verify the foregoing fact, known as the Banach-Alaoglu Theorem.

Note that we have already seen a special case of this theorem: Corollary 6.2.6 to the Helly-Bray Theorem asserts that the unit ball of $(C[a,b])^*$ is weak-* sequentially compact. Of course, in general weak-* sequential compactness does not imply weak-* compactness (this is the main point of Problem 8.14); however, when X is separable the two notions coincide, as we shall see in 8.3.6.

As in Section 6.3 (where we worked with $B[a,b]$ rather than $C[a,b]$), it will be slightly more convenient to study the space of all bounded functions defined on X rather than the space of all bounded *linear* functions defined on the space X.

8.3.1. Definition. Let X be a normed linear space; denote by $\mathcal{B}(X)$ the collection of all functions $f : X \to \Re$ with the property that

$$-\|x\| \le f(x) \le \|x\|$$

for all $x \in X$. We endow $\mathcal{B}(X)$ with the topology τ_B generated by the basis

$$\{ B(f; x_1, \ldots, x_n; \epsilon) : f \in \mathcal{B}(X), x_1, \ldots, x_n \in X \quad \text{and} \quad \epsilon > 0\},$$

where

$$B(f; x_1, \ldots, x_n; \epsilon) = \{g \in \mathcal{B}(X) : |f(x_i) - g(x_i)| < \epsilon \quad \text{for} \quad i = 1, \ldots, n\}.$$

Notice that if

$$B^* = \{x^* \in X^* : \|x^*\| \le 1\},$$

then $B^* \subseteq \mathcal{B}(X)$. Moreover, if U is any weak-* open subset of X^*, then there is a set $G \in \tau_B$ such that $G \cap B^* = U \cap B^*$; i.e., the topology that B^* inherits as a subspace of $\mathcal{B}(X)$ agrees with the weak-* topology. (See Problem 8.27.)

We will first show that $(\mathcal{B}(X), \tau_B)$ is compact; then, by showing that B^* is a τ_B-closed subset of $\mathcal{B}(X)$ we will infer that B^* is weak-* compact. The first step is this program is to prove a slightly weaker result.

8.3.2. Lemma. *Let X, $\mathcal{B}(X)$ and τ_B be as in 8.3.1; take S to be the collection of sets of the form*

$$S = \{B(f; x; \epsilon) : f \in \mathcal{B}(X), x \in X \backslash \{0\} \quad \text{and} \quad \epsilon > 0\}.$$

Then every open cover of $\mathcal{B}(X)$ by members of S admits a finite subcover.

Proof. Let $\mathcal{U}$ be an open cover of $\mathcal{B}(X)$ by members of S. For each $x \in X \backslash \{0\}$

and $S \in \mathcal{S}$ define

$$I(S; x) = \{t \in [-\|x\|, \|x\|] : \quad \text{there is a } g \in S \text{ with } \quad g(x) = t\}.$$

Observe that for each x and S, $I(S; x)$ is an open subset of $[-\|x\|, \|x\|]$ ($g \in S$ implies that $\alpha g \in S$ if $|1 - \alpha|$ is sufficiently small).

Next, for each $x \in X \backslash \{0\}$ define

$$\mathcal{S}(x) = \{B(f; x; \epsilon) : f \in \mathcal{B}(X) \quad \text{and} \quad \epsilon > 0\};$$
$$\mathcal{U}(x) = \mathcal{U} \bigcap \mathcal{S}(x);$$
$$\mathcal{I}(x) = \{I(S; x) : S \in \mathcal{U}(x)\}.$$

We first claim that there is an $x_0 \in X$ for which $\mathcal{I}(x_0)$ is an open cover of $[-\|x_0\|, \|x_0\|]$. If this is not the case, then for each x there is a number $\varphi(x) \in [-\|x\|, \|x\|]$ with the property that there is no $S \in \mathcal{U}(x)$ for which $\varphi(x) \in I(S; x)$. However, $\varphi \in \mathcal{B}(X)$ and so there is an $S \in \mathcal{U}$ for which $\varphi \in S$. Since $S \in \mathcal{S}$, we may write $S = B(f; y; \epsilon)$ for some $f \in \mathcal{B}(X), y \in X$ and $\epsilon > 0$. But then $S \in \mathcal{U}(y)$ and $\varphi(y) \in I(S; y)$, contradicting our choice of φ.

Now since $\mathcal{I}(x_0)$ is an open cover of the compact set $[-\|x_0\|, \|x_0\|]$, it follows that there are sets $\{S_1, \ldots, S_n\} \subseteq \mathcal{U}(x_0)$ with the property that

$$[-\|x_0\|, \|x_0\|] \subseteq I(S_1; x_0) \cup \cdots \cup I(S_n; x_0).$$

To complete the proof, we will argue that $\{S_1, \ldots, S_n\}$ covers $\mathcal{B}(X)$.

If $g \in \mathcal{B}(X)$, then $g(x_0) \in I(S_j; x_0)$ for some $j = 1, \ldots, n$. Since $S_j \in \mathcal{U}(x_0)$, there is an $f_j \in \mathcal{B}(X)$ and an $\epsilon_j > 0$ such that $S_j = B(f_j; x_0; \epsilon_j)$. From the foregoing it follows that there is an $h \in S_j \equiv B(f_j; x_0; \epsilon_j)$ such that $h(x_0) = g(x_0)$. But then

$$|f_j(x_0) - g(x_0)| = |f_j(x_0) - h(x_0)| < \epsilon_j,$$

showing that $g \in \cup_{j=1}^{n} S_j$, as desired.

8.3.3. Lemma. *The space $(\mathcal{B}(X), \tau_B)$ is compact.*

Proof. We suppose the contrary and let Ψ denote the class of all τ_B-open covers of $\mathcal{B}(X)$ which do not admit finite subcovers. We may partially order Ψ by inclusion; plainly if Φ is any totally ordered subfamily of Ψ, then $\bigcup \Phi$ is again a member of Ψ. Thus by Zorn's Lemma, the family Ψ contains a maximal member $\mathcal{U}$. Observe that $\mathcal{U}$ is an open cover of $\mathcal{B}(X)$, $\mathcal{U}$ has not finite subcover, and if $U \in \tau_B$, then $\mathcal{U} \cup \{U\}$ *does* admit a finite subcover of $\mathcal{B}(X)$.

With $\mathcal{S}$ as in 8.3.2, set $\mathcal{G} = \mathcal{U} \cap \mathcal{S}$. Surely no subfamily of $\mathcal{G}$ covers $\mathcal{B}(X)$, so 8.3.2 implies that $\mathcal{G}$ is *not* a cover for $\mathcal{B}(X)$. Select a function

$$f \in \mathcal{B}(X) \backslash \left(\bigcup \mathcal{G} \right)$$

and select a set $U \in \mathcal{U}$ so that $f \in U$. By the definition of τ_B we may select $x_1, \ldots, x_n \in X \backslash \{0\}$ and $\epsilon > 0$ so that

$$f \in B(f; x_1, \ldots, x_n; \epsilon)$$
$$= \bigcap_{j=1}^{n} B(f; x_j; \epsilon)$$
$$\subseteq U.$$

Since f is not in $\bigcup \mathcal{G}$, it follows that none of the sets $B(f; x_j; \epsilon)$ are in $\mathcal{U}$. Since $\mathcal{U}$ is maximal,

$$\mathcal{U} \bigcup \{B(f; x_j; \epsilon)\}$$

admits a finite subcover for each j:

$$\left\{ U_1^{(j)}, \ldots, U_{n_j}^{(j)} \right\} \bigcup \{B(f; x_j; \epsilon)\}.$$

Set $\mathcal{U}_j = \{U_1^{(j)}, \ldots, U_{n_j}^{(j)}\}$. Finally, observe that

$$U \cup \left(\bigcup_{j=1}^{n} \mathcal{U}_j \right) \supseteq B(f; x_1, \ldots, x_n; \epsilon) \cup \left(\bigcup_{j=1}^{n} \mathcal{U}_j \right)$$
$$\supseteq \left(\bigcap_{i=1}^{n} B(f; x_i; \epsilon) \cup \left(\bigcup_{j=1}^{n} \mathcal{U}_j \right) \right)$$
$$\supseteq \mathcal{B}(X),$$

showing that $\mathcal{B}(X)$ is the union of finitely many sets from $\mathcal{U}$. As this contradicts our choice of $\mathcal{U}$, we may conclude that $\mathcal{B}(X)$ is compact.

8.3.4. Banach-Alaoglu Theorem. *Let X be a normed linear space with dual space X^*; set $B^* = \{x^* \in X^* : \|x^*\| \leq 1\}$. Then B^* is weak-$*$ compact.*

Proof. Let $\mathcal{U}$ be any open cover of B^* by weak-$*$ open sets. For each $U \in \mathcal{U}$ select $G \in \tau_B$ so that $G \cap B^* = U$ and set $\mathcal{G} = \{G\}$. It suffices to show that $\mathcal{G} \bigcup \{\mathcal{B}(X) \backslash B^*\}$ is a τ_B-open cover for $\mathcal{B}(X)$, i.e., that $\mathcal{B}(X) \backslash B^*$ is in τ_B.

Let $f \in \mathcal{B}(X) \backslash B^*$. Since $f \in \mathcal{B}(X)$, $|f(x)| \leq \|x\|$ for all $x \in X$; if f were linear, this would then imply that $f \in B^*$ and thus f cannot be linear. We will exhibit an open set $U \in \tau_B$ with $f \in U$ and $U \cap B^* = \emptyset$; since $f \in \mathcal{B}(X) \backslash B^*$ was arbitrary, this will show that $\mathcal{B}(X) \backslash B^*$ is τ_B-open. There are two cases to consider.

Case 1. *There are elements $x, y \in X$ for which $f(x + y) \neq f(x) + f(y)$.*

If for some $\epsilon > 0$,

$$(B(f; x, y; \epsilon) \cup B(f; x + y; \epsilon)) \bigcap B^* = \emptyset,$$

we are through, and so for each $\epsilon > 0$ we may select $g \in B^*$ so that

$$g \in B(f; x, y; \epsilon) \cap B(f; x + y; \epsilon).$$

Then

$$|f(x + y) - f(x) - f(y)| \le |f(x + y) - g(x + y)| + |g(x) - f(x)| + \cdots$$
$$\cdots + |g(y) - f(y)|$$
$$\le 3\epsilon;$$

since ϵ was arbitrary, this shows that $f(x + y) = f(x) + f(y)$, a contradiction.

Case 2. *There is an $x \in X$ and an $\alpha \in \Re$ for which $f(\alpha x) \ne \alpha f(x)$.*
Once again, for each $\epsilon > 0$ we may select

$$g \in B(f; \alpha x; \epsilon) \cap B(f; x; \epsilon) \cap B^*.$$

Then

$$|f(\alpha x) - \alpha f(x)| \le |f(\alpha x) - g(\alpha x)| + |\alpha g(x) - \alpha f(x)| \le (1 + \alpha)\epsilon;$$

once again, since ϵ was arbitrary, this implies that $f(\alpha x) = \alpha f(x)$, a contradiction.

8.3.5. Corollary. *Let X be a normed linear space with dual space X^*. A subset C of X^* is weak-$*$ compact if and only if C is weak-$*$ closed and bounded.*

Although this is certainly not the only proof possible of the Banach-Alaoglu Theorem, it is the shortest and most direct one. The set S in 8.3.2 is called a *subbasis* for the topology τ_B since every basic open set in τ_B can be realized as the intersection of finitely many members of S. Lemma 8.3.3 then shows that a set C is compact if and only if every open cover of C by subbasis elements admits a finite cover, a result called the *Alexander Sub-basis Theorem*. This result is also intimately connected with the Tychonov Theorem dealing with products of compact sets (see Problem 8.25).

Recalling Problem 8.13, we see that weak-$*$ compactness and weak-$*$ sequential compactness need not coincide. In light of 4.2.6 and Problem 4.11, this shows that the weak-$*$ topology is not, in general, a metrizable topology. However, in the special case that X is separable, we can define a metric on B^* which is compatible with the weak-$*$ topology.

8.3.6. Theorem. *If X is a separable normed linear space, then there is a metric ϱ defined on B^* which generates the weak-$*$ topology. In particular, if X is separable, then B^* is weak-$*$ sequentially compact.*

Proof. Let $\{x_n\}$ be a countable dense subset of X and define

$$\varrho(x^*, y^*) = \sum_{n=1}^{\infty} 2^{-n} \frac{|(x^* - y^*)(x_n)|}{1 + |(x^* - y^*)(x_n)|}.$$

It is routine to verify that ϱ defines a metric on B^*; we wish to show that a set U is weak-$*$ open if and only if it is open under ϱ.

First suppose that U is weak-$*$ open and select $f \in U$; next select $\epsilon > 0$ and $\{y_1, \ldots, y_n\} \subseteq X$ so that $B(f; y_1, \ldots, y_n; \epsilon) \subseteq U$. It suffices to find $\delta > 0$ so that

$$\{g \in B^* : \varrho(f, g) < \delta\} \subseteq B(f; y_1, \ldots, y_n; \epsilon).$$

For $j = 1, \ldots, n$, select x_{m_j} so that

$$\|x_{m_j} - y_j\| < \frac{\epsilon}{3}$$

and set $N = \max\{m_1, \ldots, m_n\}$ and $C = \max_{m_j}\{1 + 2\|x_{m_j}\|\}$. Now if $\delta = 2^{-N}\epsilon(3C)^{-1}$ and $\varrho(f, g) < \delta$ then, for $j = 1, \ldots, n$,

$$
\begin{aligned}
|f(y_j) - g(y_j)| &\leq |f(y_j) - f(x_{m_j})| + |f(x_{m_j}) - g(x_{m_j})| + |g(x_{m_j}) - g(y_j)| \\
&\leq \frac{\epsilon}{3} + \frac{|f(x_{m_j}) - g(x_{m_j})|}{1 + |f(x_{m_j}) - g(x_{m_j})|} C + \frac{\epsilon}{3} \\
&\leq \frac{2\epsilon}{3} + 2^N C \varrho(f, g) \\
&\leq \epsilon,
\end{aligned}
$$

showing that $g \in B(f; x_1, \ldots, x_n; \epsilon)$, as desired.

Now let U be a ϱ-open subset of B^* and fix $f \in U$; select $\epsilon > 0$ so that $\{g \in B^* : \varrho(f, g) < \epsilon\} \subseteq U$. It suffices to find $\{x_1, \ldots, x_n\}$ and $\delta > 0$ so that $B(f; x_1, \ldots, x_n; \delta) \subseteq \{g \in B^* : \varrho(f, g) < \epsilon\}$. Choose N so large that

$$\sum_{n > N} 2^{-n} \leq \frac{\epsilon}{2}$$

and set $\delta = \epsilon/2$. Then if $g \in B(f; x_1, \ldots, x_N; \delta)$,

$$
\begin{aligned}
\varrho(f, g) &= \sum_{n=1}^{\infty} 2^{-n} \frac{|(f - g)(x_n)|}{1 + |(f - g)(x_n)|} \\
&\leq \sum_{n=1}^{N} 2^{-n} \frac{\delta}{1 + \delta} + \sum_{n > N} 2^{-n} \frac{\|x_n\| \|f - g\|}{1 + \|x_n\| \|f - g\|} \\
&\leq \delta + \frac{\epsilon}{2} \\
&= \epsilon,
\end{aligned}
$$

showing that $\{g \in B^* : \varrho(f, g)\} \supseteq B(f; x_1, \ldots, x_n; \delta)$, as desired.

PROBLEMS

8.24. Definition. A topological space (X, τ) is *Hausdorff* if given any two distinct points $x, y \in X$, there are disjoint open sets U and V with $x \in U$ and $y \in V$.
(a) Show that X^* endowed with the weak-$*$ topology is Hausdorff.

Definition. Let (X, τ) be a compact, Hausdorff topological space; denote by $C(X)$ the collection of all continous functions from X to $\Re$ with norm $\|f\| = \sup\{|f(x)| : x \in X\}$.
(b) Show that $C(X)$ is a Banach space.
(c) Let E be a Banach space. Show that there is a compact Hausdorff space X such that E is isometrically isomorphic with a closed subspace of $C(X)$.

8.25. Definition. Let $\{X_\alpha, \tau_\alpha) : \alpha \in A\}$ be a family of topological spaces where A is some arbitrary index set. The *product of* $\{X_\alpha\}$, $\prod X_\alpha$, is the collection of all functions $f : A \to \bigcup X_\alpha$ with the property that $f(\alpha) \in X_\alpha$. The *product topology* on $\prod X_\alpha$ is the topology generated by basis sets of the form

$$\left\{ g \in \prod X_\alpha : g(\alpha) \in U_\alpha \right\},$$

where $U_\alpha \in \tau_\alpha$ and $U_\alpha = X_\alpha$ for all but finitely many α.
(a) Show that if each (X_α, τ_α) is compact, then so is $\prod X_\alpha$.
(b) Deduce the Banach-Alaoglu Theorem from (a). [Part (a) is the Tychonov Theorem.]

8.26. Let $1 < p \le \infty$ and let T be a bounded linear mapping from $\mathcal{L}_1(\Re)$ to $\mathcal{L}_p(\Re)$. Show that the following are equivalent:
(a) There is an $h \in \mathcal{L}_p$ such that $T(f) = f * h$ for all $f \in \mathcal{L}_1$.
(b) $T(f) * g = T(f * g)$ for all $f, g \in \mathcal{L}_1$.

8.27. Let U be a weak-$*$ open subset of X^*. Show that there is a set $G \in \tau_B$ such that $G \cap B^* = U \cap B^*$.

8.28. Let X be a normed linear space and let T be a mapping from X to X. Show that the following are equivalent.
(a) $f(T(x) - T(y)) \ge \|x - y\|^2$ for some $f \in X^*$ with $\|f\|^2 = \|x\|^2 = f(x)$.
(b) $\|x - y\| \le \|x - y + \lambda(T(x) - T(y))\|$ for all $x, y \in X$ and all $\lambda \ge 0$.

(Mappings T of the foregoing type are said to be *accretive* and play an important role in the theory of partial differential equations.)

8.29. Let X be a Banach space and let B and B^{**} be the closed unit balls of X and X^{**}, respectively. Show that $\varsigma(X)$ is weak-$*$ dense in B^{**}.

8.30. Let X be a Banach space. Show that $\varsigma(X) = X^{**}$ if and only if the closed unit ball of X is weakly compact.

8.31. *Adjoints of linear operators.* **Definition.** Let X and Y be normed linear spaces and denote by $B(X, Y)$ the collection of all continuous linear mappings from X to Y with norm as in 3.41. For $T \in B(X, Y)$ we define the *adjoint* of T to be the mapping $T^* : Y^* \to X^*$ given by

$$T^*(y^*)(x) = y^*(T(x)).$$

(a) Verify that $T^* \in B(Y^*, X^*)$ and $\|T^*\| = \|T\|$.

(b) Verify that T^* is injective if and only if $T(X)$ is (norm) dense in Y.

(c) Suppose that T is injective and that $T(X)$ is (norm) dense in Y. Suppose that there is an $m > 0$ such that $\|T^*(y^*)\| \geq m\|y^*\|$ for all $y^* \in Y^*$. Suppose further that $T^*(Y^*)$ is not surjective.

  (i)   Show that there is an $x^* \in X^*$ such that $\|x_0^* - x^*\| > 1$ for all $x^* \in T^*(Y^*)$.

  (ii) For each n set $X_n^* = \{x^* \in T^*(Y^*) : \|x_0^* - x^*\| \leq 1\}$. Show that each X_n^* is weak-$*$ compact.

  (iii) For a finite set $S \subseteq X$, set

$$S_0 = \{x^* \in X^* : |(x^* - x_0^*)(x)| \leq 1 \quad \text{for all} \quad x \in S\}.$$

  Show that there are finite sets $\{S_0, \ldots, S_n, \ldots\}$ so that $S_n \subseteq \frac{1}{n}B^*$ (where $B^* = \{x^* \in X^* : \|x^*\| \leq 1\}$) and

$$S_0^0 \cap \cdots \cap S_{n-1}^0 \cap X_n = \emptyset.$$

(d) Suppose that T is injective and that $T(X)$ is norm dense in Y. Show that the following are equivalent.

  (i) T^* is surjective.

  (ii) T is surjective.

  (iii) There is a $m > 0$ such that $\|T^*(y^*)\| \geq m\|y^*\|$ for all $y^* \in Y^*$. [Use the Open Mapping Theorem and the Hahn-Banach Theorem to do (i) $\Rightarrow$ (ii) $\Rightarrow$ (iii). For (iii) $\Rightarrow$ *(i)*, use part (c) and the Riesz Representation Theorem for c_0^*.]

8.4. WEAK COMPACTNESS AND REFLEXIVITY.

We were able to deduce the Banach-Alaoglu theorem because the unit ball of X^* could be embedded in the function space $\mathcal{B}(X)$ in which we could exploit the properties of the product topology. While every Banach space embeds into a function space (see Problem 8.24) the embedding is not generally sufficiently regular to enable inferences about the weak topology to be made. In particular, the analogue of the Banach-Alaoglu Theorem fails if X is not a dual space. On the other hand, weakly compact sets have numerous powerful properties, especially in relation to convex sets, which weak-$*$ compact sets lack. In this section we will deduce some of these properties.

We begin by observing, however, that there is a large class of Banach spaces to which the results of the preceding section *do* apply.

8.4.1. Definition. Let X be a Banach space and let $\varsigma : X \to X^{**}$ be the natural embedding of X into X^{**}; if ς is surjective, then we say that the Banach space X is *reflexive*.

In light of the various Riesz Representation Theorems, Hilbert space and the spaces $\mathcal{L}_p$ $(1 < p < \infty)$ are reflexive spaces, while such spaces as $\mathcal{L}_1$, $\mathcal{L}_\infty$, and $C(X)$ are not. Note that since X^{**} is a dual space, it must be complete, and so reflexive spaces are necessarily Banach spaces.

If $\varsigma(X) = X^{**}$, then the *weak-* topology on $X^{**}(\equiv X)$ and the weak topology on X are the same.* In particular, the following theorem is an easy exercise.

8.4.2. Theorem. *Let X be a reflexive Banach space. Then a subset C of X is weakly compact if and only if C is weakly closed and norm bounded.*

8.4.3. Definition. Let X be a normed linear space and let C be a subset of X; the *convex hull* of C is the set

$$\mathrm{co}(C) = \bigcap \{K \subseteq X : K \quad \text{is convex and} \quad C \subseteq K\}$$

and the *closed convex hull* of C is the set

$$\overline{\mathrm{co}}(C) = \bigcap \{K \subseteq X : K \quad \text{is closed, convex and} \quad C \subseteq K\}.$$

8.4.4. Lemma. *Let X be a normed linear space and let C be a subset of X. Then:*

(i) *if C is convex, so is the norm closure of C;*

(ii) *the norm closure of $\mathrm{co}(C)$ is $\overline{\mathrm{co}}(C)$;*

(iii) *if $y \in \mathrm{co}(C)$, then there are nonnegative scalars $\{\alpha_1, \ldots, \alpha_n\}$ and vectors*

$$\{x_1, \ldots, x_n\} \subseteq C$$

such that

$$\sum_{i=1}^{n} \alpha_i = 1 \quad \text{and} \quad y = \sum_{i=1}^{n} \alpha_i x_i;$$

(iv) *if $y \in \overline{\mathrm{co}}(C)$, then for each $\epsilon > 0$ there are nonnegative scalars $\{\alpha_1, \ldots, \alpha_n\}$ and vectors $\{x_1, \ldots, x_n\} \subseteq C$ such that*

$$\sum_{i=1}^{n} \alpha_i = 1 \quad \text{and} \quad \left\| y - \sum_{i=1}^{n} \alpha_i x_i \right\| \le \epsilon.$$

Proof. (i) Let $\{x_n\}$ and $\{y_n\}$ be Cauchy sequences in C with limits x_∞ and y_∞, respectively, and let $\alpha \in [0,1]$. Then $(1-\alpha)x_n + \alpha y_n \in C$ for each n and, moreover,

$$\lim (1-\alpha)x_n + \alpha y_n = (1-\alpha)x_\infty + \alpha y_\infty,$$

showing that the norm closure of C is convex.

(ii)   By (i) the norm closure of co(C) is a closed convex set containing C, and thus $\overline{\text{co}}(C) \subseteq \text{cl}(\text{co}(C))$. On the other hand, $\overline{\text{co}}(C)$ is a closed set containing co(C) and thus $\text{cl}(\text{co}(C)) \subseteq \overline{\text{co}}(C)$.

(iii)   Denote by C' the collection of all convex combinations of elements of C; i.e., $u \in C'$ if and only if there are nonnegative scalars $\{\alpha_1, \ldots, \alpha_n\}$ and vectors $\{x_1, \ldots, x_n\} \subseteq C$ with

$$\sum_{i=1}^{n} \alpha_i = 1 \ \text{ and } \ y = \sum_{i=1}^{n} \alpha_i x_i.$$

Observe that C' is convex, for if

$$u = \sum_{i=1}^{n} \alpha_i x_i \in C'$$

and

$$v = \sum_{i=1}^{m+n} \beta_i x_i \in C'$$

and if $0 \leq t \leq 1$, then

$$(1-t)u + tv = \sum_{i=1}^{n}[(1-t)\alpha_i + t\beta_i]x_i + \sum_{i=n+1}^{m+n} t\beta_i x_i$$

and

$$\sum_{i=1}^{n}[(1-t)\alpha_i + t\beta_i] + \sum_{i=n+1}^{m+n} t\beta_i = (1-t) + t = 1,$$

showing C' is convex. Thus co(C) $\subseteq C'$.

An easy induction establishes that if K is any convex set containing C, then $K \supseteq C'$, i.e., that $C' \subseteq \text{co}(C)$.

(iv)   This is immediate from parts (ii) and (iii).

Another useful characterization of the closed convex hull is the following.

8.4.5. Proposition.   *Let X be a normed linear space and let C be a subset of X; then*

$$\overline{\text{co}}(C) = \bigcap_{\substack{\|f\| \leq 1 \\ f \in X^*}} \{x \in X : f(x) \leq \sup\{f(u) : u \in C\}\}.$$

Proof.   For $f \in X^*$ set $E(f) = \{x \in X : f(x) \leq \sup\{f(u) : u \in C\}\}$. Since f is linear and continuous, $E(f)$ is convex and closed. Since for each $y \in C$,

$f(y) \le \sup\{f(u) \,:\, u \in C\}$, it follows that

$$\text{co}(C) \subseteq \bigcap_{\|f\| \le 1} E(f) \equiv E.$$

Now suppose that there is an $x \in E$ such that x is not in $\overline{\text{co}}(C)$. Then by 8.1.6 there is an $f \in X^*$ with $\|f\| \le 1$ such that

$$f(x) > \sup\{f(u) \,:\, u \in \overline{\text{co}}(C)\}.$$

This contradicts $x \in E(f)$ and so completes the proof.

8.4.6. Corollary. *A convex subset of a normed linear space is weakly closed if and only if it is strongly closed.*

Proof. This follows immediately from the observation that the sets $E(f)$ described in the proof of 8.4.5 are weakly closed.

8.4.7. Corollary. *Let X be a reflexive Banach space. Then every bounded, strongly closed, and convex subset of X is weakly compact.*

8.4.8. Corollary. *Let X be a reflexive, separable Banach space and let $\{K_n\}$ be a sequence of closed, convex, nonempty subsets of X. Suppose in addition that*

(i) K_1 *is bounded;*

(ii) $K_{n+1} \subseteq K_n$ *for each n.*

Then $\bigcap_{n=1}^{\infty} K_n$ is nonempty.

Proof. For each n select $x_n \in K_n$. Then $\{x_n\}$ is contained in the weakly compact set K_1; as X is separable, the weak topology on K_1 is metrizable (see Problem 8.41) and thus there is a subsequence $\{x_{n'}\}$ and an element $x \in K_1$ such that $\{x_{n'}\}$ converges weakly to x. Since $\{x_{n'} \,:\, n' \ge N\} \subseteq K_N$ and K_N is weakly compact, it follows that $x \in K_N$; since N was arbitrary, the conclusion is immediate.

Corollary 8.4.8 actually holds for arbitrary reflexive space, as weak compactness and weak sequential compactness turn out to be the same. Although we are unable to deduce this here, it is possible to verify 8.4.8 for arbitrary Hilbert spaces (Problem 8.33).

Our final result in this section illustrates a surprising relationship between the norm and weakly compact sets. Results of the type below are of critical importance in the calculus of variations.

8.4.9. Theorem.   *Let X be a reflexive, separable Banach space and let K be a weakly compact, convex subset of X. Then K has an element of smallest norm.*

Proof.   Select $\{x_n\} \subseteq K$ so that $\lim \|x_n\| = \sigma \equiv \inf\{\|x\| : x \in K\}$. For each n set $K_n = \overline{\text{co}}\{x_j : j \geq n\}$. Then $\{K_n\}$ satisfies the conditions of 8.4.8 and so we may select $x \in \bigcap K_n$.

Fix $\epsilon > 0$ and choose N so large that $n \geq N$ implies that $\|x_n\| \leq \sigma + \frac{\epsilon}{2}$. Since $x \in K_N$, we may select nonnegative scalars $\{\alpha_N, \ldots, \alpha_{n+N}\}$ so that $\sum_{i=N}^{n+N} \alpha_i = 1$ and

$$\left\| x - \sum_{i=N}^{n+N} \alpha_i x_i \right\| \leq \frac{\epsilon}{2}.$$

Then

$$\|x\| \leq \left\| x - \sum_{i=N}^{n+N} \alpha_i x_i \right\| + \left\| \sum_{i=N}^{n+N} \alpha_i x_i \right\|$$
$$\leq \frac{\epsilon}{2} + \frac{\epsilon}{2} + \sigma$$
$$= \epsilon + \sigma.$$

Since ϵ was arbitrary, this shows that $\|x\| \leq \sigma$. Since the reverse inequality is immediate, this completes the proof.

PROBLEMS

8.32.   Let $\{f_n\}$ and f be functions in $C[0,1]$. Show that the following are equivalent.
(a) $\{f_n\}$ converges weakly to $f \in C[0,1]$;
(b) $\{\|f_n\|\}$ is bounded and $\lim f_n(t) = f(t)$ for each $t \in [0,1]$.

8.33.   Let $\mathcal{H}$ be a Hilbert space and let $\{K_n\}$ be a sequence of closed convex subsets of $\mathcal{H}$ satisfying 8.4.8(i) and (ii). Show that $\bigcap K_n$ is nonempty.

8.34.   Show that a Banach space X is reflexive if and only if X^* is reflexive.

8.35.   Let X be a reflexive Banach space and let Y be a closed linear subspace of X. Show that Y is reflexive.

8.36.   Let X be a reflexive Banach space and let $f \in X^*$. Show that there is an $x \in X$ such that $\|x\| = 1$ and $f(x) = \|f\|$. (It is a deep theorem of R. C. James that this property *characterizes* reflexive Banach spaces.)

8.37.   Let X be a reflexive, separable Banach space and let $\{x_n\}$ be a sequence in X. Show that the following are equivalent.
(i) $\{x_n\}$ converges weakly to x_∞.
(ii) $\{\|x_n\|\}$ is bounded and

$$\{x_\infty\} = \bigcap_{n=1}^{\infty} \overline{\text{co}}\{x_k : k \geq n\}.$$

8.38. Show that neither c nor c_0 is reflexive.

8.39. Let X be a nonreflexive Banach space. Show that $\varsigma(X)$ (the natural embedding of X into X^{**}) is first category in X^{**}.

8.40. *The Krein-Mil'man Theorem.* **Definition.** Let K be a subset of a vector space X. A nonempty set $S \subseteq K$ is an *extremal set* for K if

$$(1 - t)x + ty \in S \quad \text{implies that} \quad x, y \in S$$

where $0 \leq t \leq 1$ and $x, y \in K$. An extremal subset of K consisting of just one point is called an *extremal point* of K.
(a) Let X be a Banach space and let K be a nonempty subset of X. Show that K has at least one extremal point if any of the following hold:
(i) K is strongly compact;
(ii) K is weakly compact; or
(iii) X is a dual space and K is weak-$*$ compact.
(b) Let X be a Banach space and let K be a subset of X satisfying (i), (ii) or (iii) of part (a). Show that K is the closed convex hull of its extreme points (where the closure is taken, respectively, in the strong, weak, and weak-$*$ topologies).

8.41. Let X be a separable, reflexive Banach space. Show that X^* is separable. Conclude that the weak topology on bounded sets is metrizable.

8.42. *Uniformly convex Banach spaces.* Let X be a Banach space and define, for $\epsilon \geq 0$, the *modulus of convexity* of X to be the function $\delta(\epsilon)$ given by

$$\delta(\epsilon) = \inf \left\{ 1 - \left\| \frac{x + y}{2} \right\| : \|x\|, \|y\| \leq 1 \ \text{and} \ \|x - y\| \geq \epsilon \right\}.$$

The Banach space X is said to be *uniformly convex* if $\delta(\epsilon) > 0$ whenever $\epsilon > 0$.
(a) Show that $\mathcal{L}_p(X, \mathcal{M}, \mu)$ is uniformly convex if $1 < p < \infty$. (Use Clarkson's inequalities.)
(b) Let $\{K_n\}$ be a sequence of closed convex sets in a uniformly convex space X, and suppose that the sequence $\{K_n\}$ satisfies 8.4.8(i) and (ii). Show that

$$\bigcap_{n=1}^{\infty} K_n \neq \emptyset.$$

Appendix A

Basic Properties of the Real Numbers

The material in this appendix, although fundamental to this text, more properly belongs to undergraduate analysis. We include it here for the sake of completeness and to give the student in need of a quick review a reference having notation which is consistent with the rest of the text.

The appendix is divided into three sections. The first deals with completeness properties, the second with open and closed sets (topology), and the third with compactness.

Throughout we will assume that the real numbers ($\Re$), together with their conventional algebraic operations and order properties, are defined. (For an axiomatic construction of the real number system, the reader is referred, for example, to *Elements of Abstract and Linear Algebra* by H. Paley and P. Weichsel.) Largely as a notational convenience, we shall on occasion append two symbols, "$+\infty$" and "$-\infty$," to $\Re$, obtaining the *extended real numbers* $\Re^{\#}$:

$$\Re^{\#} = \Re \cup \{+\infty, -\infty\}.$$

These additional symbols are further defined to have the following properties:

$$-\infty \leq x \quad \text{and} \quad +\infty \geq x \quad \text{for all } x \in \Re^{\#},$$

$$x - \infty = -\infty, \quad x + \infty = +\infty, \quad x/\pm\infty = 0 \quad \text{for all } x \in \Re,$$

$$x \cdot (\pm\infty) = \text{sgn}(x) \cdot (\pm\infty) \quad \text{for all } x \in \Re^{\#}, x \neq 0,$$

$$x/0 = \text{sgn}(x) \cdot \infty \quad \text{if } x \neq 0.$$

where

$$\text{sgn}(x) = \begin{cases} 1 & \text{if } x > 0 \\ 0 & \text{if } x = 0 \\ -1 & \text{if } x < 0. \end{cases}$$

The following operations on the extended real numbers are *undefined*:

$$0 \cdot (\pm\infty) \qquad 0/0 \qquad \pm\infty/\pm\infty \qquad \infty - \infty.$$

In the real numbers it is necessary to distinguish between two types of "convergent" sequences: those which converge to a finite limit and those which diverge to either plus or minus infinity. By using the extended real numbers it is possible to analyze these two types of convergence with a single set of proofs and definitions. In a similar manner, the least upper bound axiom and its consequences can be more efficiently formulated for the extended real numbers. For these reasons, the extended real numbers are the vehicle of choice for most of the topics in this text.

A.1. COMPLETENESS The fundamental (analytic) completeness property of either $\Re$ or $\Re^{\#}$ is the *Least Upper Bound Axiom*. Before stating this axiom, we need to recall some fundamental definitions about sets of real numbers.

If S is a set of extended real numbers and if b is an extended real number satisfying

$$s \leq b \quad \text{for all} \quad s \in S,$$

then we say that b is an *upper bound* for S. If, in addition, we can choose b so that $b \neq +\infty$, then we say that S is *bounded from above*.

Similarly, if S is a set of extended real numbers and if a is an extended real number satisfying

$$s \geq a \quad \text{for all} \quad s \in S,$$

then we say that a is a *lower bound* for S. If, in addtion, we can choose a so that $a \neq -\infty$, then we say that S is *bounded from below*. Note that S is bounded from below if and only if $-S$ is bounded from above and that a is a lower bound for S if and only if $-a$ is an upper bound for $-S$. We will exploit this symmetry throughout the rest of this section.

In the *extended* real numbers, every set has both an upper and a lower bound ($+\infty$ and $-\infty$, respectively); this is one of the main advantages of $\Re$ over $\Re^{\#}$.

A.1.1. Least Upper Bound Axiom for $\Re$. Let S be a set of real numbers *which is bounded from above*. Then there is a real number b, called the *least upper bound of S*, with the following properties:

(i) b is an upper bound of S; and

(ii) if b' is any other upper bound for S, then $b' \geq b$.

A.1.2. Least Upper Bound Axiom for $\Re^{\#}$. Let S be a set of extended real numbers. Then there is an extended real number b, called the *supremum of S*, with the following properties:

(i) b is an upper bound of S; and

(ii) if b' is any other upper bound for S, then $b' \geq b$. We will write $b = \sup(S)$.

The notions of *greatest lower bound* and *infimum* can be defined in a manner similar to the above or, more conveniently, by using the symmetry between bounds for S and $-S$.

A.1.3. Definition. Let S be a set of real numbers which is bounded from below and let a denote the least upper bound of $-S$; then $-a$ is called the *greatest lower bound for S*. If S is a set of extended real numbers, then the *infimum of S*, denoted $\inf(S)$, is the extended real number given by $\inf(S) = -\sup(-S)$.

The Least Upper Bound Axiom (in either form) is sufficient to give us all of the conventional completeness properties of the real numbers. We begin our study of these properties by looking at the notions of limit superior and limit inferior.

A.1.4. Definition Let $\{x_n\}$ be a sequence of extended real numbers; we define the *limit superior of* $\{x_n\}$ to be the extended real number

$$\limsup x_n = \inf_{n \geq 1} \sup_{j \geq n} x_j$$

and the *limit inferior of* $\{x_n\}$ to be the extended real number

$$\liminf x_n = \sup_{n \geq 1} \inf_{j \geq n} x_j.$$

The (extended) real numbers $\limsup x_n$ and $\liminf x_n$ always exist. In the next three propositions we list some other important, if easily deduced, properties.

A.1.5. Proposition *Let* $\{x_n\}$ *be a sequence of extended real numbers and set*

$$s_n^- = \inf_{j \geq n} x_j \quad \text{and} \quad s_n^+ = \sup_{j \geq n} x_j.$$

Then for each n *the following hold:*

(i) $s_n^- \leq s_n^+$;

(ii) $s_{n+1}^+ \leq s_n^+$;

(iii) $s_{n+1}^- \geq s_n^-$.

Moreover,

(iv) $\liminf x_n \leq \limsup x_n$.

[A sequence satisfying A.1.5(ii) is said to be *nonincreasing*. A sequence satisfying A.1.5(iii) is said to be *nondecreasing*.]

Proof. Conclusions (i)-(iii) are obvious from the definitions, and so we prove only (iv). Define

$$s^- = \sup s_n^- (= \liminf x_n) \quad \text{and}$$
$$s^+ = \inf s_n^+ (= \limsup x_n).$$

Fix an integer n; observe that if $1 \leq k \leq n$, then by (iii)

$$s_k^- \leq s_n^- \leq s_n^+.$$

On the other hand, if $k > n$, then we may apply (ii) to obtain

$$s_k^- \leq s_k^+ \leq s_n^+,$$

and thus s_n^+ is an upper bound for $\{s_k^-\}$. This implies that $s_n^+ \geq s^-$; but n was arbitrary, and thus s^- is a lower bound for $\{s_n^+\}$, showing that $s^- \leq s^+$, as desired.

A.1.6. Lemma.   *Let $\{x_n\}$ and $\{y_n\}$ be sequences of extended real numbers. Then each of the following inequalities holds:*

(i) $\inf x_n + \inf y_n \leq \inf(x_n + y_n)$;

(ii) $\inf(x_n + y_n) \leq \sup x_n + \inf y_n$;

(iii) $\sup x_n + \inf y_n \leq \sup(x_n + y_n)$;

(iv) $\sup(x_n + y_n) \leq \sup x_n + \sup y_n$,

provided that the expressions involved are well-defined (not of the form $\infty - \infty$).

Proof. We prove only (iv), leaving the similar proofs of (i)-(iii) to the exercises. For (iv), we may assume without loss of generality that

$$\sup(x_n + y_n) > -\infty, \ \ \sup x_n < \infty \ \text{ and } \ \sup y_n < \infty,$$

for otherwise there is nothing to prove. We begin by arguing that under these assumptions, $\sup(x_n + y_n) < \infty$.

Suppose for contradiction that $\sup\{x_n + y_n\} = \infty$ and fix $R > \sup\{x_n\}$. Then there is an n_0 such that

$$x_{n_0} + y_{n_0} > R + \sup y_n.$$

Now y_{n_0} cannot be $+\infty$ (since $\sup\{y_n\} < \infty$ by assumption) and thus

$$x_{n_0} > R + \sup\{y_n\} - y_{n_0}. \tag{1}$$

If $\sup\{y_n\} = -\infty$, then $\{y_n\}$ must be the constant sequence $\{-\infty\}$, which in turn implies that $\{x_n + y_n\}$ must be the constant sequence $\{-\infty\}$. Since we have already eliminated $\sup\{x_n + y_n\} = -\infty$, it follows that $-\infty < \sup\{y_n\}$. Thus (1) implies that $\sup\{x_n\} \geq R$. Once again, this is in contradiction to our assumptions, and hence $\sup\{x_n + y_n\} < \infty$, as claimed.

Now let $\epsilon > 0$ be arbitrary. We may select an integer n_1 so that $\sup\{x_n + y_n\} - \epsilon \leq x_{n_1} + y_{n_1}$, and so

$$\sup\{x_n + y_n\} - \epsilon \leq x_{n_1} - y_{n_1}$$
$$\leq \sup\{x_n\} + \sup\{y_n\}. \tag{2}$$

Since ϵ was arbitrary, this shows that

$$\sup\{x_n + y_n\} \leq \sup\{x_n\} + \sup\{y_n\}.$$

(The last sentence of the proof above is shorthand for the following argument by contradiction:

If the conclusion of the lemma is false, then we may choose $\epsilon > 0$ so that $\epsilon < \sup\{x_n + y_n\} - (\sup\{x_n\} + \sup\{y_n\})$. Then the inequality (2) shows that

$$\sup\{x_n\} + \sup\{y_n\} < \sup\{x_n\} + \sup\{y_n\},$$

an evident contradiction.

The phrasing used in the proof is much more succinct and will be used throughout this text.)

A.1.7. Proposition. *Let* $\{x_n\}$ *and* $\{y_n\}$ *be sequences of extended real numbers. Then the following inequalities hold:*

(i) *if* $x_n \leq y_n$ *for all* n, *then both* $\liminf x_n \leq \liminf y_n$ *and* $\limsup x_n \leq \limsup y_n$;

(ii) $\liminf x_n + \liminf y_n \leq \liminf (x_n + y_n)$;

(iii) $\liminf (x_n + y_n) \leq \limsup x_n + \liminf y_n$;

(iv) $\limsup x_n + \liminf y_n \leq \limsup (x_n + y_n)$;

(v) $\limsup (x_n + y_n) \leq \limsup x_n + \limsup y_n$;

provided that the expressions involved are well-defined.

Proof.   Problem A.3.

A.1.8. Proposition. *Let* $\{x_n\}$ *be a sequence of extended real numbers and let* $\alpha > 0$. *Then*

(i) $\limsup \alpha x_n = \alpha \limsup x_n$ *and* $\liminf \alpha x_n = \alpha \liminf x_n$;

(ii) $\limsup x_n = -\liminf -x_n$.

Proof.   Both assertions are obvious from observations such as

$$\sup\{\alpha x_n\} = \alpha \sup\{x_n\} \quad \text{and} \quad \sup\{x_n\} = -\inf\{-x_n\}.$$

A.1.9. Definition.   Let $\{x_n\}$ be a sequence of extended real numbers and suppose that $\liminf x_n = \limsup x_n \equiv x_\infty$. Then we say that $\{x_n\}$ *converges to* x_∞ and write $\lim x_n = x_\infty$.

The definition of convergence above is slightly more convenient than the one given in calculus because it unifies the separate cases $|x_\infty| = \infty$ and $|x_\infty| < \infty$. It is equivalent to the conventional definitions in each of these cases.

A.1.10. Theorem.   *Let* $\{x_n\}$ *be a sequence of extended real numbers and suppose that* $\lim x_n = x_\infty$. *Then*

(i) $|x_\infty| < \infty$ *only if* $\{x_n\}$ *is bounded;*

(ii) *if* $|x_\infty| < \infty$, *then for each* $\epsilon > 0$ *there is an integer* N *such that* $|x_j - x_\infty| < \epsilon$ *whenever* $j \geq N$;

(iii) if $x_\infty = \infty$, then for each real number $R > 0$ there is an integer N such that $x_j \geq R$ if $j \geq N$;

(iv) if $x_\infty = -\infty$, then for each real number $R < 0$ there is an integer N such that $x_j \leq R$ if $j \geq N$.

Proof.   Once again, all the parts of this theorem have similar proofs, and so we verify only part (ii) and leave the others to the problems.

Define $\{s_n^-\}$ and $\{s_n^+\}$ as in A.1.5 and fix $\epsilon > 0$. Since

$$\sup\{s_n^-\} = x_\infty = \inf\{s_n^+\}$$

we may choose N_1 and N_2 so large that

$$x_\infty - \epsilon \leq s_{N_1}^- \quad \text{and} \quad x_\infty + \epsilon \geq s_{N_2}^+.$$

Setting $N = \max\{N_1, N_2\}$, it follows from A.1.5(ii) and (iii) that for $j \geq N$,

$$x_\infty - \epsilon \leq s_N^- \leq x_j \leq s_N^+ \leq x_\infty + \epsilon,$$

from which $|x_j - x_\infty| \leq \epsilon$ if $j \geq N$.

A.1.11. Definition.   Let $\{x_n\}$ be a sequence of real numbers and suppose, for each $\epsilon > 0$, there is an index N such that $|x_j - x_k| \leq \epsilon$ whenever $j, k \geq N$. Then the sequence $\{x_n\}$ is said to be a *Cauchy sequence*.

A.1.12. Theorem.   *For a sequence $\{x_n\}$ of real numbers, the following are equivalent:*

(i) $\{x_n\}$ is a Cauchy sequence;

(ii) $\{x_n\}$ converges to some real number x_∞.

Proof.   We first show that (i) $\Rightarrow$ (ii). By A.1.5(iv), it suffices to argue that $\limsup x_n \leq \liminf x_n$. Fix $\epsilon > 0$ and choose N so large that if $j, k \geq N$, then $|x_j - x_k| \leq \epsilon$. Then in particular, if $j, k \geq N$, it follows that $x_j \leq x_k + \epsilon$ and hence

$$\inf_{n \geq 1} \sup_{j \geq n} x_j \leq \sup_{j \geq N} x_j \leq x_k + \epsilon.$$

Now recall that $k \geq N$ was arbitrary, and hence

$$\limsup x_n \leq \sup_{n \geq N} \inf_{k \geq n} x_k + \epsilon = \liminf x_n + \epsilon.$$

Since ϵ was arbitrary, this shows that (i) $\Rightarrow$ (ii).

For the converse, fix $\epsilon > 0$ and applying A.1.10(ii), select an integer N so that $|x_j - x_\infty| \leq \epsilon/2$ if $j \geq N$. Then if $j, k \geq N$,

$$|x_j - x_k| \leq |x_j - x_\infty| + |x_\infty - x_k| \leq \epsilon.$$

The equivalences discussed in the preceding two theorems are fundamental; we will use all of these notions of convergence interchangeably, and so the student should feel comfortable with them all.

PROBLEMS

A.1. Complete the proof of Proposition A.1.5.

A.2. Complete the proof of Proposition A.1.6.

A.3. Prove Proposition A.1.7.

A.4. Complete the proof of Proposition A.1.10.

A.5. Find a sequence $\{x_n\}$ of real numbers with the property that

$$\liminf x_n < \limsup x_n.$$

A.6. Show by example that each of the inequalities in Proposition A.1.7 can be strict.

A.7. Let x be a real number in the interval $[0,1]$. Show that there is a sequence $\{a_n\}$ such that

$$x = \sum_{n=0}^{\infty} a_n 2^{-n} \left(\equiv \lim_{N \to \infty} \sum_{n=0}^{N} a_n 2^{-n} \right),$$

where $a_n \in \{0,1\}$ for each n.

A.8. Let x be a real number and let p be a positive integer. Show that there is a sequence $\{a_n\}$ such that

$$x = \sum_{n=0}^{\infty} a_n p^{-n},$$

where $a_n \in \{0,1,\ldots,p-1\}$ for each n.

A.9. Recall that a number q is **rational** if q can be written in the form $q = a/b$, where a and b are integers. Show that every nonempty open interval (x,y) of real numbers contains a rational number.

A.10. Let $\{x_n\}$ and $\{y_n\}$ be sequences of real numbers with $\lim x_n = x_\infty$ and $\lim y_n = y_\infty$. Show that
(a) $\lim(x_n + y_n) = \lim x_n + \lim y_n$.
(b) $\lim(x_n \wedge y_n) = \lim(x_n) \wedge \lim(y_n)$ where $a \wedge b \equiv \min\{a,b\}$.
(c) $\lim(x_n \vee y_n) = \lim(x_n) \vee \lim(y_n)$ where $a \vee b \equiv \max\{a,b\}$.
(d) $\lim(\alpha x_n) = \alpha \lim x_n$ for all real numbers α.
(e) If $\lim y_n \neq 0$, then

$$\lim \left(\frac{x_n}{y_n} \right) = \frac{\lim x_n}{\lim y_n}.$$

A.2. OPEN AND CLOSED SETS. In the preceding section we described convergence of sequences in the context of conventional ϵ's and N's. It often happens that this approach is rather cumbersome and a description in terms of open intervals is more convenient. This brief section reformulates the notions of convergence in this manner. (Unless otherwise specified, all numbers in the sequel are *real.*)

A.2.1. Definition.   A subset U of the real numbers is *open* if for each $x \in U$ there is an $\epsilon > 0$ such that the open interval

$$(x - \epsilon, x + \epsilon) \subseteq U.$$

(Alternatively, U is open if for each $x \in U$ there is an open interval I such that $x \in I$ and $I \subseteq U$.)

The following theorem is obvious from A.1.10 and 1.1.12.

A.2.2.   Theorem.   *For a sequence $\{x_n\}$ of real numbers the following are equivalent:*
   (i) $\{x_n\}$ converges to some real number x_∞;
   (ii) for each open set U with $x_\infty \in U$ there is an index N such that $x_j \in U$ if $j \geq N$.
   (iii) for each open set U with $0 \in U$ there is an index N such that $x_j - x_k \in U$ if $j, k \geq N$.

Some obvious properties of open sets are contained in the following.

A.2.3. Propostion
   (i) Both $\emptyset$ and $\Re$ are open sets.
   (ii) if $\{U_1, \ldots, U_n\}$ is a finite family of open sets, then

$$\cap_{i=1}^n U_i$$

   is an open set.
   (iii) If A is an arbitrary index set and if $\{U_\alpha : \alpha \in A\}$ is a family of open sets indexed by A, then

$$\cup_{\alpha \in A} U_\alpha$$

   is an open set.

Proof.   All of these assertions are more or less obvious; we prove (ii) by way of example. Let $\{U_1, \ldots, U_n\}$ be a finite collection of open sets and let $x \in \cap_k U_k$.

For each j select $\epsilon_j > 0$ so that

$$(x - \epsilon_j, x + \epsilon_j) \subseteq U_j$$

and set $\epsilon = \min\{\epsilon_1, \ldots, \epsilon_n\}$. Then since

$$(x - \epsilon, x + \epsilon) \subseteq (x - \epsilon_j, x + \epsilon_j) \subseteq U_j$$

for $j = 1, \ldots, n$, it follows that

$$(x - \epsilon_j, x + \epsilon_j) \subseteq \cap_{j=1}^n U_j.$$

Clearly, open intervals are open sets, and equally clearly, every open set is the union of open intervals. Less transparent is the following observation.

A.2.4. Theorem. *Let U be an open subset of the real numbers. Then there is a countable family $\{I_n\}$ of disjoint open intervals such that $U = \cup_n I_n$.*

Proof. If $U = \emptyset$, there is nothing to prove so we suppose that U is nonempty. For an arbitrary element $x \in U$ define

$$I(x) = \bigcup \{I : I \text{ is an open interval } x \in I \text{ and } I \subseteq U\}.$$

In other words, for each x, $I(x)$ is the largest open interval about x which is contained in U.

Suppose now that we have two such intervals $I(x)$ and $I(y)$. If $I(x) \cap I(y) \neq \emptyset$, then $I(x) \cup I(y)$ is an open interval about x which is contained in U. Since $I(x)$ is the largest such interval, it follows that $I(x) \cup I(y) \subseteq I(x)$ and hence that $I(x) \cup I(y) = I(x)$. Similarly, $I(x) \cup I(y) = I(y)$, and thus any two of the intervals $\{I(x)\}$ are either disjoint or equal.

The preceding paragraph shows that U decomposes into the union of disjoint open intervals. If we could show that any collection $\{I_\alpha\}$ of disjoint open intervals is at most countable, we would be through. But any nonempty open interval I_α must contain a rational number q_α (see Problem A.9). Since the intervals are disjoint, $q_\alpha \neq q_\beta$ if $\alpha \neq \beta$. Consequently, the mapping $I_\alpha \mapsto q_\alpha$ is an injection from $\{I_\alpha\}$ to a subset of the rational numbers. As the rational numbers are countable, this completes the proof.

A.2.5. Definition. Let $S \subseteq \Re$ and let $x \in \Re$. Suppose that for every open set U containing x, $S \cap U \backslash \{x\} \neq \emptyset$. Then x is a *cluster point* of the set S.

The notion of a cluster point slightly generalizes the notion of a sequence with a limit, as the following proposition illustrates.

A.2.6. Proposition. *Let $S \subseteq \Re$ and let $x \in \Re$. Then the following are equivalent.*
(i) *x is a cluster point of S;*
(ii) *there is a sequence $\{x_n\} \subseteq S$ such that $\lim x_n = x$ and $x_n \neq x$ for all n.*

Proof To see that (i) implies (ii), select x_n so that

$$x_n \in S \cap \left(x - \frac{1}{n}, x + \frac{1}{n}\right) \setminus \{x\}.$$

For the converse, if s is not a cluster of S, then there is an $\epsilon > 0$ such that

$$S \cap (x - \epsilon, x + \epsilon) \setminus \{x\} = \emptyset.$$

Corresponding to this ϵ, there is an integer N such that $|x_j - x| \leq \epsilon$ if $j \geq N$. But then

$$x_j \in S \cap (x - \epsilon, x + \epsilon) \setminus \{x\},$$

a contradiction.

A.2.7. Corollary. *If x is a cluster point of the sequence $\{x_n\}$, then there is a subsequence $\{x_{n_k}\}$ such that*

$$\lim_k x_{n_k} = x.$$

A.2.8. Definition. A set $S \subseteq \Re$ is *closed* if $\Re \setminus S$ is open.

Note that the interval $[a, b)$ is neither open nor closed. Sets are called "closed" because they contain all of their cluster points.

A.2.9. Theorem. *For a set $S \subseteq \Re$ the following are equivalent:*
(i) *S is closed;*
(ii) *S contains all of its cluster points.*

Proof. To see that (i) implies (ii), suppose that S is not closed and let x be a cluster of of S. If $x \notin S$, then x is in the open set S^c, and thus there is an open interval I about x with $I \subseteq S^c$, i.e.,

$$S \cap I \setminus \{x\} = \emptyset,$$

a contradiction.
For the converse, if S is not closed, then there is an element $x \in S^c$ with the property that $I \not\subseteq S^c$ whenever I is an open interval about x. But this implies that $S \cap I \setminus \{x\} \neq \emptyset$ for all open intervals I about x. By (ii), this implies that $x \in S$, again a contradiction.

PROBLEMS

A.11. *Prove or find a counterexample:* If C is a closed set, then there is a countable family $\{I_n\}$ of disjoint closed intervals with $C = \cup_n I_n$.

A.12. *Open and closed sets in $\Re^\#$*. **Definition.** A set $F \subseteq \Re^\#$ is *closed* if whenever $\{x_n\} \subseteq F$ and $\lim x_n = x_\infty$, then $x_\infty \in F$. A set $G \subseteq \Re^\#$ is *open* if $\Re^\# \backslash G$ is closed.

(a) $G \subseteq \Re^\#$ is open if and only if $G \cap \Re$ is an open subset of $\Re$.

(b) $F \subseteq \Re^\#$ is closed if and only if $F \cap \Re$ is a closed subset of $\Re$.

A.13. Let $S \subseteq \Re$; the *closure of S*, denoted $cl(S)$, is the set

$$cl(S) = \bigcap \{F : F \supseteq S \text{ and } F \text{ is closed}\};$$

the *interior of S*, denoted $int(S)$, is the set

$$int(S) = \bigcup \{U : U \subseteq S \text{ and } U \text{ is open}\};$$

the *boundary of S*, denoted ∂S, is the set

$$\partial S = cl(S) \backslash int(S).$$

(a) Find a set S for which $\emptyset \neq int(S) \subsetneq S \subsetneq cl(S)$.

(b) For any set S show that $cl(S) = S \cup \partial S$, $int(S) = S \backslash \partial S$, and $\Re = int(S) \cup \partial S \cup int(\Re \backslash S)$.

A.14. Show that

(a) both $\emptyset$ and $\Re$ are closed sets.

(b) If $\{F_1, \ldots, F_n\}$ is a finite collection of closed sets, then $\cup_i F_i$ is again a closed set.

(c) If $\{F_\alpha\}$ is an arbitrary family of closed sets, show that $\cap_\alpha F_\alpha$ is again a closed set.

(d) Suppose that $S \subseteq \Re$ is both open and closed. Show that either $S = \emptyset$ or else $S = \Re$.

A.15. Find a family $\{F_n\}$ of closed sets satisfying

(a) $F_n \subseteq F_{n+1}$ for all n; and

(b) $\cup_n F_n$ is not closed.

A.16. *Open and closed sets in $\Re \times \Re$*. **Definition.** A set $U \subseteq \Re \times \Re$ is *open* if, for each $(x, y) \in U$ there are open sets U_x and U_y contained in $\Re$ such that $x \in U_x$, $y \in U_y$ and, $U_x \times U_y \subseteq U$. A set $F \subseteq \Re \times \Re$ is *closed* if $\Re \times \Re \backslash F$ is open.

(a) State and prove the analogue of A.2.3 for open sets in $\Re \times \Re$.

(b) Show that $U \subseteq \Re \times \Re$ is open if and only if for each $(x, y) \in U$ there is an $\epsilon > 0$ such that the disk

$$\left\{ (u, v) : (x - u)^2 + (y - v)^2 \leq \epsilon^2 \right\} \subseteq U.$$

Definition. A sequence $\{(x_n, y_n)\} \subseteq \Re \times \Re$ *converges* to (x_∞, y_∞) if, for each $\epsilon > 0$ there is an N such that $j \geq N$ implies that

$$(x_j - x_\infty)^2 + (y_j - y_\infty)^2 \leq \epsilon^2.$$

The sequence $\{(x_n, y_n)\}$ is *Cauchy* if for each $\epsilon > 0$ there is an N such that $j, k \geq N$ implies that

$$(x_j - x_k)^2 + (y_j - y_k)^2 \leq \epsilon^2.$$

(c) Show that every Cauchy sequence in $\Re \times \Re$ converges to some $(x_\infty, y_\infty) \in \Re \times \Re$.

A.17. Let A and B be disjoint closed sets of real numbers. Show that there are disjoint open sets U and V with $A \subseteq U$ and $B \subseteq V$.

A.18. Show that every uncountable set of real numbers has a cluster point.

A.19. A set $S \subseteq \Re$ is *nowhere dense* if $int(S) = \emptyset$. Show that $\Re$ cannot be written as a countable union of nowhere dense sets.

A.3. COMPACT SETS   It will be very useful to determine when a given sequence of real numbers (or extended real numbers) has a convergent subsequence. The answer to this question is found in the study of compact sets.

A.3.1. Definition.   Let S be a collection of real numbers and let $\{U_\alpha\}_{\alpha \in A}$ be a family of open sets.

(i) We call $\{U_\alpha\}$ an *open cover for* S if $S \subseteq \bigcup_\alpha U_\alpha$.

(ii) We say that a cover $\{U_\alpha\}$ *admits a finite subcover* if there is a finite collection $\{U_{\alpha_1}, \ldots, U_{\alpha_n}\}$ with the property that

$$S \subseteq \bigcup_{i=1}^{n} U_{\alpha_i}.$$

(iii) We call the set S *compact* if every open cover admits a finite subcover.

The fundamental result about compact sets of real numbers is contained in the following theorem. The technique used to prove this theorem is an important one; our proof of the Differentiation Theorem in Chapter 5 is based, in part, on these ideas.

A.3.2. Theorem.   *If $a \leq b$ are real numbers, then the interval $[a, b]$ is compact.*

Proof.   Let $\{U_\alpha\}$ be an open cover for $[a, b]$ and define

$$P = \{t \in [a, b] : \{U_\alpha\} \text{ admits a finite subcover on } [a, t]\}.$$

Observe that $a \in P$, so $P \neq \emptyset$. Further, if $a \leq t_1 \leq t_2$ and $t_2 \in P$, then $t_1 \in P$. Notice that it will suffice to show that $b \in P$. To do this we establish two claims about the set P.

Claim 1. *If* $t \in P$ *and* $t < b$, *then there is an* $\epsilon > 0$ *such that* $t + \epsilon \in P$.

To prove Claim 1, select $\{U_{\alpha_1}, \ldots, U_{\alpha_n}\}$ so that

$$[a, t] \subseteq \bigcup_{i=1}^{n} U_{\alpha_i}.$$

Select an index j so that $t \in U_{\alpha_j}$. Next select $\epsilon > 0$ so small that $t + \epsilon < b$ and

$$(t - 2\epsilon, t + 2\epsilon) \subseteq U_{\alpha_j}.$$

Then $t + \epsilon \in [a, b]$ and

$$[a, t + \epsilon] \subseteq [a, t] \cup [t, t + \epsilon] \subseteq \bigcup_{i=1}^{n} U_{\alpha_j},$$

showing $t + \epsilon \in P$.

Claim 2. *If* $\sigma = \sup P$, *then* $\sigma \in P$.

To verify Claim 2, observe that $a \leq \sigma \leq b$ and so there is a set U_{α_0} so that $\sigma \in U_{\alpha_0}$. By Claim 1, $\sigma > a$ and so for $\epsilon > 0$ sufficiently small, $\sigma - \epsilon > a$ and $(\sigma - 2\epsilon, \sigma + 2\epsilon) \subseteq U_{\alpha_0}$. Since $\sigma - \epsilon$ is not an upper bound for P, it follows that $\sigma - \epsilon \in P$ and so there is a finite collection $\{U_{\alpha_1}, \ldots, U_{\alpha_n}\}$ so that $[a, \sigma - \epsilon] \subseteq \bigcup_{i=1}^{n} U_{\alpha_i}$. But then

$$[a, \sigma] = [a, \sigma - \epsilon] \cup [\sigma - \epsilon, \sigma] \subseteq \bigcup_{i=0}^{n} U_{\alpha_i},$$

showing that $\sigma \in P$.

To complete the argument, observe that if $\sigma < b$, then by Claim 1 there is an $\epsilon > 0$ so that $\sigma + \epsilon \in P$, contradicting our choice of σ. Thus $\sigma = b$ and $b \in P$, as desired.

With the Theorem A.3.2 in hand, we can completely characterize the the compact subsets of the reals.

A.3.3. Theorem. *For a subset C of the real numbers, the following are equivalent:*

(i) C is compact.

(ii) C is closed and bounded.

Proof. We first verify that (i) implies (ii). Observe that if x is cluster point of the compact set C, then $x \in C$, for if this is not the case, then we may set

$$U_n = \left(-\infty, x - \frac{1}{n}\right) \cup \left(x + \frac{1}{n}, \infty\right),$$

so that $C \subseteq (-\infty, x) \cup (x, \infty) = \cup U_n$. Then $\{U_n\}$ is an open cover of the compact set C and so there is an integer N so that

$$C \subseteq \bigcup_{i=1}^{N} U_n = \left(-\infty, x - \frac{1}{N}\right) \cup \left(x + \frac{1}{N}, \infty\right).$$

Since this contradicts x being a cluster point of C, it follows that $x \in C$.

Because of this observation, C^c contains no cluster points of C. Thus, for each $y \in C^c$ there is an open set G_y so that $C \cap G_y \backslash \{y\} = \emptyset$; since $y \in C^c$, this implies that $C \cap G_y = \emptyset$, so that $G_y \subseteq C^c$. Thus C^c is the union of the open sets $\{G_y : y \in C^c\}$, implying that C^c is open and hence that C is closed.

To see that C must be bounded, notice that $\{(-n, n) : n \in \mathcal{Z}\}$ is an open cover for C and so there is an N so that $C \subseteq \bigcup_{i=1}^{N}(-i, i) = (-N, N)$.

For the converse, suppose that C is closed and bounded and let $\{U_\alpha\}$ be an open cover of C. Select an interval $[a, b]$ so that $C \subseteq [a, b]$. Since $\{U_\alpha\} \cup \{C^c\}$ is an open cover for $[a, b]$ there is a finite collection $\{U_{\alpha_1}, \ldots, U_{\alpha_n}\}$ such that

$$C \subseteq [a, b] \subseteq C^c \cup \bigcup_{i=1}^{n} U_{\alpha_i},$$

and thus $C \subseteq \bigcup_{i=1}^{n} U_{\alpha_i}$. Since $\{U_\alpha\}$ admits a finite subcover, this shows that C is compact.

A.3.4. Corollary. *If $\{x_n\}$ is a bounded sequence of real numbers, then $\{x_n\}$ has a convergent subsequence.*

Proof. If $\{x_n\}$ has only finitely many distinct terms, then we may extract a constant subsequence from $\{x_n\}$; thus we suppose that the sequence $\{x_n\}$ assumes infinitely many distinct values.

It suffices to show that $\{x_n\}$ has a cluster point. Choose a closed interval $[a, b]$ so that $\{x_n\} \subseteq [a, b]$ and suppose that $\{x_n\}$ has no cluster points. Then for each $t \in [a, b]$ we can find an open set U_t such that $U_t \cap \{x_n\} \backslash \{t\} = \emptyset$. The family $\{U_t\}$ is an open cover of the compact set $[a, b]$ and so there is a finite collection $\{U_{t_1}, \ldots, U_{t_p}\}$ so that

$$\{x_n\} \subseteq \bigcup_{i=1}^{p} U_{t_i}.$$

Since $U_{t_i} \cap \{x_n\}$ is at most the singleton t_i, this contradicts our initial assumption that $\{x_n\}$ assumes infinitely many distinct values.

We conclude this section with a theorem due to Lebesgue which we will use in our discussions of the integral.

A.3.5. Theorem. *Let C be a compact set and let $\{U_\alpha\}$ be an open cover of C. Then there is a number $\delta > 0$ with the following property:*

*If $s, t \in C$ and $|s - t| < \delta$, then there is a set U_{α_0} with **both** s and t in U_{α_0}.*

Proof. We suppose that the conclusion of the theorem is not true and proceed by contradiction. Then for each $n > 0$ there are numbers t_n and s_n with $|t_n - s_n| \leq 1/n$ and with the property that t_n and s_n are never in the same U_α. Since C is compact, there are subsequences $\{t_{n'}\}$ and $\{s_{n'}\}$ and an element $x \in C$ such that

$$\lim t_{n'} = \lim s_{n'} = x$$

($\{t_{n'}\}$ and $\{s_{n'}\}$ have the same limit since $\lim |t_{n'} - s_{n'}| = 0$.) Since $x \in C$ there is an open set U_{α_0} with $x \in U_{\alpha_0}$. We may select N so large that $n' \geq N$ implies that $t_{n'} \in U_{\alpha_0}$ and $s_{n'} \in U_{\alpha_0}$, in contradiction to our choice of the sequences $\{t_n\}$ and $\{s_n\}$.

The number δ given in the Theorem A.3.5 is called the *Lebesgue number* of the open cover $\{U_\alpha\}$.

PROBLEMS

A.20. For a set $C \subseteq \Re$ show that the following are equivalent:
  (a) C is compact.
  (b) Every infinite subset of C has a cluster point in C;
  (c) Every family $\{F_\alpha\}$ of closed subsets of C having the finite intersection property has nonempty intersection. (A family $\{F_\alpha\}$ has the *finite intersection property* if every finite subfamily has nonempty intersection.)
  (d) Every sequence in C has a subsequence which converges to a limit in C.

A.21. Show that $cl(S)$ is compact if and only if S is bounded.

A.22. Given any set X and a family of open sets τ definied on X, we say that $C \subseteq X$ is compact if every open cover of C by members of τ admits a finite subcover.
  (a) Show that $\Re^\#$ is compact with τ as defined in Problem A.12.
  (b) Show that $C \subseteq \Re \times \Re$ is compact if and only if C is closed and bounded (see Problem A.16). (*Hint:* First show that $[a, b] \times [c, d]$ is compact.)

A.23. If C_1 and C_2 are compact subsets of $\Re$, then each of the following sets is also compact.

$$C_1 + C_2 = \{x + y : x \in C_1 \text{ and } y \in C_2\}$$

$$C_1 \cdot C_2 = \{x \cdot y : x \in C_1 \text{ and } y \in C_2\}$$

$$C_1/C_2 = \{x/y : x \in C_1 \text{ and } y \in C_2\} \text{ (provided that } 0 \notin C_2)$$

A.24. Let C be a compact subset of $\Re$ and let $\epsilon > 0$. Show that there is a finite set $\{x_1, \ldots, x_n\} \subset C$ such that for any $t \in C$, there is an x_j with $|t - x_j| \leq \epsilon$.

Appendix B
The Axiom of Choice

The search for increased rigor in mathematical argument which began in the late eighteenth and early nineteenth centuries eventually led to the development of "axiom schemes," (hopefully) consistent sets of basic principles from which all mathematics could be deduced.[*] Generally, these schemes contain some version of the "Axiom of Choice." This axiom merits some special attention in this book for two reasons. First, although it is accepted by most mathematicians, it remains rather controversial. Second, the axiom has some unusual consequences which are sometimes useful in proving analysis theorems. (In fact, it is at least in part this utility of the axiom which has lead to its rather grudging acceptance.) In this brief appendix, we will discuss the axiom and some of its equivalent formulations and give two examples of its use. (We will not prove that the various formulations are equivalent. The interested student can consult, for example, Suppes' book.)

To see how the Axiom of Choice arises, consider the following set-theoretic problem. Suppose that X and Y are sets and $f : X \to Y$ is a surjective function. Does there exist a function $g : Y \to X$ such that $g(f(x)) = x \ \forall \ x \in X$? Since f is surjective, for each $y \in Y$ the set

$$f^{-1}(y) = \{x \in X \, : \, f(x) = y\}$$

is nonempty. Thus the function g will exist if we can choose $g(y) \in f^{-1}(y)$ simultaneously for each $y \in Y$. This is exactly the content of the Axiom of Choice; more precisely:

> *Axiom of Choice.* Let A be an arbitrary nonempty index set and let $\{E_\alpha : \alpha \in A\}$ be a family of nonempty sets indexed by A. Then there is a function
> $$c : A \to \bigcup_{\alpha \in A} E_\alpha$$
> such that $c(\alpha) \in E_\alpha$ for each $\alpha \in A$.

The axiom seems reasonable since, for each fixed α we can certainly make the desired choice. The problem is that the axiom asserts the existence of the choice function c without giving enough information to tell us exactly how

---

[*] For a historical survey of some of the various axiom systems, see P. Suppes' *Axiomatic Set Theory*, Van Nostrand, Princeton, NJ (1960).

to find it (by applying a finite number of rules). Moreover, this is the *only* formal axiom of set theory which is nonconstructive in this manner. This nonconstructive character of the axiom is especially disturbing today when computer algorithms are of increasing and vital importance to all aspects of mathematics. For these reasons, the use of the axiom should be avoided where possible and explictly noted where it is used.

The Axiom of Choice is used in four places in this text. In Problem 2.24 it is used to construct a nonmeasurable set; it is used to prove that every Hilbert space contains a complete orthonormal set; it is used in the proof of the Hahn-Banach Theorem; finally, it is used in the proof of the Banach-Alaoglu Theorem. Each of these uses is essential in some manner, although it is often possible to deduce a special case without appeal to the axiom (as in, for example, 3.4.12 where a complete orthonormal set is explicitly constructed for a separable Hilbert space).

Usually, the Axiom of Choice is not directly used but rather some equivalent form; these equivalent forms all involve statements about various orderings on sets. The ideas are stated in the next definition.

B.1. Definition. Let A be a set and let $\preceq$ be a relation on X.[*] Then $\preceq$ is
   (i) *reflexive* if $a \preceq a$ for all $a \in A$;
   (ii) *antisymmetric* if $a \preceq b$ and $b \preceq a$ implies that $a = b$ for all $a, b \in A$;
   (iii) *transitive* if $a \preceq b$ and $b \preceq c$ implies that $a \preceq c$ for all $a, b, c \in A$;
   (iv) *symmetric* if $a \preceq b$ implies that $b \preceq a$ for all $a, b \in A$;
   (v) an *equivalence relation* if it is reflexive, symmetric, and transitive;
   (vi) is a *partial order* if it is reflexive, antisymmetric, and transitive. In this case, the pair $(A, \preceq)$ is called a *partially ordered set.*

B.2. Examples
   (i) If $A = \Re$ and $x \preceq y \leftrightarrow x \leq y$, then $\preceq$ is a partial order but not an equivalence relation;
   (ii) if $A = Z$ and

$$x \preceq y \quad \leftrightarrow \quad y - x \text{ is divisible by 3,}$$

   then $\preceq$ is an equivalence relation but not a partial order;
   (iii) if X is any nonempty set and $A = P(X) = \{E : E \subseteq X\}$, the power set of X, and if $E \preceq F \leftrightarrow E \subseteq F$, then $\preceq$ is a partial order but not an equivalence relation. Note that, unlike example (ii), not all sets are comparable under $\preceq$; i.e., it is possible for both of the statements $E \npreceq F$ and $F \npreceq E$ to be true;
   (iv) if $A = C[0, 1]$, then $f \preceq g \leftrightarrow f(x) \leq g(x)$ for all x is a partial order. Once

---

[*] Formally, a relation R is a subset of $A \times A$ and we say that $x \preceq y \leftrightarrow (x, y) \in R$.

again, not all functions are comparable under this partial order. However, given any pair of functions f and g, there is a function $h \in C[0,1]$ (namely, $f \vee g$) such that both $f \preceq h$ and $g \preceq h$.

The examples above show that partial orders can be a little less regular than one might expect. Some further properties of partial orders meriting attention are:

B.3. Definition. Let $(X, \preceq)$ be a partially ordered set and let $C \subseteq A$. Then
(i) an element $x \in X$ is an *upper bound* for C if $c \preceq x$ for all $c \in C$;
(ii) an element $x \in X$ is a *lower bound* for C if $x \preceq c$ for all $c \in C$;
(iii) an element $x \in X$ is *maximal* if $y \preceq x$ for all $y \in X$;
(iv) an element $x \in X$ is *minimal* if $x \preceq y$ for all $y \in X$;
(v) X is *totally ordered* if either $x \preceq y$ or $y \preceq x$ for all $x, y \in X$ (sometimes this is called *linearly ordered*);
(vi) if $(C, \preceq)$ is totally ordered, then C is called at *chain* in $(X, \preceq)$;
(vii) $(X, \preceq)$ is *well ordered* if it is totally ordered and if every subset $E \subseteq X$ has a lower bound $x \in E$.

Note that the natural numbers are well-ordered by their usual order, whereas the real numbers are not.

The following theorem contains the essential facts relating the above to the Axiom of Choice.

B.4. Theorem. *The following are equivalent:*
(i) The Axiom of Choice.
(ii) (Zorn's Lemma) If $(X, \preceq)$ is a partially ordered set in which every chain has an upper bound, then X has a maximal element.
(iii) (Zermelo's Theorem) Every set can be well-ordered.

We use only Zorn's Lemma in this text; Zermelo's Theorem is the basis for "transfinite induction." There are two other commonly used and equivalent formulations of the Axiom of Choice which we will not discuss, namely Tukey's Lemma and the Hausdorff Maximality Principle.

We conclude by giving two examples of the use of Zorn's Lemma. The first is a rather standard proof that every vector space has a basis.

B.5. Definition. Let V be a vector space and let $E \subseteq V$; the set E is *linearly independent* if whenever $\{v_1, \ldots, v_n\}$ is a finite collection from E and $\{\alpha_1, \ldots, \alpha_n\}$ is a finite collection of real numbers which are not all zero, then $\sum_{i=1}^{n} \alpha_i v_i \neq 0$.

B.6. Theorem. *If V is a nontrivial (i.e., $V \neq \{0\}$) vector space, then there is a linearly independent set $\mathcal{B} \subseteq V$ with the property that for each $v \in V$ there are vectors $\{v_1, \ldots, v_n\} \subseteq \mathcal{B}$ and scalars $\{\alpha_1, \ldots, \alpha_n\}$ such that $v = \sum_{i=1}^{n} \alpha_i v_i$. Moreover, the representation is unique.*

Proof. Since V is nontrivial, we may select $p \in V$ such that $p \neq 0$. Note that the set $\{p\}$ is a linearly independent subset of V. Now take A to be the set

$$A = \{E \subseteq V : E \text{ is linearly independent}\}$$

and define a partial order $\preceq$ on E by saying that $E_1 \preceq E_2 \leftrightarrow E_1 \subseteq E_2$. We will use Zorn's Lemma to show that A has a maximal element. Note that $A \neq \emptyset$.

Thus let $\mathcal{C}$ be any chain in A and take $E_0 = \bigcup \{E : E \in \mathcal{C}\}$. Certainly, $E_0 \supseteq E$ for all $E \in A$; if E_0 is linearly independent, then E_0 is an upper bound for the chain $\mathcal{C}$ and Zorn's Lemma applies to show that A has a maximal element.

To see that E_0 is linearly independent, let $\{e_1, \ldots, e_n\} \subseteq E_0$ and let $\{\alpha_1, \ldots, \alpha_n\}$ be a collection of scalars for which

$$\sum_{i=1}^{n} \alpha_i e_i = 0. \tag{1}$$

For each i we may select $E_i \in \mathcal{C}$ such that $e_i \in E_i$, $i = 1, \ldots, n$. Now $\{E_1, \ldots, E_n\}$ is a finite and totally ordered collection, and thus, by reindexing if needed, we may suppose that $E_1 \subseteq E_2 \subseteq \ldots \subseteq E_n$ (finiteness of the collection is essential at this point). Thus $\{e_1, \ldots, e_n\} \subseteq E_n$. As E_n is linearly indeddpendent by assumption, (1) shows that each of the scalars α_i, $i = 1, \ldots, n$ must be zero. As we started with an arbitrary collection $\{e_1, \ldots, e_n\} \subseteq E_0$, this shows that E_0 is linearly independent, as desired.

Thus by Zorn's Lemma, the set A contains a maximal member E_∞. We next show that E_∞ must be the set $\mathcal{B}$ of the conclusion of the theorem. If not, then there is an element $v_0 \in V$ such that v_0 cannot be represented as a finite linear combination of members of the set E_∞. We claim that the set $E_\infty \cup \{v_0\}$ is linearly independent. To see this, suppose that we have vectors $\{v_1, \ldots, v_n\} \subseteq E_\infty$ and scalars $\{\alpha_0, \ldots, \alpha_n\}$ not all zero for which $\sum_{i=0}^{n} \alpha_i v_i = 0$. If $\alpha_0 = 0$, this contradicts E_∞ linearly independent, and thus $\alpha_0 \neq 0$. But then we may solve for v_0:

$$v_0 = \sum_{i=1}^{n} \frac{\alpha_i}{\alpha_0} v_i,$$

contradicting our choice of v_0. Thus the set $E_\infty \cup \{v_0\}$ is linearly independent. But this is in contradiction to the set E_∞ being maximal in A!

Uniqueness of the representation is left as an easy exercise for the student.

Our next example is a little more challenging. The technique involved is a primitive version of the one used to prove the Hahn-Banach Theorem in Chapter 8.

B.7. Theorem. *Let V be a vector space and let X be a subspace of V; let f be a linear mapping from X to the reals. Then there is a linear mapping g from V to the reals with the property that $f(x) = g(x)$ for all $x \in X$.*

Proof. Take A to be the set of all order pairs (h, E) such that
(1) E is a subspace of V and $X \subseteq E$;
(2) $h : E \to \Re$ is linear and $h(x) = f(x)$ for all $x \in X$.
Next define an ordering $\preceq$ on A by saying that $(h_1, E_1) \preceq (h_2, E_2)$ if and only if
(3) $E_1 \subseteq E_2$; and
(4) $h_1(x) = h_2(x)$ for all $x \in E_1$.
It is easily checked that $\preceq$ is antisymmetric and reflexive. To see that $\preceq$ is transitive as well, suppose that $(h_1, E_1) \preceq (h_2, E_2)$ and $(h_2, E_2) \preceq (h_3, E_3)$. Then $E_1 \subseteq E_2 \subseteq E_3$, so $E_1 \subseteq E_3$. In addition, if $x \in E_1 \subseteq E_2$, then $h_1(x) = h_2(x) = h_3(x)$, and so $(h_1, E_1) \preceq (h_3, E_3)$, as desired. In particular, $(A, \preceq)$ is a partially ordered set.

Next suppose that $C = \{(E_\alpha, h_\alpha)\}$ is a chain in A. Set $E_0 = \cup_\alpha E_\alpha$. Notice that if $x \in E_\alpha \cap E_\beta$, then, since C is totally ordered, either $E_\alpha \subseteq E_\beta$ or else $E_\beta \subseteq E_\alpha$. In either case, $h_\alpha(x) = h_\beta(x)$. Thus if $x \in E_0$ and if $h_0(x) = \sup\{h_\alpha(x)\}$, then $h_0(X) = h_\alpha(x)$ for all α for which h_α is defined.

Notice that the above implies that h_0 is linear: If $x, y \in E_0$ and if $s, t \in \Re$, then (since C is totally ordered) there is an index α for which $x, y \in E_\alpha$, and so

$$\begin{aligned} h_0(sx + ty) &= h_\alpha(sx + ty) \\ &= sh_\alpha(x) + th_\alpha(y) \\ &= sh_0(x) + th_0(y). \end{aligned}$$

In particular, (E_0, h_0) is an upper bound for the chain C.

Thus by Zorn's Lemma, A contains a maxmimal member (E_∞, h_∞). If $E_\infty = V$, then surely (E_∞, h_∞) is the desired extension of f to all of V, and so suppose for contradiction that $E_\infty \neq V$. Then there is a $y \in V \backslash E_\infty$. Set $\hat{V} = sp(E_\infty \cup \{y\})$. If $x \in \hat{V}$, then there is a vector $u \in E_\infty$ and a scalar $t \in \Re$ so that $x = u + ty$. Moreover, this representation is unique, for if $u + ty = v + sy$, then $u - v = (s - t)y$. Since $y \notin E_\infty$, this implies that $s = t$ and hence that $u = v$.

Thus we may define $\hat{g}$ on $\hat{V}$ by $\hat{g}(x) = \hat{g}(u+ty) = h_\infty(u) + t$. Note that $\hat{g}$ is linear, $\hat{g}$ agrees with h_∞ on E_∞, and that $\hat{V} \underset{\neq}{\supseteq} E_\infty$. This contradicts the maximality of E_∞ and so completes the proof.

PROBLEMS

B.1. Show that if $(X, \preceq)$ is a partially ordered set with the property that every chain has a lower bound, then X has a minimal element.

B.2. Let $(E, \preceq)$ be a partially ordered set and let $f : E \to E$ be a mapping for which $x \preceq y \Rightarrow f(x) \preceq f(y)$. Show that there is an $x \in X$ for which $f(x) = x$.

B.3. Let $\mathcal{U}$ be a set and define a relation $\preceq$ on $\mathcal{P}(\mathcal{U})$ by $A \preceq B$ if and only if there is an injection $f : A \to B$. If $A \preceq B$ and $B \preceq A$, show that there is a bijection $f : A \to B$.

B.4.   Let X and Y be Banach spaces and let $P : X \to Y$ be a mapping which satisfies:

      (i) P sends open sets in X to open sets in Y;

     (ii) $\{(x, P(x)) : x \in X\}$ is closed in $X \times Y$; and

   (iii) $\|P(x) - P(y)\| \geq \|x - y\|$ for all $x, y \in X$.

Show that P is surjective.

B.5.   Let $(X, \preceq)$ be a partially ordered set; suppose for each $x, y \in X$ that the set $\{u \in X : x \preceq u \text{ and } y \preceq u\}$ has a minimal element $x \vee y$ and the set $\{u \in X : u \preceq x \text{ and } u \preceq y\}$ has a maximal element $x \wedge y$. Then $(X, \preceq)$ is called a *lattice*. A set $I \subseteq X$ is called an *ideal* if

$$x \in I \text{ and } y \in I \Rightarrow x \vee y \in I$$

and

$$x \in X \text{ and } y \in I \Rightarrow x \wedge y \in I.$$

Suppose that X has a maximum and at least one other element. Show that X contains an ideal $I \neq X$.

Appendix C
Selected References

In this section we include a brief survey of some of the more important references in integration theory. This list is by no means exhaustive, our purpose being merely to provide the student some historical references and with a starting place for researching a topic. The student desiring an exhaustive historical survey is referred to the notes in [21].

[1]   Ahlfors, L., *Complex Analysis*, McGraw-Hill, New York, 1953.

[2]   Alaoglu, L., *Weak topologies of normed linear spaces*, Ann. of Math. (2) **41**, 252-267 (1940).

[3]   Alexandroff, P., and Hopf, H., *Topologie*, Springer, Berlin, 1935.

[4]   Arzela, C., *Sulle funzioni di linee*, Mem. Accad. Sci. Ist. Bologna Cl. Fis. Mat. (5) **5**, 55-74 (1895).

[5]   Ascoli, G., *Le curve limiti di una varieta data di curve*, Atti della R. Accad. dei Lincei Memorie della Cl. Sci. Fis. Mat. Nat. (3) **18**, 521-586 (1883-1884).

[6]   Asplund, E., and Bungart, L., *A First Course in Integration*, Holt, Rinehart and Winston, New York, 1966.

[7]   Baire, R., *Sur les fonctions des variables réelles*, Ann. Mat. Pura Appl. **3**, 1-122 (1899).

[8]   Banach, S., *Théorie des opérations linéaires*, Monografje Matematyczne, Warsaw, 1932.

[9]   Banach, S., Steinhaus, H., *Sur le principe de la condensation de singularites*, Fund. Math. **9**, 50-61 (1927).

[10] Bartle, R., *The Elements of Real Analysis*, Wiley, New York, 1976.

[11] Borel, E., *Leçons sur la théorie des fonctions*, Gauthier-Villars, Paris, 1898.

[12] Bourbaki, N., *Eléments de mathématique, Livre VI, Integration*, Hermann et Cie, Act. Sci. et Ind. **1175**, Paris, 1952.

[13] Bray, H., *Elementary properties of the Stieltjes integral*, Ann. of Math. (2) **20**,

177-186 (1918-1919).

[14] Brouwer, L. E. J., *Uber ein eindeutige, stetige Transformationen von Flachen in sich*, Math. Ann. **69**, 176-180 (1910).

[15] Cantor, G., *Uber unendliche, lineare Punktmannigfaltigkeiten*, Math. Ann. **22**, 453-458 (1884).

[16] Caratheodory, C., *Vorlesungen uber reelle Funktionen*, Teubner, Leipzig, 1927.

[17] Cauchy, A., *Oeuvres*, ser. I and II, Gauthier-Villars, Paris, 1900.

[18] Clarkson, J., *Uniformly convex spaces*, Trans. Amer. Math. Soc. **40**, 396-414 (1936).

[19] Daniell, P., *A general form of the integral*, Ann. of Math. **19**, 279-294 (1919).

[20] Dini, U., *Fondamenti per la teorica delle funzioni di variabili reali*, Pisa (1878).

[21] Dunford, N., Schwartz, J., *Linear Operators, Part I: General Theory*, Wiley-Interscience, New York, 1957.

[22] Eberlein, W., *Weak compactness in Banach spaces, I*, Proc. Nat. Acad. Sci. U.S.A. **36**, 51-53 (1947).

[23] Egorov, D., *Sur les suites des fonctions mesurables*, Compt. Rend. Acad. Sci. Paris **152**, 244-246 (1911).

[24] Fatou, P., *Séries trigonométriques et séries de Taylor*, Acta Math. **30**, 335-340 (1906).

[25] Fischer, E., *Sur la convergence en maoyenne*, C.R. Acad. Sci. Paris **144**, 1022-1024 (1907).

[26] Friedrichs, K., *On Clarkson's inequalities*, Comm. Pure Appl. Math. **23**, 603-607 (1970).

[27] Fubini, F., *Sugli integrali multipli*, Rend. Accad. Nazl. Linceir (Rome) **24**, 204-206 (1907).

[28] Goldberg, R., *Methods of Real Analysis*, Ginn-Blaisdel, Waltham, MA, 1964.

[29] Hahn, H., *Uber lineare Gleichungssysteme in linearen Raumen*, J. Reine Angew. Math. **157**, 214-229 (1927).

[30] Helley, E., *Uber lineare Funktionaloperationen*, S.-B. K. Akad. Wiss. Wein Math.-Naturwiss. Kl. **121**, IIa, 265-297 (1912).

[31] Helley, E., *Uber Systeme linearer Gleichungen mit unendlich vielen Unbekannten*, Monatsh. fur Math. u. Phys. **31**, 60-91 (1921).

[32] Hewitt, E., Stromberg K., *Real and Abstract Analysis*, Springer-Verlag, Berlin, 1965.

[33] Hilbert, D., *Grundzuge einer allgemeinen Theorie der linearen Integralgleichungen, I-VI*, Teubner, Leipzig, 1912.

[34] Hobson, E., *On some fundamental properties of Lebesgue integrals in a two-dimensional domain*, Proc. London Math. Soc. **8**, 22-39 (1910).

[35] Hölder, E., *Uber einen Mittelwertsatz*, Nachr. Akad. Wiss. Gottingen, Math.-Phys. Kl. 1889, 38-47 (1889).

[36] Jordan, G., *Sur la série de Fourier*, C.R. Acad. Sci. Paris **92**, 228-230 (1881).

[37] Kelley, J., *General Topology*, D. van Nostrand, New York, 1955.

[38] Krein, M., Mil'man, D., *On extreme points of regularly convex sets*, Studia Math. **9**, 133-138 (1940).

[39] Lebesgue, H., *Leçons sur l'integration et la recherche des fonctions primitives*, Gauthier-Villars, Paris, 1904.

[40] Lusin, N., *Sur les propriétés des fonctions mesurables*, Compt. Rend. Acad. Sci. Paris **150**, 1688-1690 (1912).

[41] McShane, E., *Linear functionals on certain Banach spaces*, Proc. Amer. Math. Soc. **1**, 402-408 (1950).

[42] Minkowski, H., *Gesammelte Abhandlungen*, Teubner, Leipzig, 1911.

[43] Nikodym. O., *Sur les fonctions d'ensembles*, Compte Rendus du I Congres des Math. des Pays Slaves, Warsaw, 304-313 (1929).

[44] Radon, J., *Theorie und Anwendungen der absolut additiven Mengenfunktionen*, S.-B. Akad. Wiss. Wien **122** , 1295-1438 (1913).

[45] Riesz, F., *Sur les systèmes orthogonaux de fonctions*, C.R. Acad. Sci. Paris **144**, 615-619 (1907).

[46] Riesz, F., *Untersuchungen uber Systeme integrierbarer Funktionen*, Math. Ann. **69**, 449-497 (190.8).

[47] Riesz, F. and Sz.-Nagy, B., *Leçons d'analyse fonctionelle*, Akademiai Kiado, Budapest, 1952.

[48] Royden, H. L., *Real Analysis*, Macmillan, Toronto, 1963.

[49] Rudin, W., *Functional Analysis*, Mcgraw-Hill, New York, 1973.

[50] Rudin, W., *Real and Complex Analysis*, McGraw-Hill, New York, 1966.

[51] Stone, M., *Notes on Integration, I-IV*, Proc. Nat. Acad. Sci. U.S.A. **34**, 336-342, 447-445, 483-490, **35**, 50-58 (1948-1949).

[52] Titchmarsh, E., *The Theory of Functions*, Clarendon Press, Oxford, 1932.

[53] Tonelli, L., *Sull'integrazione per parti*, Rend. Accad. Nazl. Lincei (Rome) **18**, 246-253 (1909).

[54] Weierstrass, K., *Mathematische Werke*, Mayer und Muller, 1894.

[55] Widder, D., *The Laplace Transform*, Princeton Univ. Press, Princeton, NJ, 1941.

[56] Willard, S., *General Topology*, Addison-Wesley, Reading, MA, 1970.

List of Symbols

Index